Advances in Fuzzy Logic and Artificial Neural Networks

Advances in Fuzzy Logic and Artificial Neural Networks

Guest Editor

Francisco Rodrigues Lima-Junior

Basel • Beijing • Wuhan • Barcelona • Belgrade • Novi Sad • Cluj • Manchester

Guest Editor
Francisco Rodrigues Lima-Junior
Federal Technological University of Paraná
Curitiba
Brazil

Editorial Office
MDPI AG
Grosspeteranlage 5
4052 Basel, Switzerland

This is a reprint of the Special Issue, published open access by the journal *Mathematics* (ISSN 2227-7390), freely accessible at: ttps://mdpi.com/si/128382.

For citation purposes, cite each article independently as indicated on the article page online and as indicated below:

Lastname, A.A.; Lastname, B.B. Article Title. *Journal Name* **Year**, *Volume Number*, Page Range.

ISBN 978-3-7258-2941-5 (Hbk)
ISBN 978-3-7258-2942-2 (PDF)
https://doi.org/10.3390/books978-3-7258-2942-2

© 2024 by the authors. Articles in this book are Open Access and distributed under the Creative Commons Attribution (CC BY) license. The book as a whole is distributed by MDPI under the terms and conditions of the Creative Commons Attribution-NonCommercial-NoDerivs (CC BY-NC-ND) license (https://creativecommons.org/licenses/by-nc-nd/4.0/).

Contents

About the Editor vii

Francisco Rodrigues Lima-Junior
Advances in Fuzzy Logic and Artificial Neural Networks
Reprinted from: *Mathematics* 2024, 12, 3949, https://doi.org/10.3390/math12243949 1

Vitor Amado de Oliveira Bobel, Tiago F. A. C. Sigahi, Izabela Simon Rampasso, Gustavo Hermínio Salati Marcondes de Moraes, Lucas Veiga Ávila, Walter Leal Filho and Rosley Anholon
Analysis of the Level of Adoption of Business Continuity Practices by Brazilian Industries: An Exploratory Study Using Fuzzy TOPSIS
Reprinted from: *Mathematics* 2022, 10, 4041, https://doi.org/10.3390/math10214041 4

Malinka Ivanova and Mariana Durcheva
M-Polar Fuzzy Graphs and Deep Learning for the Design of Analog Amplifiers
Reprinted from: *Mathematics* 2023, 11, 1001, https://doi.org/10.3390/math11041001 21

Mohammad Hijji, Hikmat Yar, Fath U Min Ullah, Mohammed M. Alwakeel, Rafika Harrabi, Fahad Aradah, et al.
FADS: An Intelligent Fatigue and Age Detection System
Reprinted from: *Mathematics* 2023, 11, 1174, https://doi.org/10.3390/math11051174 37

Eugenia I. Toki, Giorgos Tatsis, Vasileios A. Tatsis, Konstantinos Plachouras, Jenny Pange and Ioannis G. Tsoulos
Applying Neural Networks on Biometric Datasets for Screening Speech and Language Deficiencies in Child Communication
Reprinted from: *Mathematics* 2023, 11, 1643, https://doi.org/10.3390/math11071643 53

Francisco Rodrigues Lima-Junior, Mery Ellen Brandt de Oliveira and Carlos Henrique Lopes Resende
An Overview of Applications of Hesitant Fuzzy Linguistic Term Sets in Supply Chain Management: The State of the Art and Future Directions
Reprinted from: *Mathematics* 2023, 11, 2814, https://doi.org/10.3390/math11132814 68

Saleem Abdullah, Alaa O. Almagrabi and Nawab Ali
A New Method for Commercial-Scale Water Purification Selection Using Linguistic Neural Networks
Reprinted from: *Mathematics* 2023, 11, 2972, https://doi.org/10.3390/math11132972 108

Simona Hašková, Petr Šuleř and Róbert Kuchár
A Fuzzy Multi-Criteria Evaluation System for Share Price Prediction: A Tesla Case Study
Reprinted from: *Mathematics* 2023, 11, 3033, https://doi.org/10.3390/math11133033 128

Predrag S. Stanimirović, Miroslav Ćirić, Spyridon D. Mourtas, Pavle Brzaković and Darjan Karabašević
Simulations and Bisimulations between Weighted Finite Automata Based on Time-Varying Models over Real Numbers
Reprinted from: *Mathematics* 2024, 12, 2110, https://doi.org/10.3390/math12132110 145

Gabriel Marín Díaz, Raquel Gómez Medina and José Alberto Aijón Jiménez
Integrating Fuzzy C-Means Clustering and Explainable AI for Robust Galaxy Classification
Reprinted from: *Mathematics* 2024, 12, 2797, https://doi.org/10.3390/math12182797 172

Nikolay Hinov and Bogdan Gilev
Neural Network-Based Design of a Buck Zero-Voltage-Switching Quasi-Resonant DC–DC Converter
Reprinted from: *Mathematics* **2024**, *12*, 3305, https://doi.org/10.3390/math12213305 . . . . **200**

About the Editor

Francisco Rodrigues Lima-Junior

Francisco Rodrigues Lima-Junior is a professor in the Department of Management and Economics and the Graduate Program in Administration at the Federal University of Technology – Paraná (UTFPR). He is also a research productivity fellow with the National Council for Technological Development (CNPq), Brazil. He has authored numerous works on fuzzy logic and neural networks, published in high-impact journals, with over 2900 citations on Google Scholar and 1200 on the Web of Science. He serves as a Guest Editor for the journals *Mathematics* and *Symmetry* (MDPI) and a reviewer for journals from prominent publishing groups such as Elsevier, Springer, IEEE, and Taylor & Francis. He coordinates the research group "Decision Making in Operations Management" at UTFPR and is a member of the "Production Performance Management" research group at the University of São Paulo (USP).

Editorial

Advances in Fuzzy Logic and Artificial Neural Networks

Francisco Rodrigues Lima-Junior

Postgraduate Program in Administration, Federal Technological University of Paraná, Curitiba 80230-901, Brazil; frjunior@utfpr.edu.br

1. Introduction

Fuzzy logic and artificial neural networks are among the most prominent AI approaches, recognized for their importance across various domains. Since its introduction by Professor Lotfi Zadeh in 1965, fuzzy logic has been the subject of numerous theoretical studies and practical applications. Over the years, several extensions of fuzzy logic have been developed to better model different uncertainty phenomena. Consequently, fuzzy logic has proven to be a robust framework for handling subjective, imprecise, ambiguous, or incomplete information. It finds frequent use in control systems, decision-making processes, expert systems, and recommendation engines. Despite the extensive body of research, fuzzy logic remains a fertile and relevant study area that attracts significant academic interest.

Similarly, artificial neural networks have evolved remarkably since their inception in the mid-20th century. Advances in network architectures, training algorithms, development tools, and computational power have significantly expanded their application. Artificial neural networks excel due to their learning capabilities, memory, fault tolerance, and distributed processing. These advantages explain their widespread use in tasks such as value prediction, pattern recognition, clustering, anomaly detection, natural language processing, and image generation.

Given these AI techniques' vast applicability and importance, I am honored to introduce this Special Issue in *Mathematics* on "Advances in Fuzzy Logic and Artificial Neural Networks." This Special Issue features 10 papers selected through a rigorous blind review process. In total, 34 authors from 16 countries contributed to these publications. The distribution of countries of the authors is illustrated in Figure 1, derived from Table 1. Notably, the majority of contributors are from Greece and Brazil.

Citation: Lima-Junior, F.R. Advances in Fuzzy Logic and Artificial Neural Networks. *Mathematics* **2024**, *12*, 3949. https://doi.org/10.3390/math12243949

Received: 11 December 2024
Revised: 13 December 2024
Accepted: 13 December 2024
Published: 16 December 2024

Copyright: © 2024 by the author. Licensee MDPI, Basel, Switzerland. This article is an open access article distributed under the terms and conditions of the Creative Commons Attribution (CC BY) license (https://creativecommons.org/licenses/by/4.0/).

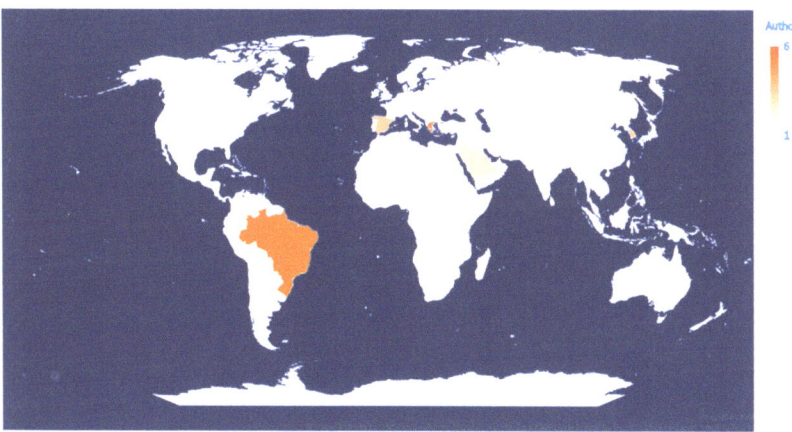

Figure 1. Map highlighting the countries of the authors of the articles published in this Special Issue.

Table 1. Countries of the authors of the articles published in this Special Issue.

Country	Count
Greece	6
Brazil	5
Spain	3
The Republic of Korea	3
The Czech Republic	3
Serbia	2
Pakistan	2
Saudi Arabia	2
Portugal	1
Russia	1
Norway	1
Bulgaria	1
Israel	1
Chile	1
Germany	1
The United Kingdom	1
Total	34

2. Brief Overview of the Contributions to the Special Issue

This Special Issue combines groundbreaking research that explores AI approaches across diverse contexts, showcasing interdisciplinary solutions to complex challenges.

In "Integrating Fuzzy C-Means Clustering and Explainable AI for Robust Galaxy Classification", the authors combine fuzzy C-means clustering with explainable AI methods, such as SHAP and LIME, to enhance the accuracy of galaxy classification while addressing data uncertainty from the Galaxy Zoo project. This method not only improves classification performance but also suggests potential applications for environmental management. Similarly, in "Simulations and Bisimulations between Weighted Finite Automata Based on Time-Varying Models over Real Numbers", the use of dynamic neural systems to solve simulation and bisimulation problems in weighted finite automata, coupled with advanced mathematical techniques, highlights the evolving role of AI in tackling complex problems.

The application of fuzzy logic extends to financial forecasting in "A Fuzzy Multi-Criteria Evaluation System for Share Price Prediction", where fuzzy systems outperform neural networks in predicting Tesla stock trends. The article "Analysis of the Level of Adoption of Business Continuity Practices by Brazilian Industries" applies Fuzzy TOPSIS to assess business continuity practices, providing insights into the resilience of companies across various sectors. In addition, "An Overview of Applications of Hesitant Fuzzy Linguistic Term Sets in Supply Chain Management" offers a comprehensive review of HFLTS techniques, identifying emerging trends and opportunities in supply chain optimization.

This Special Issue also includes innovative applications of neural networks, such as "Neural Network-Based Design of a Buck Zero-Voltage-Switching Quasi-Resonant DC–DC Converter", which focuses on energy-efficient converter design. In "M-Polar Fuzzy Graphs and Deep Learning for the Design of Analog Amplifiers", a hybrid methodology combining deep learning and fuzzy graphs is presented to optimize analog amplifier design.

Healthcare applications are also a key focus. "Applying Neural Networks on Biometric Datasets for Screening Speech and Language Deficiencies in Child Communication" utilizes neural networks to identify speech deficiencies in children, aiding in early clinical diagnoses. "A New Method for Commercial-Scale Water Purification Selection Using Linguistic Neural Networks" proposes a hierarchical linguistic neural network for selecting the most effective water purification methods. Lastly, "FADS: An Intelligent Fatigue and Age Detection System" introduces an AI-powered system to monitor fatigue and age, with potential applications in healthcare monitoring.

This compilation underscores the transformative potential of fuzzy logic and neural networks, driving advances across diverse domains such as science, industry, and healthcare. These studies highlight AI's versatility and open new avenues for research and practical application in addressing some of today's most pressing challenges.

Conflicts of Interest: The author declares no conflicts of interest.

List of Contributions

1. Díaz, G.M.; Medina, R.G.; Jiménez, J.A.A. Integrating fuzzy C-means clustering and explainable AI for robust galaxy classification. *Mathematics* **2024**, *12*, 2797. https://doi.org/10.3390/math12182797.
2. Stanimirović, P.S.; Ćirić, M.; Mourtas, S.D.; Brzaković, P.; Karabašević, D. Simulations and bisimulations between weighted finite automata based on time-varying models over real numbers. *Mathematics* **2024**, *12*, 2110. https://doi.org/10.3390/math12132110.
3. Hašková, S.; Šuleř, P.; Kuchár, R. A fuzzy multi-criteria evaluation system for share price prediction: A Tesla case study. *Mathematics* **2023**, *11*, 3033. https://doi.org/10.3390/math11133033.
4. Bobel, V.A.d.O.; Sigahi, T.F.A.C.; Rampasso, I.S.; Marcondes de Moraes, G.H.S.; Ávila, L.V.; Filho, W.L.; Anholon, R. Analysis of the level of adoption of business continuity practices by Brazilian industries: An exploratory study using fuzzy TOPSIS. *Mathematics* **2022**, *10*, 4041. https://doi.org/10.3390/math10214041.
5. Lima-Junior, F.R.; de Oliveira, M.E.B.; Resende, C.H.L. An overview of applications of hesitant fuzzy linguistic term sets in supply chain management: The state of the art and future directions. *Mathematics* **2023**, *11*, 2814. https://doi.org/10.3390/math11132814.
6. Hinov, N.; Gilev, B. Neural network-based design of a buck zero-voltage-switching quasi-resonant DC–DC converter. *Mathematics* **2024**, *12*, 3305. https://doi.org/10.3390/math12213305.
7. Ivanova, M.; Durcheva, M. M-polar fuzzy graphs and deep learning for the design of analog amplifiers. *Mathematics* **2023**, *11*, 1001. https://doi.org/10.3390/math11041001.
8. Toki, E.I.; Tatsis, G.; Tatsis, V.A.; Plachouras, K.; Pange, J.; Tsoulos, I.G. Applying neural networks on biometric datasets for screening speech and language deficiencies in child communication. *Mathematics* **2023**, *11*, 1643. https://doi.org/10.3390/math11071643.
9. Abdullah, S.; Almagrabi, A.O.; Ali, N. A new method for commercial-scale water purification selection using linguistic neural networks. *Mathematics* **2023**, *11*, 2972. https://doi.org/10.3390/math11132972.
10. Hijji, M.; Yar, H.; Ullah, F.U.M.; Alwakeel, M.M.; Harrabi, R.; Aradah, F.; Cheikh, F.A.; Muhammad, K.; Sajjad, M. FADS: An intelligent fatigue and age detection system. *Mathematics* **2023**, *11*, 1174. https://doi.org/10.3390/math11051174.

Short Biography of Author

Francisco Rodrigues Lima Junior is a professor in the Department of Management and Economics and the Graduate Program in Administration at the Federal University of Technology—Paraná (UTFPR). He is also a research productivity fellow with the National Council for Technological Development (CNPq), Brazil. He has authored numerous works on fuzzy logic and neural networks, published in high-impact journals, with over 2900 citations on Google Scholar and 1200 on the Web of Science. He serves as a guest editor for the journals Mathematics and Symmetry (MDPI) and a reviewer for journals from prominent publishing groups such as Elsevier, Springer, IEEE, and Taylor & Francis. He coordinates the research group "Decision Making in Operations Management" at UTFPR and is a member of the "Production Performance Management" research group at the University of São Paulo (USP).

Disclaimer/Publisher's Note: The statements, opinions and data contained in all publications are solely those of the individual author(s) and contributor(s) and not of MDPI and/or the editor(s). MDPI and/or the editor(s) disclaim responsibility for any injury to people or property resulting from any ideas, methods, instructions or products referred to in the content.

Article

Analysis of the Level of Adoption of Business Continuity Practices by Brazilian Industries: An Exploratory Study Using Fuzzy TOPSIS

Vitor Amado de Oliveira Bobel [1], Tiago F. A. C. Sigahi [1,*], Izabela Simon Rampasso [2], Gustavo Hermínio Salati Marcondes de Moraes [3,4], Lucas Veiga Ávila [5], Walter Leal Filho [6,7] and Rosley Anholon [1]

[1] School of Mechanical Engineering, State University of Campinas, Campinas 13083-872, Brazil
[2] Departamento de Ingeniería Industrial, Universidad Católica del Norte, Antofagasta 1270755, Chile
[3] School of Applied Sciences, State University of Campinas, Limeira 13484-350, Brazil
[4] School of Management Sciences, North-West University, Vanderbijlpark 1174, South Africa
[5] Center for Social Sciences and Humanities, Federal University of Santa Maria, Santa Maria 97105-900, Brazil
[6] Faculty of Life Sciences, Hamburg University of Applied Sciences, 20099 Hamburg, Germany
[7] School of Science and the Environment, Manchester Metropolitan University, Manchester M15 6BH, UK
* Correspondence: sigahi@unicamp.br

Abstract: The COVID-19 outbreak caused several negative effects in industries of all sizes and in all parts of the world, leading academic and practitioners to ask whether organizations could have been better prepared to face disruptive situations. This paper aims to analyze business continuity practices performed by Brazilian industries. A survey was conducted with academics who work in the field of organizational resilience and business continuity and are familiar with the reality of Brazilian companies in the industrial sector. The participants assessed 16 practices (P) proposed by the ISO 22301:2020, considering two categories: large industries (LI) and small and medium-sized industries (SMI). Data analysis was performed using Hierarchical Cluster Analysis, frequency analysis, Fuzzy TOPSIS and sensitivity analysis. For LIs, P4 (leaders conduct periodic critical analyses of practices) was considered the practice with the best application rate, while for SMIs, P2 (understand stakeholders' needs and expectations, and use information in business continuity management) was chosen. In all scenarios tested for LIs and SMIs, P8 (well-structured systematic processes to analyze the impact of abnormal situations on their business and the potential risks of a disruption) and P16 (periodic audits of their business continuity management activities to identify opportunities for improvement, and information record) are in the bottom quartile. When compared to LIs in the Brazilian context, SMIs exhibit more profound deficiencies in terms of applying business continuity practices. The findings of this study can be of great value to assist managers in improving organizational resilience. Organizations should be better prepared to face future disruptive events, whether biological, social, technological, or economic.

Keywords: business continuity; organizational resilience; management system standards; Fuzzy TOPSIS; ISO 22301; Brazil

MSC: 03B52; 03E72

1. Introduction

The COVID-19 pandemic has disrupted not only people's daily lives, but also the global economic system [1,2], highlighting the need for companies to pay greater attention to business continuity management [3–5]; that is, to their capacity "to continue the delivery of products and services within acceptable time frames at predefined capacity during a disruption" [6]. Such macroeconomic impacts have implied that businesses had to adapt to a new reality, dealing with revenue losses, demand fluctuations, lockdowns, organizational changes, and a variety of other restrictive situations and conditions imposed on their activities as a result of the economic losses and health crisis [6–8].

Examples of how the COVID-19 pandemic has negatively impacted various economic sectors abound. The aviation sector was one of the most affected with losses in aviation-supported jobs (reduction from 46 million to 41.7 million) and direct aviation jobs (43% reduction), including airlines, airports, manufacturers and air traffic management [9–11]. Another sector directly impacted by the pandemic was healthcare, where organizations faced constant challenges due to a lack of qualified professionals, inputs, resources, working conditions, and growing demand [12]. Although these effects have become more visible, many other industries can be mentioned such as manufacturing [13], energy [14], agriculture [15] and food [16], among others.

The pandemic's effects on businesses had a number of consequences for both large and small industries, entrepreneurs and employees. The study conducted by Gautam [17], for example, revealed that more than 20 million informal sector workers in Bangladesh were unable to carry out their activities due to the implications caused by the pandemic, triggering significant social and economic disruptions in this country. Even in developed countries such as the United States (US), 76.89% of entrepreneurs and business partners claimed that the pandemic had an impact on their business, with 31.93% forced to close temporarily and 4.20% permanently [9]. Large enterprises such as Brooks Brothers and Virgin Atlantic went bankrupt due to COVID-19 [17,18]. It was also noted that family businesses and startups were especially affected by the pandemic [19,20]. Studies conducted with 5800 small businesses in the US showed a reduction of about 40% of jobs, leading thousands of them to drastic cost cutting, additional bank loans or bankruptcy [21–23].

Despite the fact that government responses to the pandemic varied by country, causing more severe effects in some regions, Collins et al. [24] asserted that years of organizational optimization in industries were lost, and critical systems involved in the production of goods and services revealed a lack of resilience, redundancy, investments in diversification, and adaptive capacity. This has disrupted supply chains, which are overly interdependent in today's globalized world [24,25]. Because countries dealt with the pandemic inequitably, the effects and failures in supply chains were felt throughout the crisis; thus, even if a country has achieved positive results in combating COVID-19, shortages of certain goods and services will still be felt if the pandemic continues to affect economic activities in different parts of the world [26,27].

All of these negative effects felt by industries of all sizes and in all parts of the world led academics and practitioners to ask whether organizations could have been better prepared to face disruptive situations by implementing business continuity and resilience concepts and approaches [4,28]. The debate on this topic has gained traction, and studies have revealed that there is still enormous room for improvement and application of business continuity concepts and practices, and that, in general, organizations and managers around the world have neglected them, which could have mitigated the impacts of the pandemic [29,30].

Given the context presented, ISO 22301:2020 [6] is characterized as an important, globally recognized standard that provides security and organizational resilience guidelines, with the goal of establishing a management system that enables business continuity following a disruption. In this exploratory study, ISO 22301:2020 is used as a framework analysis with the purpose of examining the level of adoption of business continuity principles by Brazilian industries. This study is important for the dissemination of knowledge about this important management tool, as well as for the benefit of managers who can improve organizational resilience; organizations that can be better prepared for future disruptive events; and countries whose economic development is highly dependent on the continuity of their businesses.

The remainder of this paper is organized as follows. Section 2 provides the background on security, resilience and business continuity management systems (BCMS) considering the structure of the ISO 22301:2020. In addition, business continuity management is discussed in relation to firm size. Section 3 presents the methods, including the structuring of the survey, Hierarchical Cluster Analysis (HCA), frequency analysis, Fuzzy TOPSIS (Technique for Order Preference by Similarity to Ideal Solution) and sensitivity analysis. Section 4 contains discussions that take into account the results obtained for large, medium and small industries in an integrated manner. Finally, Section 5 presents the conclusions, limitations and suggestions for future studies.

2. Background

2.1. Security and Resilience: BCMS from the Perspective of the ISO 22301:2020

The ISO 22301:2020 standard defines the structure, best practices, and requirements for implementing, maintaining, and improving a BCMS. This management system guides organizations in developing the requirements for supporting in disruption contexts, while also assessing whether the impacts are acceptable or not; in addition, they contribute to reducing the likelihood of negative impacts occurring or making recovery possible when they occur [6].

According to ISO [6], when implementing a BCMS it is important that organizations comprehend the external and internal issues that can contribute to disruptive situations. These issues may differ depending on the organization's goals, products, services, and the degree and types of risk it is willing to take. As a result, when establishing a BCMS, organizations need to identify the relevant stakeholders and define the requirements to be applied in the interaction with them.

The first requirements described in ISO 22301:2020 for the implementation of a BCMS are legal and regulatory obligations, for which a continuous process to identify them should be put in place, with the goal of ensuring their products and/or services are law-compliant. In addition, when establishing a BCMS, the standard recommends that its scope be defined in accordance with its missions, goals, and obligations, as well as the parts of the organization in which this management system must be applied [6].

ISO 22301:2020 emphasizes that organizations' top management must demonstrate engagement and leadership in the implementation of the BCMS, establishing policies to ensure commitment to system requirements and providing resources to achieve the defined goals [6]. Thus, when planning the implementation of the BCMS, the organization must identify the risks and opportunities that should be addressed based on its scope of application in order to prevent and reduce unwanted effects and ensure that the intended results are achieved. Furthermore, actions to assess system effectiveness must be defined, including what will be done, what resources will be required, when the actions will be completed, and how the results will be evaluated [6].

According to ISO 22301:2020, it is important that the organization determine what skills, training, and experiences are required for employees to perform the activities associated with business continuity. It is also critical to educate employees on business continuity policies and the BCMS principles [6].

Finally, ISO 22301:2020 recommends that organizations implement and maintain systematic processes for analyzing the impact on business and assessing the risks of disruption to operations. These analyses must be conducted on a regular basis in order to identify and choose business continuity strategies that account for solutions before, during and after a disruption or crisis. Thus, these strategies must comprise one or more actions that meet the requirements to continue, recover and protect the organization's activities or that reduce the probability and time of disruption, while constantly reassessing the BCMS parameters to achieve continuous improvement [6].

2.2. Business Continuity Management and Firm Size

The COVID-19 pandemic had a wide-ranging impact on businesses of all sizes, sectors and activities. As discussed by Anholon et al. [8], Margherita and Heikkilä [4] and Graham and Loke [31], the lack of structured business continuity management practices and systems at the beginning of this crisis exacerbated these impacts, emphasizing the importance of better preparation in the face of situations characterized by high uncertainty, complexity, and potentially disruptive disruption. Despite the absence of well-established BCMS, organizations were forced to reinvent themselves and consequently apply business continuity management concepts, even if instinctively and remedially, in order to survive the challenges posed by the pandemic [6,32,33].

Kraus et al. [23] investigated how small family businesses in five European countries (Austria, Germany, Italy, Liechtenstein and Switzerland) dealt with the pandemic and what risk mitigation strategies were adopted. As an example, a German appliance manufacturer reduced its own production and working hours to minimum levels in order to deal with logistical problems and low production due to the decrease in the availability of components, thus prioritizing its survival. Another example was an Austrian dairy company, which focused its resources on production to meet the increased demand for cheese during the pandemic, while implementing sanitary measures and reorganizing its production lines, despite potential losses in other product or service markets. These authors conclude that typical characteristics of family business models contributed to the survival of the companies studied, particularly the emphasis on long-term survival over immediate results and shareholder interests [23]. Other studies on small and family businesses support the importance of business continuity and risk mitigation strategies for survival in times of crisis [34–36].

Regarding the business continuity strategies used by large and medium-sized industries, authors such as Margherita and Heikkilä [4] and Papadopoulos et al. [33] stated that, in general, they relied on innovation and the use of new technologies, both to establish new forms of work (e.g., home office) and to develop new products and services from the application mainly of artificial intelligence and data science. The application of new technologies by large companies has also extended to the development of new supply chain management strategies [37,38].

Margherita and Heikkilä [4] have structured the business resilience responses and measures of various large and medium-sized industries into five categories: Operations and Value Systems, Customer Experience and Support, Human Resources and Workforce, Leadership and Change Management, and Community and Social Engagement. The main measures in the category of 'Operations and Value Systems' were directed at dealing with supply chain and logistics issues, with a focus on improving connectivity and digital integration not only within the company but also with suppliers and key business partners, as well as actions to prioritize essential inputs and restructure the business model. The Customer Experience and Support actions focused on rearranging store and office space, reducing human contact in the purchasing and sales processes, and increasing investment in digital marketing. Measures to make work more flexible, adapt the work environment and implement remote work were included in the category Human Resources and Workforce. Finally, Leadership and Change Management actions focused on creating positive scenarios to maintain customer and stakeholder trust in the organization, whereas Community and Social Engagement actions included campaigns, donations and investments in COVID-19 mitigation [4].

It is important to note that the literature concurs that logistical disruption was one of the most common problems faced by industries worldwide during the pandemic [4,22,32]. The reasons for this, according to Sharma et al. [7], are the high interdependence of supply chains in a globalized world and the low capacity for resilience and organizational flexibility. These authors mentioned Samsung as an example of a company that had diversified its industrial parks in several countries and, as a result, was able to more effectively cope with production interruptions and the differences of each country in the COVID-19 confrontation, adjusting production to the constraints and determinations of each location [7].

Overall, the pandemic has made researchers, managers and businesses all over the world aware of the importance of BCMS. Even when business continuity concepts and practices were used in an unplanned and unstructured manner, and they proved to be critical for survival and recovery during the pandemic. In a context of growing complexity and uncertainty [39], it is certain that new disruptive events will emerge, whether biological,

3. Materials and Methods

3.1. Survey Structuring and Research Instrument Development

With the purpose of examining the level of adoption of business continuity principles by Brazilian industries, a survey was conducted with academics who work in the field of organizational resilience and business continuity and are familiar with the reality of Brazilian companies in the industrial sector. It is important to mention that this study was approved by the Research Ethics Committee of the university (CAEE: 50579021.8.0000.5404).

The survey addressed 16 practices related to the BCMS proposed by ISO 56002:2020 (Table 1).

social, technological, or economic [40]. In this sense, regardless of the geographical region, economic sector, or size of the company, advancing knowledge about the structuring and implementation of BCMS is required [7,30,32].

Table 1. The business continuity management practices that comprised the survey.

Code	Description
P1	Organizations analyze internal and external issues that may jeopardize business continuity management results on a regular basis
P2	Organizations seek to understand stakeholders' needs and expectations, and then use that information in activities and decisions related to business continuity management
P3	Organizations define the clear scope related to the activities to be performed within the business continuity management
P4	Organizations' top management conduct periodic critical analyses in relation to activities associated with business continuity management and ensure the necessary resources so that they occur in the best possible way
P5	Organizations define the roles and responsibilities associated with business continuity management correctly and ensure that employees understand them
P6	Organizations have employees qualified to work on business continuity management activities; to that end, they must strive to develop such competencies through education, training, and/or experience
P7	Organizations define the goals associated with business continuity management in a consistent, measurable manner and communicate them to all parties involved; they also develop strategies for achieving those goals, including what will be done, what resources will be needed, who will be involved, how long the plan will be, and how the results will be analyzed
P8	Organizations have well-structured systematic processes in place to analyze the impact of abnormal situations on their business and the potential risks of a disruption, including a pre-defined maximum period during which non-resumption of activities will become unacceptable and the minimum capacity required for recovery
P9	Organizations have structured risk assessment and intervention processes in place to help them make decisions concerning business continuity management
P10	Organizations have well-structured documentation control systems to support business continuity management
P11	Organizations define strategies and solutions for business continuity in a way that is appropriate to their reality, taking into account size and financial capacity, protecting priority activities, reducing the likelihood of disruption, and ensuring the provision of necessary resources in disruptive situations
P12	Concerning operational plans to manage the organization during disruption, organizations define the purpose, scope, objectives, roles and responsibilities of employees, as well as the operation and coordination of each team, allowing for a structured and well-defined response during disruption

Table 1. Cont.

Code	Description
P13	Organizations have structured plans to assess the appropriate moment for the resumption of their operations
P14	Organizations conduct periodic tests to assess efficiency and validate their business continuity strategies and solutions, reflecting on existing improvement opportunities; they also assess the ability of suppliers and partners to maintain operations during disruptions, as well as compliance with legal and regulatory requirements and industry best practices
P15	Organizations are clear about how they measure business continuity management performance, taking into account key factors such as relevant indicators, monitoring frequency and responsible employees
P16	Organizations conduct periodic audits of their business continuity management activities to identify opportunities for improvement; they also record information in any format so that it can assist decision making and be used as a lesson learned in the future

Source: Elaborated by the authors based on ISO 22301:2020 [6].

Respondents were asked about the adoption of each practice considering two categories: large industries (LI) and small and medium-sized industries (SMI). Respondents should rate each practice for each category using the following scale: Not applied (NA); Applied superficially (AS); Applied reasonably (AR); Applied properly (AP); or Applied in a well-structured way (AW).

The first section of the survey consisted of questions aiming at characterizing the sample, including the respondent's research area, whether he/she conducted master/doctoral student supervision activities in the area of organizational resilience and business continuity management, and academic experience in the field. In the second part of the survey, the 16 practices elaborated based on the ISO 22301:2020 were presented, and the respondents assessed their level of adoption by Brazilian LIs and SMIs.

Considering the research purpose, the recommendations of Apostolopoulos and Liargovas [41] were followed for sampling, in which a non-probabilistic and judgmental procedure was adopted to select participants with conceptual and practical knowledge qualified to participate in the research. The respondents were selected based on the analysis of their curriculums registered on the main professional platform for researchers in Brazil, i.e., the Lattes platform, which is validated by the National Council for Scientific and Technological Development. Only those with experience in the field were invited to participate in the survey, based on their professional background.

3.2. Data Analysis

The data analysis was performed using Hierarchical Cluster Analysis (HCA), frequency analysis, Fuzzy TOPSIS (Technique for Order Preference by Similarity to Ideal Solution), and sensitivity analysis.

The HCA enabled classifying data into groups that are most similar to each other. According to Nielsen [42], the HCA works by progressively grouping the data in order to obtain a class in which the data can be grouped based on their similarity at each step of the algorithm. A binary tree of clusters or dendrogram is generated as a result of this calculation, with its "root" containing all of the data to be treated and each partition allowing for a new classification, with the option to truncate this process at each new partition iteration. In this study, the HCA was used to ascertain how respondents were classified based on their educational level, experience, and knowledge on organizational resilience and business continuity management. The percentage indicated by respondents for each of the 16 business continuity management practices evaluated in each category (i.e., LI and SMI) was analyzed using frequency analysis.

The Fuzzy TOPSIS application was based on the adapted version proposed by Chen [43], which is a method widely used in a variety of academic researches [44]. The Fuzzy TOPSIS is performed by combining the TOPSIS developed by Hwang and Yoon [45], which was initially used to support multi-criteria decision making [46,47], and fuzzy logic, through which Chen [43] proposed the application of fuzzy numbers for representing linguistic variables. The use of fuzzy logic in conjunction with TOPSIS allowed for the consideration of uncertainties presented in the answers of the respondents and their classification, as well as the generation of a ranking of the analyzed business continuity practices.

In this study, the Fuzzy TOPSIS was used to rank the business continuity practices presented in Table 1 based on expert opinions about the level of adoption in Brazilian LIs and SMIs. According to Chen's [43] methodological procedures, the practices served as alternatives and the respondents as criteria with weights based on their educational background, professional experience, and subject-matter expertise. As in Chen's [43] application, this study used the triangular fuzzy numbers. As explained by Pedrycz [48] and Klir and Yuan [49], despite its simplicity, it can be useful in cases where variations in shape have little impact on the analysis.

The fuzzy version of (a) the scales used (Figure 1a) and the levels for grouping respondents (Figure 1b) are shown in Figure 1.

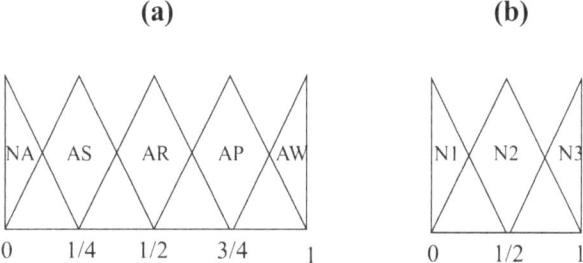

NA [0.0; 0.0; 1/4]   AS [0.0; 1/4; 1/2]   AR [1/4; 1/2; 3/4]   AP [1/2; 3/4; 1.0]   AW [3/4; 1.0; 1.0]

N1 [0.0; 0.0; 1/2]   N2 [0.0; 1/2; 1.0]   N3 [1/2; 1.0; 1.0]

Figure 1. (a) Fuzzy values of the evaluation scale of practices; (b) Fuzzy values of the levels for allocation of respondents based on education, experience, and knowledge.

Following Chen [43], based on the fuzzy numbers the responses were organized in a matrix $\tilde{G}$ (1) containing the scores in their fuzzy triangular form. In the sequence, the vector $\tilde{E}$ (2) was formed, where w_j is the respondent's level (based on education, experience, and knowledge) in its fuzzy triangular form, representing the fuzzy weights of the respondents.

$$\tilde{G} = \begin{bmatrix} \tilde{x}_{11} & \cdots & \tilde{x}_{1m} \\ \vdots & \ddots & \vdots \\ \tilde{x}_{n1} & \cdots & \tilde{x}_{nm} \end{bmatrix}; \tilde{x}_{ij} = [a_{ij}, \ b_{ij}, \ c_{ij}] \qquad (1)$$

$$\tilde{E} = [\tilde{w}_1, \tilde{w}_2, \ldots \tilde{w}_n]; \tilde{w}_j = [w_1, w_2, w_3] \qquad (2)$$

In the next step, the matrix $\tilde{G}$ was normalized based on the highest score value, obtaining the matrix $\mathcal{R}$ (3) as follows:

$$\mathcal{R} = [\tilde{r}_{ij}]_{mxn}; \tilde{r}_{ij} = \left[\frac{a_{ij}}{C_j^*}, \frac{b_{ij}}{C_j^*}, \frac{c_{ij}}{C_j^*}\right] \to C_j^* = \max(i)c_{ij} \qquad (3)$$

The matrix $\mathcal{R}$ was then weighted by the vector $\widetilde{E}$, generating the matrix υ (4):

$$\upsilon = [\widetilde{v}_{ij}]_{mxn} \to i = 1, 2, \ldots, m; j = 1, 2, \ldots, n \to \widetilde{v}_{ij} = \widetilde{r}_{ij}\, (.)\, \widetilde{w}_j \qquad (4)$$

Next, the Positive Ideal Solution (unit vector) (5) and Negative Ideal Solution (null vector) (6) were used to calculate the distances $d(\widetilde{a}, \widetilde{b})$ (7) related to each element of the matrix υ using the following equations:

$$A^* = [\widetilde{v}_1^*, \widetilde{v}_2^*, \widetilde{v}_3^*]\, ,\ \text{where}\ \widetilde{v}_j^* = [1,1,1] \qquad (5)$$

$$A^- = [\widetilde{v}_1^-, \widetilde{v}_2^-, \widetilde{v}_3^-]\, ,\ \text{where}\ \widetilde{v}_j^* = [0,0,0] \qquad (6)$$

$$d\left(\widetilde{a}, \widetilde{b}\right) = \sqrt{\frac{1}{3}\left[(a_1-b_1)^2 + (a_2-b_2)^2 + (a_3-b_3)^2\right]} \qquad (7)$$

The total positive (d_i^*) (8) and negative (d_i^-) (9) distances in relation to each alternative were obtained through the sum of the partial distances as follows:

$$d_i^* = \sum_{j=1}^{n} d\left(\widetilde{v}_{ij}, \widetilde{v}_j^*\right) \qquad (8)$$

$$d_i^- = \sum_{j=1}^{n} d\left(\widetilde{v}_{ij}, \widetilde{v}_j^-\right) \qquad (9)$$

Then, the last step of the method was calculating the proximity coefficient (CC_i) (10), which allowed the structuring of the ranking of the alternatives (practices).

$$CC_i = \frac{d_i^-}{d_i^* + d_i^-} \qquad (10)$$

Finally, in order to conduct the sensitivity analysis, various scenarios were examined, each one adjusting for the exclusion of a group of respondents as defined by the HCA and evaluating how each one influenced the ordering of the business continuity management practices.

4. Results and Discussion
4.1. Hierarchical Cluster Analysis

The following criteria were used to assess the respondents' educational level, experience and knowledge; years of experience in the field of innovation; conducting research directly related to innovation management; and teaching, training, and supervision activities for qualified human resources in the field. These criteria were applied to each of the 22 respondents, and the HCA allowed them to be classified into seven groups, as shown in the dendrogram (Figure 2).

The experts were then assigned to levels 1 (Groups 3 and 6), 2 (Groups 2, 4 and 5) and 3 (Group 1) (Table 2). This classification was based on the analysis of experts' characteristics in each group. The experts were graded as 1 or 2 based on their data about years of experience, research area in which they work and whether or not they supervise master's and doctorate students. For respondents with up to 15 years of experience, a grade of 1 was assigned, and a grade of 2 was assigned to those with more than 15 years of experience. For those who did not emphasize research directly related to business continuity, a grade of 1 was assigned, and a grade of 2 was assigned to those who did. In terms of graduate student supervision, those who did not perform it received a grade of 1, while those who did received a grade of 2. All of the N3 respondents were graded 2 in all the indicators. N2 respondents were graded 2 in two of the three indicators. N1 respondents presented at least two indicators with a grade of 1.

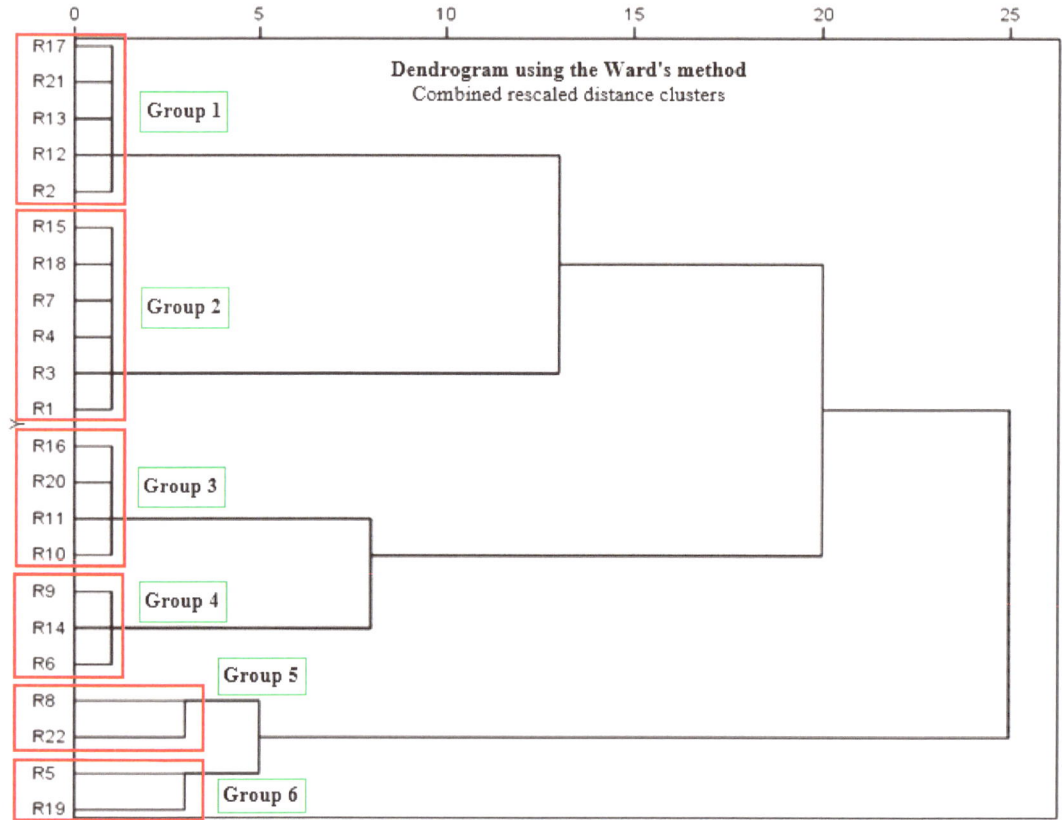

Figure 2. Dendrogram from the Hierarchical Cluster Analysis.

Table 2. Allocation of the respondents based on educational level, experience and knowledge.

Level	Respondents
N1	R5, R10, R11, R16, R19 and R20
N2	R1, R3, R4, R6, R7, R8, R9, R14, R15, R18 and R22
N3	R2, R12, R13, R17 and R21

4.2. Observed Frequency Analysis

For each of the 16 practices evaluated (P1 to P16), which illustrate essential topics of BCMS proposed by the ISO 22301:2020, the frequency of responses was calculated corresponding to the scale options: Not applied (NA); Applied superficially (AS); Applied reasonably (AR); Applied properly (AP); and Applied in a well-structured way (AW). This procedure was performed for the categories of large industries (LI) and small and medium-sized industries (SMI). Following an overview of data frequencies, specific considerations by level (N1, N2, N3) are presented.

4.2.1. Data Analysis for Large Industries (LI)

The global frequency of data referring to LIs is presented in Table 3.

Table 3. Global frequency on the application of each practice in LIs.

Practices	Level of Application *				
	NA	AS	AR	AP	AW
P1	0.000	0.000	0.318	0.500	0.182
P2	0.000	0.000	0.409	0.455	0.136
P3	0.000	0.000	0.227	0.591	0.182
P4	0.000	0.045	0.318	0.500	0.136
P5	0.000	0.091	0.409	0.409	0.091
P6	0.000	0.136	0.409	0.318	0.136
P7	0.000	0.091	0.318	0.545	0.045
P8	0.000	0.227	0.455	0.273	0.045
P9	0.000	0.136	0.455	0.364	0.045
P10	0.000	0.136	0.409	0.227	0.227
P11	0.000	0.136	0.227	0.500	0.136
P12	0.000	0.136	0.409	0.318	0.136
P13	0.000	0.182	0.455	0.318	0.045
P14	0.091	0.409	0.182	0.273	0.045
P15	0.000	0.273	0.318	0.364	0.045
P16	0.045	0.273	0.273	0.273	0.136

* Source: Not applied (NA); Applied superficially (AS); Applied reasonably (AR); Applied properly (AP); Applied in a well-structured way (AW).

In the scenario of LIs in Brazil, it is worth noting that none of the respondents selected the option Not applied (NA) for 14 out of the 16 practices presented (from P1 to P13, and P15). Furthermore, none of the respondents selected the option Applied superficially (AS) for P1 (analyze internal and external issues that may jeopardize business continuity management results on a regular basis), P2 (understand stakeholders' needs and expectations, and use information in business continuity management) and P3 (define a clear scope for BCMS), indicating that these three practices are at least reasonably applied in the Brazilian LIs according to the experts' perceptions.

With the exception of P14 (periodic tests to assess the business continuity strategies and the ability of suppliers and partners to maintain operations and comply with the law during disruptions), it is possible to notice that most of the answers were concentrated in the options 'Applied reasonably' (AR) and 'Applied properly' (AP). P14 was the worst rated practice at the two lowest adoption levels, i.e., not applied (9.1%) and applied superficially (40.9%).

It is worth noting that P1, P4 (leaders conduct periodic critical analyses of practices) and P11 (define strategies and solutions for business continuity taking into account their reality) reached 50% agreement in the "Applied properly" option. The only cases that showed more than 50% agreement in the responses were P3 and P7 (define the goals associated with the BCMS in a consistent, measurable manner and communicate them to all parties involved), considered properly applied by 59.1% and 54.5% of the participants, respectively.

Finally, P10 (organizations have well-structured documentation control systems to support business continuity management) was the only practice considered to be applied in a well-structured way by at least 20% of the participants.

4.2.2. Data Analysis for Small and Medium-Sized Industries (SMI)

The global frequency of data referring to SMIs is presented in Table 4.

In the scenario of SMIs in Brazil, none of the practices was considered to be applied in a well-structured way. Furthermore, 9 out of the 16 practices presented (P1, P7–10, P13–16) were not evaluated as properly applied by none of the respondents, and only P3 (define a clear scope for BCMS) reached more than 10% in this level of adoption.

Table 4. Global frequency on the application of each practice in SMIs.

Practices	Level of Application *				
	NA	AS	AR	AP	AW
P1	0.045	0.636	0.318	0.000	0.000
P2	0.045	0.682	0.227	0.045	0.000
P3	0.091	0.455	0.318	0.136	0.000
P4	0.182	0.545	0.227	0.045	0.000
P5	0.182	0.591	0.136	0.091	0.000
P6	0.273	0.409	0.273	0.045	0.000
P7	0.091	0.727	0.182	0.000	0.000
P8	0.364	0.545	0.091	0.000	0.000
P9	0.409	0.409	0.182	0.000	0.000
P10	0.409	0.364	0.227	0.000	0.000
P11	0.227	0.409	0.318	0.045	0.000
P12	0.182	0.545	0.227	0.045	0.000
P13	0.273	0.455	0.273	0.000	0.000
P14	0.455	0.409	0.136	0.000	0.000
P15	0.318	0.500	0.182	0.000	0.000
P16	0.591	0.273	0.136	0.000	0.000

* Source: Not applied (NA); Applied superficially (AS); Applied reasonably (AR); Applied properly (AP); Applied in a well-structured way (AW).

It is possible to observe that most of the answers were concentrated in the option Applied superficially (AS), with only three exceptions, that is P10 (organizations have well-structured documentation control systems to support business continuity management), P14 (periodic tests to assess the business continuity strategies and the ability of suppliers and partners to maintain operations and comply with the law during disruptions) and P16 (periodic audits of their business continuity management activities to identify opportunities for improvement, and information record), which presented the highest frequency in the lowest level of adoption (Not applied) with 40.9%, 45.5% and 59.1%, respectively.

The highest levels of agreement were observed for P1 (analyze internal and external issues that may jeopardize business continuity management results on a regular basis), P2 (understand stakeholders' needs and expectations, and use information in business continuity management) and P7 (define the goals associated with the BCMS in a consistent, measurable manner and communicate them to all parties involved), all on the option Applied superficially (AS), reaching 63.6%, 68.2% and 72.7%, respectively.

4.3. Comparative Ordering Analysis via Fuzzy TOPSIS

Following the procedures proposed by Chen [43] and described in Section 3.2, the proximity coefficients (CC_i) were calculated and based on them the practices were ordered considering LIs and SMIs. In addition, sensitivity analysis was used to verify the influence of each group of respondents in the ordering of practices.

4.3.1. Data Analysis for Large Industries (LI)

The results of the ordering of practices via Fuzzy TOPSIS and sensitivity analysis for LIs is presented in Table 5. The details for calculating the CC_i related to LIs for all groups are presented in Supplementary Materials (from Table S1 to Table S7).

Table 5. The ordering of practices via Fuzzy TOPSIS and sensitivity analysis for LIs.

Practices	CC_i	All Groups	Group Excluded for Sensitivity Analysis					
			G1	G2	G3	G4	G5	G6
P1	0.4365	2nd	2nd	2nd	2nd	2nd	2nd	2nd
P2	0.4228	4th	3rd	5th	3rd	4th	3rd	4th
P3	0.4453	1st	1st	1st	1st	1st	1st	1st
P4	0.4240	3rd	5th	3rd	4th	3rd	4th	3rd
P5	0.4009	7th	9th	6th	7th	7th	7th	7th
P6	0.3796	10th	6th	14th	12th	9th	11th	10th
P7	0.4145	5th	10th	4th	5th	5th	5th	5th
P8	0.3626	14th	13th	13th	15th	15th	15th	14th
P9	0.3773	11th	8th	12th	11th	12th	14th	11th
P10	0.3914	9th	11th	9th	8th	8th	8th	9th
P11	0.4087	6th	4th	7th	6th	6th	6th	6th
P12	0.3920	8th	7th	8th	9th	10th	9th	8th
P13	0.3712	13th	12th	11th	13th	14th	13th	13th
P14	0.3351	16th	16th	15th	16th	16th	16th	16th
P15	0.3749	12th	15th	10th	10th	11th	10th	12th
P16	0.3624	15th	14th	16th	14th	13th	12th	15th

For Brazilian LIs, the first place, that is, the business continuity management practice with the highest level of application in the perception of the specialists, was P3 (define a clear scope for BCMS), while the last place was P14 (periodic tests to assess the business continuity strategies and the ability of suppliers and partners to maintain operations and comply with the law during disruptions).

The sensitivity analysis reveals that P3 and P1 (analyze internal and external issues that may jeopardize business continuity management results on a regular basis) were ranked as first and second, respectively, in all scenarios, demonstrating a robustness in the evaluation of these practices in the scenario of the Brazilian LIs.

It is also worth noting that the last place in the ranking changes in the scenario in which G2 is removed from the calculation, causing P16 (periodic audits of their business continuity management activities to identify opportunities for improvement, and information record) to assume this position.

4.3.2. Data Analysis for Small and Medium-Sized Industries (SMI)

Finally, the results of the ordering of practices via Fuzzy TOPSIS and sensitivity analysis for SMIs is presented in Table 6. The details for calculating the CC_i related to SMIs for all groups are presented in Supplementary Materials (from Table S8 to Table S14).

When all groups of respondents are considered in the scenario of Brazilian SMIs, once more P3 stands out, being ranked in first place. The same occurs for P1, which takes second place. In turn, P16 appears last.

The sensitivity analysis reveals that none of the practices maintains the same position in all scenarios, with the only exception being P16, which was ranked last regardless of the group removed in the calculation. Considering the first position, it is worth noting that G2 is the only group that influences the classification of P3, which moves to second place, giving way to P1.

Table 6. The ordering of practices via Fuzzy TOPSIS and sensitivity analysis for SMIs.

Practices	CC_i	All Groups	Group Excluded for Sensitivity Analysis					
			G1	G2	G3	G4	G5	G6
P1	0.3660	2nd	2nd	1st	2nd	3rd	3rd	2nd
P2	0.3632	3rd	3rd	3rd	3rd	2nd	2nd	3rd
P3	0.3756	1st	1st	2nd	1st	1st	1st	1st
P4	0.3371	6th	7th	5th	7th	4th	5th	6th
P5	0.3405	4th	4th	6th	5th	5th	8th	4th
P6	0.3296	9th	10th	7th	9th	7th	9th	7th
P7	0.3364	7th	8th	4th	6th	9th	6th	8th
P8	0.2842	14th	12th	13th	14th	15th	14th	14th
P9	0.2762	15th	14th	14th	15th	14th	15th	15th
P10	0.2922	12th	13th	11th	12th	13th	12th	12th
P11	0.3324	8th	6th	9th	8th	6th	7th	9th
P12	0.3393	5th	5th	8th	4th	8th	4th	5th
P13	0.3223	10th	9th	10th	10th	10th	10th	10th
P14	0.2845	13th	15th	15th	13th	12th	13th	13th
P15	0.3042	11th	11th	12th	11th	11th	11th	11th
P16	0.2481	16th	16th	16th	16th	16th	16th	16th

4.4. Considerations on the Level of Expertise of Participants and Industry Categories

In the case of Brazilian LIs, when considering the allocation of the respondents by level based on educational level, experience and knowledge (N1, N2 and N3) generated by the HCA (see Section 4.1), it was observed that respondents allocated in N3 rarely indicated that a business continuity management practice is applied in a well-structured way. For SMIs, it was noted that the vast majority of responses for most of the practices examined were in the lower ranges of the scale for the three levels of respondents.

The consideration of the level of expertise of participants in the comparative ordering analysis via Fuzzy TOPSIS revealed that, for LIs, the main deficiencies are observed in P14, P16 and P8, whereas the more well-established are P3, P1 and P4; and for SMIs, the main deficiencies are observed in P16, P9 and P8, whereas the more well-established are P3, P1 e P2, although they are applied superficially or reasonably. These findings are summarized in Table 7.

Table 7. Most deficient and well-established practices comparatively defined for LIs and SMIs.

Categories	Business Continuity Practices	
	Most Deficient	Most Well-Established
LI	P14	P4
SMI	P9	P2
Both	P8 and P16	P3 and P1

In general, the findings suggest that Brazilian industries, regardless of their size, should pay special attention to business continuity practices related to P8 (well-structured systematic processes to analyze the impact of abnormal situations on their business and the potential risks of a disruption) and P16 (periodic audits of their business continuity management activities to identify opportunities for improvement, and information record). When the categories are differentiated, P14 (periodic tests to assess the business continuity strategies and the ability of suppliers and partners to maintain operations and comply with the law during disruptions) inspires greater attention for LIs and P9 (structured risk assessment and intervention processes to help them make decisions concerning business continuity management) for SMIs.

As documented in the literature [4,8,31], the COVID-19 pandemic evidenced that business continuity management practices are necessary for the survival of companies of all sizes worldwide. In this sense, both LIs and SMIs must improve risks analyses (P8) and audit processes (P16). To support them in this process, besides ISO 22301, other standards of ISO can be useful, more specifically, ISO 31000 [50], which establishes guidelines for risk management, and ISO 19011 [51], which provides guidance for companies to audit their management systems. Regarding the deficiencies, ISO 31000 can also be of great value particularly for SMIs (P9). For LIs (P14), in addition to ISO 22301, the literature presents some useful contributions. An example is the study of Sadeghi et al. [52], in which the authors verified the importance of collaboration and rewards to enhance business continuity management practices throughout the supply chain.

5. Conclusions

The aim of this study was to analyze the level of adoption of business continuity practices by Brazilian industries, based on the structure of the ISO 22301:2020 and the opinion of experts on the subject. For LIs, P4 (leaders conduct periodic critical analyses of practices) was considered the practice with the best application rate, while for SMIs, P2 (understand stakeholders' needs and expectations, and use information in business continuity management) was chosen. In all scenarios tested for LIs and SMIs, P8 (well-structured systematic processes to analyze the impact of abnormal situations on their business and the potential risks of a disruption) and P16 (periodic audits of their business continuity management activities to identify opportunities for improvement, and information record) are in the bottom quartile.

Despite the recent publication of ISO 22301:2020, the recommended business continuity management practices have had some adherence in the reality of Brazilian LIs, while SMIs remain in more deficient situations. In relation to the practical implications of the presented findings, they may be useful for Brazilian managers who want to develop a structured BCMS and/or consolidate existing practices in their organizations. The most deficient practices evidenced in this research should receive greater attention in the development of such management systems. In this regard, it is critical to highlight the systemic impact that improvements in organizational resilience capacity can have, because industry recovery is a critical factor in the economic development of all countries in a post-pandemic world.

Regarding the theoretical contribution of this study, the main novelty of it is the use of the combination of Hierarchical Cluster Analysis, frequency analysis, Fuzzy TOPSIS and sensitivity analysis to analyze the context of business continuity practices performed by Brazilian industries. These methodological procedures can be used by researchers to evaluate other realities and compare them to the results presented here. Furthermore, the findings of this paper indicate the main deficiencies of SMIs and LIs concerning the analyzed practices, and they can be used as the foundation for future studies aimed at developing guidelines for companies to overcome their weaknesses and become more resilient.

The limitations of the study are related to the specific context studied and the methods utilized. Additional research is needed to delve deeper into business continuity management practices in the Brazilian context, using qualitative methods and expanding the sample of consulted experts. Another avenue for research is to investigate other contexts, including both developing and developed countries. The findings of these studies can be compared, which can be particularly valuable to managers, as well as for the resilience of LIs and SMIs.

Finally, it is worth noting that ISO 22301:2020 serves as an important reference for industries seeking to implement organizational resilience and business continuity practices, and researchers can be excellent partners in better preparing organizations for the emergence of new disruptive events of any nature.

Supplementary Materials: The following supporting information can be downloaded at: https://www.mdpi.com/article/10.3390/math10214041/s1, Table S1: Matrix $\tilde{G}$ containing the scores assigned by respondents in the fuzzy triangular form for LIs; Table S2: Vector $\tilde{E}$ representing the fuzzy weights of the respondents for LIs; Table S3: Matrix $\mathcal{R}$ containing the scores assigned by respondents in the fuzzy triangular form normalized based on the highest score value for LIs; Table S4: Matrix v resulted from matrix $\mathcal{R}$ weighted by the vector $\tilde{E}$ for LIs; Table S5: Distance to the Positive Ideal Solution for LIs; Table

S6: Distance to the Negative Ideal Solution for LIs; Table S7: Total positive and negative distances and CC_i for LIs; Table S8: Matrix $\tilde{G}$ containing the scores assigned by respondents in the fuzzy triangular form for SMIs; Table S9: Vector $\tilde{E}$ representing the fuzzy weights of the respondents for SMIs; Table S10: Matrix $\mathcal{R}$ containing the scores assigned by respondents in the fuzzy triangular form normalized based on the highest score value for SMIs; Table S11: Matrix υ resulted from matrix $\mathcal{R}$ weighted by the vector $\tilde{E}$ for SMIs; Table S12: Distance to the Positive Ideal Solution for SMIs; Table S13: Distance to the Negative Ideal Solution for SMIs; Table S14: Total positive and negative distances and CC_i for SMIs.

Author Contributions: Conceptualization, V.A.d.O.B. and R.A.; methodology, V.A.d.O.B. and R.A.; validation, V.A.d.O.B. and R.A.; formal analysis, V.A.d.O.B., T.F.A.C.S. and R.A.; investigation, V.A.d.O.B. and R.A.; resources, T.F.A.C.S., G.H.S.M.d.M., L.V.Á., I.S.R., W.L.F. and R.A.; writing—original draft preparation, V.A.d.O.B., T.F.A.C.S. and R.A.; writing—review and editing, T.F.A.C.S., G.H.S.M.d.M., L.V.Á., I.S.R., W.L.F. and R.A.; supervision, R.A.; project administration, R.A. All authors have read and agreed to the published version of the manuscript.

Funding: The authors are grateful for the support of the National Council for Scientific and Technological Development (CNPq/Brazil) under the grants n° 304145/2021-1, n° 303924/2021-7, and PIBIC-CNPq.

Data Availability Statement: Not applicable.

Conflicts of Interest: The authors declare no conflict of interest. The funders had no role in the design of the study; in the collection, analyses, or interpretation of data; in the writing of the manuscript; or in the decision to publish the results.

References

1. Ibn-Mohammed, T.; Mustapha, K.B.; Godsell, J.; Adamu, Z.; Babatunde, K.A.; Akintade, D.D.; Acquaye, A.; Fujii, H.; Ndiaye, M.M.; Yamoah, F.A.; et al. A critical review of the impacts of COVID-19 on the global economy and ecosystems and opportunities for circular economy strategies. *Resour. Conserv. Recycl.* **2021**, *164*, 105169. [CrossRef] [PubMed]
2. Jawad, M.; Maroof, Z.; Naz, M. Impact of pandemic COVID-19 on global economies (a seven-scenario analysis). *Manag. Decis. Econ.* **2021**, *42*, 1897–1908. [CrossRef] [PubMed]
3. Chen, J.; Huang, J.; Su, W.; Štreimikienė, D.; Baležentis, T. The challenges of COVID-19 control policies for sustainable development of business: Evidence from service industries. *Technol. Soc.* **2021**, *66*, 101643. [CrossRef] [PubMed]
4. Margherita, A.; Heikkilä, M. Business continuity in the COVID-19 emergency: A framework of actions undertaken by world-leading companies. *Bus. Horiz.* **2021**, *64*, 683–695. [CrossRef] [PubMed]
5. Schmid, B.; Raju, E.; Jensen, P.K.M. COVID-19 and business continuity-learning from the private sector and humanitarian actors in Kenya. *Prog. Disaster Sci.* **2021**, *11*, 100181. [CrossRef]
6. ISO 22313:2020; Security and Resilience—Business Continuity Management Systems—Guidance on the Use of ISO 22301. International Organization for Standardization: Geneva, Switzerland, 2020.
7. Sharma, P.; Leung, T.Y.; Kingshott, R.P.J.; Davcik, N.S.; Cardinali, S. Managing uncertainty during a global pandemic: An international business perspective. *J. Bus. Res.* **2020**, *116*, 188–192. [CrossRef]
8. Anholon, R.; Silva, D.; Souza Pinto, J.; Rampasso, I.S.; Domingos, M.L.C.; Dias, J.H.O. COVID-19 and the administrative concepts neglected: Reflections for leaders to enhance organizational development. *Kybernetes* **2021**, *50*, 1654–1660. [CrossRef]
9. Yue, P.; Korkmaz, A.G.; Yin, Z.; Zhou, H. Household-owned Businesses' Vulnerability to the COVID-19 Pandemic. *Emerg. Mark. Financ. Trade* **2021**, *57*, 1662–1674. [CrossRef]
10. International Air Transport Association. *Global Outlook for Air Transport*; International Air Transport Association: Montreal, QU, Canada, 2022.
11. Air Transport Action Group. *Covid-19 Impact on Aviation*; Air Transport Action Group: Geneva, Switzerland, 2021.
12. Groenewold, M.R.; Burrer, S.L.; Ahmed, F.; Uzicanin, A.; Free, H.; Luckhaupt, S.E. Increases in Health-Related Workplace Absenteeism Among Workers in Essential Critical Infrastructure Occupations During the COVID-19 Pandemic—United States, March–April 2020. *MMWR. Morb. Mortal. Wkly. Rep.* **2020**, *69*, 853–858. [CrossRef]
13. Sahoo, P. Ashwani COVID-19 and Indian Economy: Impact on Growth, Manufacturing, Trade and MSME Sector. *Glob. Bus. Rev.* **2020**, *21*, 1159–1183. [CrossRef]
14. Madurai Elavarasan, R.; Shafiullah, G.; Raju, K.; Mudgal, V.; Arif, M.T.; Jamal, T.; Subramanian, S.; Sriraja Balaguru, V.S.; Reddy, K.S.; Subramaniam, U. COVID-19: Impact analysis and recommendations for power sector operation. *Appl. Energy* **2020**, *279*, 115739. [CrossRef] [PubMed]
15. Siche, R. What is the impact of COVID-19 disease on agriculture? *Sci. Agropecu.* **2020**, *11*, 3–6. [CrossRef]
16. Singh, S.; Kumar, R.; Panchal, R.; Tiwari, M.K. Impact of COVID-19 on logistics systems and disruptions in food supply chain. *Int. J. Prod. Res.* **2021**, *59*, 1993–2008. [CrossRef]
17. Gautam, S.; Setu, S.; Khan, M.G.Q.; Khan, M.B. Analysis of the health, economic and environmental impacts of COVID-19: The Bangladesh perspective. *Geosyst. Geoenviron.* **2022**, *1*, 100011. [CrossRef]

18. Maheshwari, S.; Friedman, V. Brooks Brothers, Founded in 1818, Files for Bankruptcy. *New York Times*. 8 September 2020. Available online: https://www.nytimes.com/2020/07/08/business/brooks-brothers-chapter-11-bankruptcy.html (accessed on 25 September 2022).
19. Rushe, D. Virgin Atlantic Files for Bankruptcy Protection as Covid Continues to Hurt Airlines. *The Guard.* 4 August 2020. Available online: https://www.theguardian.com/business/2020/aug/04/virgin-atlantic-files-for-bankruptcy-as-covid-continues-to-hurt-airlines (accessed on 25 September 2022).
20. Kuckertz, A.; Brändle, L.; Gaudig, A.; Hinderer, S.; Morales Reyes, C.A.; Prochotta, A.; Steinbrink, K.M.; Berger, E.S.C. Startups in times of crisis–A rapid response to the COVID-19 pandemic. *J. Bus. Ventur. Insights* **2020**, *13*, e00169. [CrossRef]
21. Belitski, M.; Guenther, C.; Kritikos, A.S.; Thurik, R. Economic effects of the COVID-19 pandemic on entrepreneurship and small businesses. *Small Bus. Econ.* **2022**, *58*, 593–609. [CrossRef]
22. Bartik, A.W.; Bertrand, M.; Cullen, Z.; Glaeser, E.L.; Luca, M.; Stanton, C. The impact of COVID-19 on small business outcomes and expectations. *Proc. Natl. Acad. Sci. USA* **2020**, *117*, 17656–17666. [CrossRef]
23. Kraus, S.; Clauss, T.; Breier, M.; Gast, J.; Zardini, A.; Tiberius, V. The economics of COVID-19: Initial empirical evidence on how family firms in five European countries cope with the corona crisis. *Int. J. Entrep. Behav. Res.* **2020**, *26*, 1067–1092. [CrossRef]
24. Collins, A.; Florin, M.-V.; Renn, O. COVID-19 risk governance: Drivers, responses and lessons to be learned. *J. Risk Res.* **2020**, *23*, 1073–1082. [CrossRef]
25. Ivanov, D. Predicting the impacts of epidemic outbreaks on global supply chains: A simulation-based analysis on the coronavirus outbreak (COVID-19/SARS-CoV-2) case. *Transp. Res. Part E Logist. Transp. Rev.* **2020**, *136*, 101922. [CrossRef]
26. Ivanov, D.; Dolgui, A. Viability of intertwined supply networks: Extending the supply chain resilience angles towards survivability. A position paper motivated by COVID-19 outbreak. *Int. J. Prod. Res.* **2020**, *58*, 2904–2915. [CrossRef]
27. Vo, T.A.; Mazur, M.; Thai, A. The impact of COVID-19 economic crisis on the speed of adjustment toward target leverage ratio: An international analysis. *Financ. Res. Lett.* **2022**, *45*, 102157. [CrossRef] [PubMed]
28. Gebhardt, M.; Spieske, A.; Kopyto, M.; Birkel, H. Increasing global supply chains' resilience after the COVID-19 pandemic: Empirical results from a Delphi study. *J. Bus. Res.* **2022**, *150*, 59–72. [CrossRef]
29. Hadjielias, E.; Christofi, M.; Tarba, S. Contextualizing small business resilience during the COVID-19 pandemic: Evidence from small business owner-managers. *Small Bus. Econ.* **2022**. [CrossRef]
30. Beninger, S.; Francis, J.N.P. Resources for business resilience in a COVID-19 world: A community-centric approach. *Bus. Horiz.* **2022**, *65*, 227–238. [CrossRef]
31. Graham, J.; Loke, H. Theory rewritten: Business Continuity and Crisis Management in the wake of the Covid-19 pandemic. *J. Contingencies Cris. Manag.* **2022**, *30*, 231–233. [CrossRef]
32. Han, H.; Qian, Y. Did Enterprises' Innovation Ability Increase During the COVID-19 Pandemic? Evidence From Chinese Listed Companies. *Asian Econ. Lett.* **2020**, *1*, 18072. [CrossRef]
33. Papadopoulos, T.; Baltas, K.N.; Balta, M.E. The use of digital technologies by small and medium enterprises during COVID-19: Implications for theory and practice. *Int. J. Inf. Manag.* **2020**, *55*, 102192. [CrossRef]
34. Suder, M.; Kusa, R.; Duda, J.; Dunska, M. How small printing firms alleviate impact of pandemic crisis? Identifying configurations of successful strategies with fuzzy-set qualitative comparative analysis. *Entrep. Bus. Econ. Rev.* **2022**, *10*, 61–80. [CrossRef]
35. Nguyen, V.K.; Pyke, J.; Gamage, A.; de Lacy, T.; Lindsay-Smith, G. Factors influencing business recovery from compound disasters: Evidence from Australian micro and small tourism businesses. *J. Hosp. Tour. Manag.* **2022**, *53*, 1–9. [CrossRef]
36. Santos, E.; Tavares, V.; Tavares, F.O.; Ratten, V. How is risk different in family and non-family businesses?–A comparative statistical analysis during the COVID-19 pandemic. *J. Fam. Bus. Manag.* **2021**. [CrossRef]
37. Bag, S.; Dhamija, P.; Luthra, S.; Huisingh, D. How big data analytics can help manufacturing companies strengthen supply chain resilience in the context of the COVID-19 pandemic. *Int. J. Logist. Manag.* **2021**. [CrossRef]
38. Nisar, Q.A.; Haider, S.; Ameer, I.; Hussain, M.S.; Gill, S.S.; Usama, A. Sustainable supply chain management performance in post COVID-19 era in an emerging economy: A big data perspective. *Int. J. Emerg. Mark.* **2022**. [CrossRef]
39. Taskan, B.; Junça-Silva, A.; Caetano, A. Clarifying the conceptual map of VUCA: A systematic review. *Int. J. Organ. Anal.* **2022**. [CrossRef]
40. Donthu, N.; Gustafsson, A. Effects of COVID-19 on business and research. *J. Bus. Res.* **2020**, *117*, 284–289. [CrossRef] [PubMed]
41. Apostolopoulos, N.; Liargovas, P. Regional parameters and solar energy enterprises. *Int. J. Energy Sect. Manag.* **2016**, *10*, 19–37. [CrossRef]
42. Nielsen, F. Hierarchical Clustering. In *Introduction to HPC with MPI for Data Science*; Nielsen, F., Ed.; Springer: New York, NY, USA, 2016; pp. 195–211.
43. Chen, C.-T. Extensions of the TOPSIS for group decision-making under fuzzy environment. *Fuzzy Sets Syst.* **2000**, *114*, 1–9. [CrossRef]
44. Lima Junior, F.R.; Carpinetti, L.C.R. A comparison between TOPSIS and Fuzzy-TOPSIS methods to support multicriteria decision making for supplier selection. *Gest. Prod.* **2015**, *22*, 17–34. [CrossRef]
45. Hwang, C.-L.; Yoon, K. *Multiple Attribute Decision Making*; Lecture Notes in Economics and Mathematical Systems; Springer: Berlin/Heidelberg, Germany, 1981; Volume 186, ISBN 978-3-540-10558-9.
46. Akram, M.; Shumaiza; Arshad, M. Bipolar fuzzy TOPSIS and bipolar fuzzy ELECTRE-I methods to diagnosis. *Comput. Appl. Math.* **2020**, *39*, 1–21. [CrossRef]

47. Tominaga, L.K.D.G.; Martins, V.W.B.; Rampasso, I.S.; Anholon, R.; Silva, D.; Pinto, J.S.; Leal Filho, W.; Lima Junior, F.R. Critical analysis of engineering education focused on sustainability in supply chain management: An overview of Brazilian higher education institutions. *Int. J. Sustain. High. Educ.* **2021**, *22*, 380–403. [CrossRef]
48. Pedrycz, W. Why triangular membership functions? *Fuzzy Sets Syst.* **1994**, *64*, 21–30. [CrossRef]
49. Klir, G.J.; Yuan, B. *Fuzzy Sets and Fuzzy Logic: Theory and Applications*; Prentice Hall: Upper Saddle River, NJ, USA, 1995.
50. ISO. *ISO 31000-Gestão de Riscos-Diretrizes (Risks Management-Guidelines)*; ABNT: Rio de Janeiro, Brazil, 2018.
51. *ISO NBR ISO 19011*; Diretrizes Para Auditoria de Sistemas de Gestão [Management Systems Audit Guidelines]. ISO: Geneva, Switzerland, 2018.
52. Sadeghi R., J.K.; Azadegan, A.; Ojha, D.; Ogden, J.A. Benefiting from supplier business continuity: The role of supplier monitoring and buyer power. *Ind. Mark. Manag.* **2022**, *106*, 432–443. [CrossRef]

Article

M-Polar Fuzzy Graphs and Deep Learning for the Design of Analog Amplifiers

Malinka Ivanova [1,*] and Mariana Durcheva [1,2]

[1] Department of Informatics, Faculty of Applied Mathematics and Informatics, Technical University of Sofia, 1797 Sofia, Bulgaria
[2] Department of Mathematics, Shamoon College of Engineering, Ashdod 77245, Israel
* Correspondence: m_ivanova@tu-sofia.bg

Abstract: The design of analog circuits is a complex and repetitive process aimed at finding the best design variant. It is characterized by uncertainty and multivariate approaches. The designer has to make different choices to satisfy a predefined specification with required parameters. This paper proposes a method for facilitating the design of analog amplifiers based on m-polar fuzzy graphs theory and deep learning. M-polar fuzzy graphs are used because of their flexibility and the possibility to model different real-life multi-attribute problems. Deep learning is applied to solve a regression task and to predict the membership functions of the m-polar fuzzy graph vertices (the solutions), taking on the role of domain experts. The performance of the learner is high since the obtained errors are very small: Root Mean Squared Error is from 0.0032 to 0.0187, Absolute Error is from 0.022 to 0.098 and Relative Error is between 0.27% and 1.57%. The proposed method is verified through the design of three amplifiers: summing amplifier, subtracting amplifier, and summing/subtracting amplifier. The method can be used for improving the design process of electronic circuits with the possibility of automating some tasks.

Keywords: m-polar fuzzy graph; machine learning; deep learning; analog amplifier; design process; automation

MSC: 05C72; 68T07; 94C15

1. Introduction

The design process of analog circuits is characterized by its complexity and uncertainty, but also by its multivariate approaches. According to one predefined specification with initial parameters, it is possible to have multiple solutions that satisfy this assignment. It is obvious that the designer must find the best design by entering one iterative procedure, and making one or another decision. In support of the designer, a wide variety of methods and methodologies are proposed—from the use of standard matrix theory [1] to contemporary machine learning approaches [2].

The topic related to machine learning-driven design of analog circuits is under extensive investigation, because of the desire for facilitating the designer and automating as many engineering tasks as possible. This could reduce the design time, increase the design quality and decrease the effort of both a beginner and experienced engineer.

Some works show the applicability of evolutionary computing in analog circuit design and possibilities for time controlling [3], achieving efficient and flexible design [4], and design optimization [5]. Bayesian optimization techniques are also under investigation, pointing out another different possible approach for improving the circuit design process regarding design speed and accuracy [6,7].

Deep learning (DL) as a part of machine learning is based on architectures of artificial neural networks (ANNs) and is used for solving classification or regression tasks in different domains and industries, including electronic circuit design [8] and analysis [9]. This popularity of supervised DL is due to its capability to perform predictions and classifications with high accuracy. This is possible because of the flexibility of ANNs construction and DL algorithm parameters adjustment.

Nowadays, the theory of fuzzy graphs gets its popularity due to its wide applications in many areas of real life. The concept of a fuzzy graph was initially introduced by Kauffman [10], but the development of the fuzzy graphs theory is due to Rosenfeld [11] and Yeh and Bang [12]. The fuzzy graphs theory has been developing in different directions, such as fuzzy tolerance graphs [13], fuzzy threshold graphs [14], interval-valued fuzzy graphs [15–17], fuzzy k-competition graphs and p-competition fuzzy graphs [18], $m-$step fuzzy competition graphs [19], and hesitant fuzzy graphs [20]. In 1994, Zhang [21,22] introduced the concept of bipolar fuzzy sets as a generalization of fuzzy sets. This theory was developed by Akram [23] to present bipolar fuzzy graphs. In 2014, Chen et al. [24] discussed the notion of m-polar fuzzy sets as a generalization of bipolar fuzzy sets. Ghorai and Pal [25] introduced the m-polar fuzzy graphs as a generalization of bipolar fuzzy graphs and defined different operations. The theory of m-polar fuzzy graphs was further developed in [26–29]. A detailed explanation could be found in two contemporary monographs [30,31].

This work is the first to use m-polar fuzzy graphs theory with the benefits of deep learning for the design of analog amplifiers, but we consider the idea to combine the advantages of fuzzy graphs and deep learning to be very promising. Graphs theory is used, taking into account its powerful capability for solving different real-life problems. Modelling of such problems often involves multi-polar information, including uncertainty and process limits, so attracting m-polar fuzzy graphs for describing such problems is very useful. Deep learning predictions are utilized in the role of the domain experts to point out the membership functions of m-polar fuzzy graph vertices. The regression task is solved with high performance.

The aim of the paper is to present a method for designing analog amplifiers by discussing some basic concepts regarding m-polar fuzzy graphs and considering the advantages of deep learning. The following is a summary of the research work's main contribution:

1. The activity for designing analog amplifiers is defined as a multi-attribute problem, which is solved here using the m-polar fuzzy graphs and deep learning.

2. For the verification of the proposed method, three electronic circuits are designed: an inverting summing amplifier, a subtracting amplifier (differential amplifier), and a summing/subtracting amplifier with an operational amplifier.

3. The predictive models are experimentally evaluated using data sets collected considering the functional and electrical behavior of the examined circuits.

The paper is organized as follows: In Section 2, a review of the basic concept of fuzzy graphs theory and m-polar fuzzy graphs as well as different types of product operations for m-polar fuzzy graphs are presented. Section 3 is dedicated to discussing the usage of ANNs in solving circuit design problems. Section 4 considers the proposed method. In Section 5, the presented method for analog circuit design is demonstrated and verified in the real-life application of analog amplifier design. The conclusion is driven in the last section.

2. Fuzzy Graphs Theory

2.1. Basic Concepts of Fuzzy Graphs

A graph is a pair $G = (V, E)$ of a nonempty set of vertices V (or nodes) and a set of edges E. Each edge has either one or two vertices associated with it, called its *endpoints*. An edge is said to connect its endpoints. A graph is called *simple* if it has no loops and no multiple edges.

Definition 1 [32]. *A fuzzy graph $G = (V, \sigma, \mu)$ is a triple consisting of a nonempty set V together with a pair of functions $\sigma : V \to [0, 1]$ and $\mu : E \to [0.1]$ such that for all $x, y \in V$, $\mu(xy) \leq \sigma(x) \wedge \sigma(y)$, where $\wedge$ stands for the minimum.*

Definition 2 [33]. *A fuzzy graph $G = (V, \sigma, \mu)$ is complete if $\mu(xy) = \sigma(x) \wedge \sigma(y)$ for all $x, y \in V$.*

Definition 3 [24]. *An m-polar fuzzy set (or a $[0,1]^m$-set) on X is mapping $M : X \to [0,1]^m$.*

The $[0,1]^m$ (m-power of $[0,1]$) is considered to be a poset with point-wise order $\leq$ (m is natural), where $\leq$ is defined by $x \leq y \Leftrightarrow P_i(x) \leq P_i(y)$ for each $i = 1, 2, \ldots, m$, $x, y \in [0,1]^m$, and $P_i : [0,1]^m \to [0,1]$ is the i-th projection mapping ($i = 1, 2, \ldots, m$). Here, $1 = (1, 1, \ldots, 1)$ is the greatest value and $0 = (0, 0, \ldots, 0)$ is the smallest value in $[0,1]^m$.

Definition 4 [24]. *Let σ be an m-polar fuzzy set on a set V. An m-polar fuzzy relation on σ is an m-polar fuzzy set μ of $V \times V$ such that $\mu(xy) \leq \sigma(x) \wedge \sigma(y)$ for all $x, y \in V$, i.e., for each $i = 1, 2, \ldots, m$, for all $x, y \in V$: $P_i \circ \mu(xy) \leq P_i \circ \sigma(x) \wedge P_i \circ \sigma(y)$.*

Definition 5 [30]. *An m-polar fuzzy graph $G = (V, \sigma, \mu)$ is a triple consisting of a nonempty set V together with a pair of functions $\sigma : V \to [0,1]^m$ and $\mu : E = V \times V \to [0,1]^m$, where σ is an m-polar fuzzy set on the set of vertices V and μ is an m-polar fuzzy relation in V such that for all $x, y \in V$, $\mu(xy) \leq \sigma(x) \wedge \sigma(y)$, where $\wedge$ stands for minimum.*

It can be noted that $\mu(x, y) = 0$ for all $x, y \in V \times V - E$.

2.2. Products in m-Polar Fuzzy Graphs

2.2.1. Direct (Tensor) Product

Definition 6 [30]. *Let $G_1 = (\sigma_1, \mu_1)$ of $G_1^* = (V_1, E_1)$ and $G_2 = (\sigma_2, \mu_2)$ of $G_2^* = (V_2, E_2)$ be two m-polar fuzzy graphs. The direct product of G_1 and G_2 is denoted by $G_1 \times G_2$ and is defined as a pair $G_1 \times G_2 = (\sigma_1 \times \sigma_2, \mu_1 \times \mu_2)$, such that for each $i = 1, 2, \ldots, m$:*

$$P_i \circ (\sigma_1 \wedge \sigma_2)(x_1, x_2) = P_i \circ \sigma_1(x_1) \wedge P_i \circ \sigma_2(x_2) \text{ for all } (x_1, x_2) \in V_1 \times V_2$$

$$P_i \circ (\mu_1 \wedge \mu_2)((x_1, x_2)(y_1, y_2)) = P_i \circ \mu_1(x_1 y_1) \wedge P_i \circ \mu_2(x_2 y_2)$$

for all $x_1 y_1 \in E_1, x_2 y_2 \in E_2$.

Proposition 1 [30]. *The direct product of two m-polar fuzzy graphs is an m-polar fuzzy graph.*

2.2.2. Semi-Strong Product

Definition 7 [31]. *Let $G_1 = (\sigma_1, \mu_1)$ of $G_1^* = (V_1, E_1)$ and $G_2 = (\sigma_2, \mu_2)$ of $G_2^* = (V_2, E_2)$ be two m-polar fuzzy graphs. The semi-strong product of G_1 and G_2 (denoted by $G_1 \boxdot G_2$) is defined as a graph $G_1 \boxdot G_2 = (\sigma_1 \boxdot \sigma_2, \mu_1 \boxdot \mu_2)$ of $G^* = (V_1 \times V_2, E)$ (here $E = \{(x_1, y_1)(x_1, y_2) \mid x_1 \in V_1, y_1 y_2 \in E_2\} \cup \{(x_1, y_1)(x_2, y_2) \mid x_1 x_2 \in E_1, y_1 y_2 \in E_2\}$) such that for each $i = 1, 2, \ldots, m$:*

$P_i \circ (\sigma_1 \boxdot \sigma_2)(x, y) = P_i \circ \sigma_1(x) \wedge P_i \circ \sigma_2(y)$ *for all $(x, y) \in V_1 \times V_2$;*
$P_i \circ (\mu_1 \boxdot \mu_2)(x_1, x_2)(x_1, y_2) = P_i \circ \sigma_1(x_1) \wedge P_i \circ \mu_2(x_2 y_2)$ *for all $x_1 \in V_1$ and $x_2 y_2 \in E_2$;*
$P_i \circ (\mu_1 \boxdot \mu_2)(x_1, x_2)(y_1, y_2) = P_i \circ \mu_1(x_1 y_1) \wedge P_i \circ \mu_2(x_2 y_2)$ *for all $x_1 x_2 \in E_1$ and $x_2 y_2 \in E_2$;*

Proposition 2 [26]. *The semi-strong product of two m-polar fuzzy graphs is an m-polar fuzzy graph.*

2.2.3. Strong Product

Definition 8 [31]. *Let $G_1 = (\sigma_1, \mu_1)$ of $G_1^* = (V_1, E_1)$ and $G_2 = (\sigma_2, \mu_2)$ of $G_2^* = (V_2, E_2)$ be two m-polar fuzzy graphs. The strong product of G_1 and G_2 (denoted by $G_1 \otimes G_2$) is defined as a graph $G_1 \otimes G_2 = (\sigma_1 \otimes \sigma_2, \mu_1 \otimes \mu_2)$ of $G^* = (V_1 \times V_2, E)$ where*

$$E = \{(x_1,y_1)(x_1,y_2) \mid x_1 \in V_1, y_1y_2 \in E_2\} \cup \{(x_1,y)(x_2,y) \mid x_1x_2 \in E_1, y \in V_2\}$$
$$\cup \{(x_1,y_1)(x_2,y_2) \mid x_1x_2 \in E_1, y_1y_2 \in E_2\}$$

such that for each $i = 1, 2, \ldots, m$:
$P_i \circ (\sigma_1 \otimes \sigma_2)(x,y) = P_i \circ \sigma_1(x) \wedge P_i \circ \sigma_2(y)$ for all $(x,y) \in V_1 \times V_2$;
$P_i \circ (\mu_1 \otimes \mu_2)(x_1,x_2)(x_1,y_2) = P_i \circ \sigma_1(x_1) \wedge P_i \circ \mu_2(x_2y_2)$ for all $x_1 \in V_1$ and $x_2y_2 \in E_2$;
$P_i \circ (\mu_1 \otimes \mu_2)(x_1,x_2)(y_1,x_2) = P_i \circ \mu_1(x_1y_1) \wedge P_i \circ \sigma_2(x_2)$ for all $x_1y_1 \in E_1$ and $x_2 \in V_2$;
$P_i \circ (\mu_1 \otimes \mu_2)(x_1,x_2)(y_1,y_2) = P_i \circ \mu_1(x_1y_1) \wedge P_i \circ \mu_2(x_2y_2)$ for all $x_1y_1 \in E_1$ and $x_2y_2 \in E_2$.

Proposition 3 [31]. *The strong product of two m-polar fuzzy graphs is an m-polar fuzzy graph.*

2.2.4. Lexicographic Product

Definition 9 [30]. *Let $G_1 = (\sigma_1, \mu_1)$ of $G_1^* = (V_1, E_1)$ and $G_2 = (\sigma_2, \mu_2)$ of $G_2^* = (V_2, E_2)$ be two m-polar fuzzy graphs. The lexicographic product of G_1 and G_2 (denoted by $G_1 \bullet G_2$) is defined as a pair $G_1 \bullet G_2 = (\sigma_1 \bullet \sigma_2, \mu_1 \bullet \mu_2)$ such that for each $i = 1, 2, \ldots, m$:*

$P_i \circ (\sigma_1 \bullet \sigma_2)(x_1,x_2) = P_i \circ \sigma_1(x_1) \wedge P_i \circ \sigma_2(x_2)$ for all $(x_1,x_2) \in V_1 \times V_2$;
$P_i \circ (\mu_1 \bullet \mu_2)((x_1,x_2)(x_1,y_2)) = P_i \circ \sigma_1(x_1) \wedge P_i \circ \mu_2(x_2y_2)$ for all $x_1 \in V_1$ and $x_2y_2 \in E_2$;
$P_i \circ (\mu_1 \bullet \mu_2)(x_1,x_2)(y_1,y_2) = P_i \circ \mu_1(x_1y_1) \wedge P_i \circ \mu_2(x_2y_2)$ for all $x_1y_1 \in E_1$ and $x_2y_2 \in E_2$;

Proposition 4 [30]. *The lexicographic product of two m-polar fuzzy graphs is an m-polar fuzzy graph.*

Some applications of m-polar fuzzy graphs for facilitating the decision-making process in a wide variety of studied domains are discussed in [31,34,35]. It is obvious that the presented methods and techniques have the potential to deal with multi-attribute, multi-criteria, and multi-objective problems in uncertain and fuzzy environments. This is the reason such an approach is chosen for application in the field of electronic circuit design.

3. Deep Learning and Applications in Electronic Circuit Design

Deep Learning (DL) comprises multiple methods and techniques, based on ANNs, which are utilized for different purposes—from studying a process, event, or facts to analyzing, predicting, or optimizing some parameters. Contemporary surveys summarize approaches, types, and architectures of deep learning algorithms, as well as discuss their applications in the context of supervised, unsupervised, and reinforcement learning [36,37]. The advantages of DL, such as universal usage, robustness, workability with different data types, and scalability are also described. The cases in which DL is suitable for usage are explained to show its usefulness in the unavailability of domain experts, the impossibility to gather the expertise and the complexity of the problem.

In the automation of electronic circuit design, a few papers are devoted to the advantages of ANNs utilization in support of the designers' tasks.

Dieste-Velasco et al. propose a methodology for assisting the analytical design of analog circuits as it combines the statistical technique factorial design of experiments and ANNs [38]. Such an approach allows the behavioral modelling of circuits to be done with high accuracy.

The work of Devi et al. is focused on the automated design of analog circuits through the usage of ANN-based supervised learning [39]. The method is verified in the design of two different analog circuits and the obtained results are characterized by high accuracy and small mean squared error.

Wang et al. applied deep learning to solve the sizing problem of analog circuits [40]. Two ANNs architectures are proposed, Recurrent Neural Network (RNN) and DL, which are proven to possess the capability to predict the transistor size with high accuracy.

More sophisticated solutions for automating the design process of complex analog systems [41] and for improving the circuit synthesis [42] point out the possibility of reducing the design process resources and decreasing the designer's effort.

Budak et al. deal with the speed of the design of integrated analog circuits proposing an efficient method for sizing [43]. The method is verified in the design of analog amplifiers and a comparator, and the results show its benefit.

It seems that DL possesses a big potential for the implementation of a supportive, automated, and smart design process that facilitates decision-making and problem-solving. All these advantages and possibilities are considered when choosing the DL for creating predictive models and speeding up the circuit design.

4. Proposed Method

The design of analog amplifiers could be defined as a multi-attribute problem, which is solved here through the utilization of the m-polar fuzzy graphs theory and deep learning. The developed method for amplifier design is presented in Figure 1 and includes three stages:

- In the first stage, a dataset is prepared according to a predefined specification regarding the designed amplifier. All possible variants of the designed electronic circuit are found and membership functions of attributes are predicted through a deep learning algorithm.
- The second stage points out the suitable design solutions, considering the requested parameters, and after obtaining the membership values of vertices and edges, an m-polar fuzzy graph is constructed.
- In the third stage, the most suitable solutions are prioritized, finding the best one, according to the user's specifications and certain requirements.

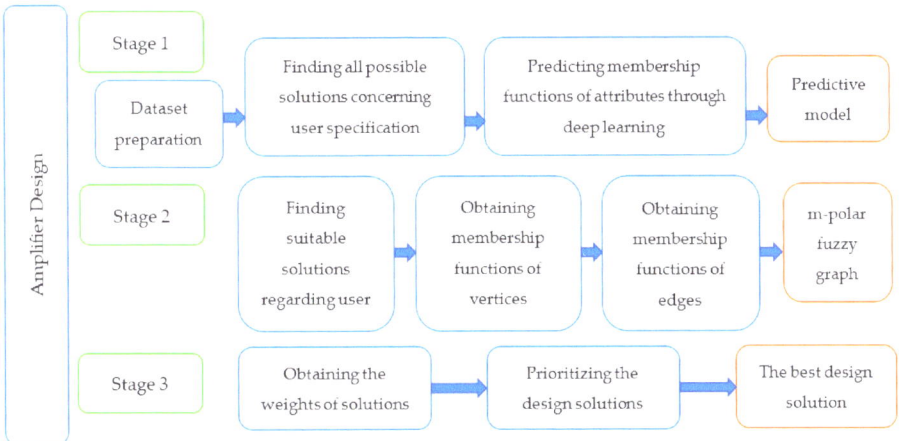

Figure 1. Proposed method.

5. Experimentation and Results

The problem related to finding the most useful designs of an electronic circuit is examined as a multi-attribute problem for solving. The typical attributes of a given circuit that better explain its electrical and functional behavior are chosen, as they have to be adjusted considering an input specification. For the verification of the proposed method, comprising m-polar fuzzy graphs theory and DL, three electronic circuits are designed: an inverting summing amplifier, a subtracting amplifier (differential amplifier), and a summing/subtracting amplifier with an operational amplifier.

5.1. Design of Inverting Summing Amplifier

Summing amplifiers are very often implemented through operational amplifiers with negative feedback and topology shown in Figure 2. They are used for solving equations such as [44]:

$$v_{out} = -(k_1 v_1 + k_2 v_2 + \ldots + k_n v_n) \quad (1)$$

where scaling coefficients $k_1, k_2, \ldots, k_n$ are defined as the ratio between the feedback resistor R_F and the input resistors $R_1, R_2, \ldots, R_n$, $v_1, v_2, \ldots, v_n$ are input voltage signals, and v_{out} is the output voltage.

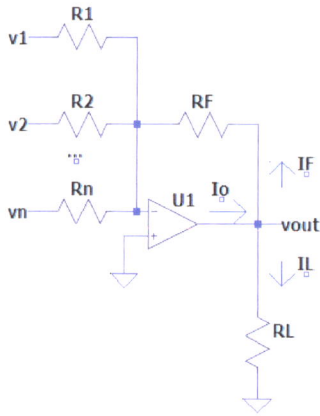

Figure 2. Inverting summing amplifier [44].

One of the challenges for designers is how to satisfy Equation (1), taking into account the possible ranges of input values $v_1, v_2, \ldots, v_n$ and output signal v_{out} and considering the allowed range of power dissipation P_D typical for the used operational amplifier for a given load R_L.

Power dissipation has to possess appropriate values for the normal functioning of the operational amplifier. Otherwise, the operational amplifier will be damaged and the realized analog circuit will not work as expected or will stop working. The amplifier power dissipation P_D is found by knowing the supply voltage V_S, maximal quiescent current I_Q, output current I_o, load current I_L, and output voltage v_{out} [45].

The design process of summing amplifiers Is related to finding the values of input resistors $R_1, R_2, \ldots, R_n$, taking into account the given scaling coefficients $k_1, k_2, \ldots, k_n$ in Equation (1), the output voltage v_{out} at given input voltages $v_1, v_2, \ldots, v_n$, and at certain values of the feedback resistor R_F and the load resistor R_L.

Let us suppose that the used operational amplifier is OPA 322 [46] and the equation to be solved when designing an inverting summing amplifier, taking into consideration the allowed power dissipation is:

$$v_{out} = -(3v_1 + 6v_2) \quad (2)$$

During the first stage of the proposed method, a dataset in the form of Table 1 is prepared, considering a predefined specification. The investigated range of R_F is from 60 kΩ to 6 kΩ, v_1 and v_2 have values from 0.01 V to 0.5 V, and R_L takes values from 0.5 kΩ to 10 kΩ. During the design process multiple solutions $S_1, S_2, \ldots, S_n$ are possible, which must first be found. All solutions possess common attributes $A_1 = R_1$, $A_2 = R_2$, $A_3 = v_1$, $A_4 = v_2$,

$A_5 = v_{out}$, $A_6 = R_F$, $A_7 = R_L$, $A_8 = P_D$, e.g., $S_1 = f(A_1, A_2, A_3, A_4, A_5, A_6, A_7, A_8)$, $S_2 = f(A_1, A_2, A_3, A_4, A_5, A_6, A_7, A_8), \ldots, S_n = f(A_1, A_2, A_3, A_4, A_5, A_6, A_7, A_8)$.

Table 1. Prepared dataset with possible solutions.

S	R_F,kΩ	R_1,kΩ	R_2,kΩ	v_1,V	v_2,V	v_{out},V	R_L,kΩ	P_D,mW	$\sigma(P_D)$
S_1	60	20	10	0.01	0.01	0.09	10	10.511	0.999
S_2	60	20	10	0.05	0.01	0.21	10	10.590	0.992
S_3	60	20	10	0.1	0.01	0.36	10	10.684	0.983
S_4	60	20	10	0.15	0.01	0.51	10	10.772	0.975
S_5	60	20	10	0.2	0.01	0.66	10	10.855	0.968
...	...	...	...	...	...	...	...	...	...

All obtained solutions are assigned membership values for each attribute $\sigma^{S_i}(A_j)$ corresponding to the membership functions. The created dataset is presented in Table 1 and the last column includes the membership values of power dissipation P_D. The membership values of other attributes are not shown, but they are predicted in a similar way through a deep learning algorithm with a regression task. The ANN includes two hidden layers with 50 neurons at each layer. The utilized activation function is Rectifier. The dataset contains 13,090 records, of which 70% are used for training and 30% for testing.

The achieved accuracy of the predictive model is assessed considering standard metrics such as Root mean squared error (RMSE) = 0.0098, Absolute error (AE) = 0.0053, and Relative error (RE) = 0.98%. Figure 3 presents the prediction chart and the linear regression result.

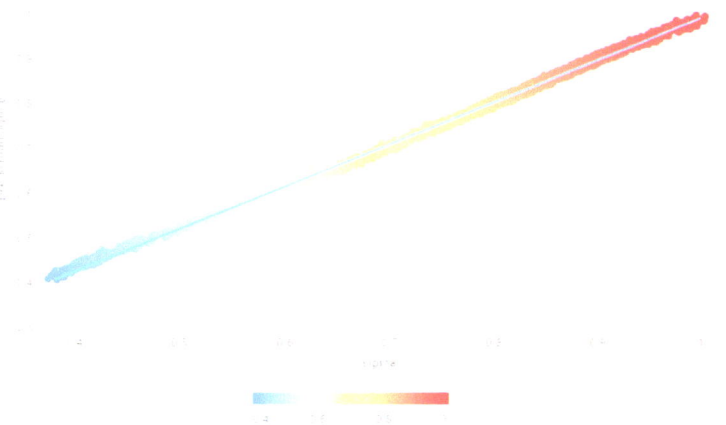

Figure 3. Prediction chart of the power dissipation P_D membership values (sigma) at inverting summing amplifier design.

In the second stage, the designer is looking for solutions with maximal membership function of power dissipation P_D, e.g., minimal values of power dissipation P_D and allowable values of the rest of the examined attributes. It is found that these requirements are satisfied by 51 of 13,090 designs: $SS_1, SS_2, .., SS_{51}$. The membership functions of vertices (solutions) of the fuzzy graph are predicted through deep learning as ten random solutions are shown in Figure 4. For further exploration, five solutions which are the most appropriate, are considered. Thus, effort and resources are reduced, selecting and examining the closest to the user specification solutions.

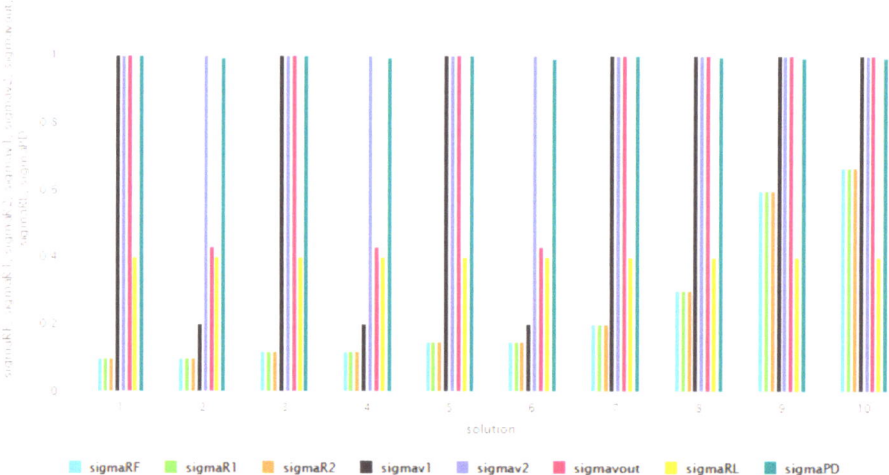

Figure 4. Membership values of ten random solutions at design of inverting summing amplifier.

An m-polar fuzzy graph is constructed with five vertices (the most suitable five solutions with minimal power dissipation P_D–SS_1, SS_2, ..., SS_5 and edges between them (Figure 5). The membership functions of the edges (SS_i, SS_k), with respect to the attributes $A_1 \div A_8$ are found according to the following equation [31] (Table 2):

$$\mu(SS_i, SS_k) = \sigma^{SS_i}(A_j) \wedge \sigma^{SS_k}(A_j) \qquad (3)$$

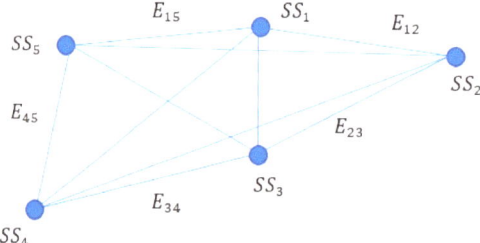

Figure 5. M-polar fuzzy graph at design of inverting summing amplifier.

Table 2. Membership values of the edges.

$SS_i \otimes SS_k$	A_1	A_2	A_3	A_4	A_5	A_6	A_7	A_8
$E_{12} = SS_1 \otimes SS_2$	0.1	0.1	0.1	1	1	1	0.4	0.999
$E_{23} = SS_2 \otimes SS_3$	0.12	0.12	0.12	1	1	1	0.4	0.999
$E_{34} = SS_3 \otimes SS_4$	0.1	0.1	0.1	1	1	1	0.4	0.999
$E_{45} = SS_4 \otimes SS_5$	0.1	0.1	0.1	1	1	1	0.444	0.999
$E_{15} = SS_1 \otimes SS_5$	0.1	0.1	0.1	1	1	1	0.4	0.999
...	...	...	...	...	...	...	...	...

In the third stage, the membership values of the solutions considering the membership values of every attribute, are calculated through the weight function:

$$w(S_i) = \frac{\sum \sigma^{SS_i}(A_j)}{n} \qquad (4)$$

where n is the number of attributes.

Figure 6 gives information on the calculated weights of every candidate for the best solution. It seems that the SS_3 is the best-found design considering the specified requirements. The best design could be used for amplifier prototyping with priorities in comparison to the other four designs.

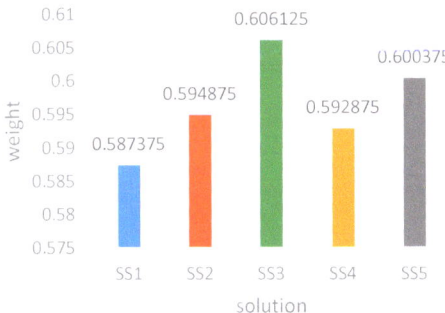

Figure 6. Calculated weights of the solutions at design of the inverting summing amplifier.

5.2. Design of Subtracting Amplifier (Differential Amplifier)

The function realized from the subtracting amplifier (differential input amplifier), presented in Figure 7, is described by the following equation [44]:

$$v_{out} = k(v_2 - v_1) \qquad (5)$$

where $k = \frac{R_F}{R_1}$, e.g., the output voltage is proportional to or equal to the difference between the input voltages.

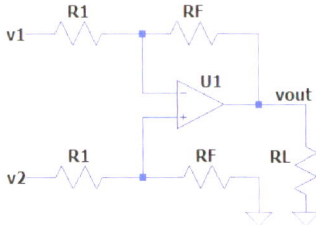

Figure 7. Subtracting amplifier [44].

Let us suppose that the specification requires the subtracting amplifier to realize the following function at minimal power dissipation P_D:

$$v_{out} = 3(v_2 - v_1) \qquad (6)$$

The suggestion is that the used operational amplifier is OPA 322 [46] and the explored range of R_F is from 60 kΩ to 3 kΩ, v_1 and v_2 have values from $0.01 V$ to $0.5 V$, R_L takes values from 0.5 kΩ to 10 kΩ. The design solutions are characterized by seven attributes:

$A_1 = R_1, A_2 = v_1, A_3 = v_2, A_4 = v_{out}, A_5 = R_F, A_6 = R_L, A_7 = P_D$,
e.g., $S_1 = f(A_1, A_2, A_3, A_4, A_5, A_6, A_7)$, $S_2 = f(A_1, A_2, A_3, A_4, A_5, A_6, A_7)$, ..., $S_n = f(A_1, A_2, A_3, A_4, A_5, A_6, A_7)$.

The prepared dataset is similar to Table 1. The membership values of the rest of the attributes are also predicted through a deep-learning algorithm. The learning records are 5710 and the prediction chart of the membership values of power dissipation (sigma) is presented in Figure 8. The created predictive model is characterized by very small errors: RMSE = 0.0032, AE = 0.0022, RE = 0.27%.

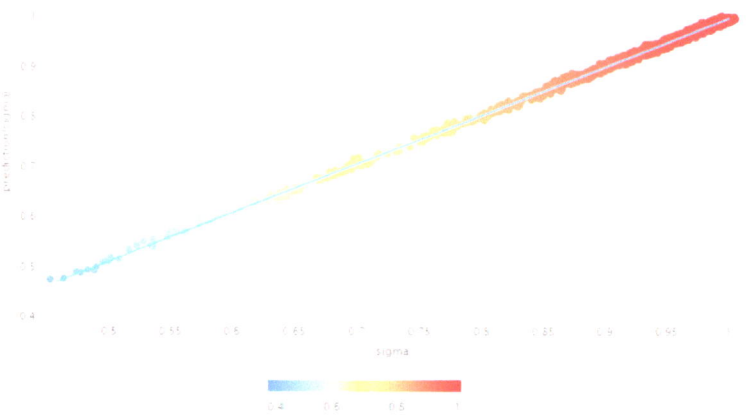

Figure 8. The prediction chart of the membership values of power dissipation (sigma) at subtracting amplifier design.

Figure 9 depicts the membership values of ten randomly chosen designs.

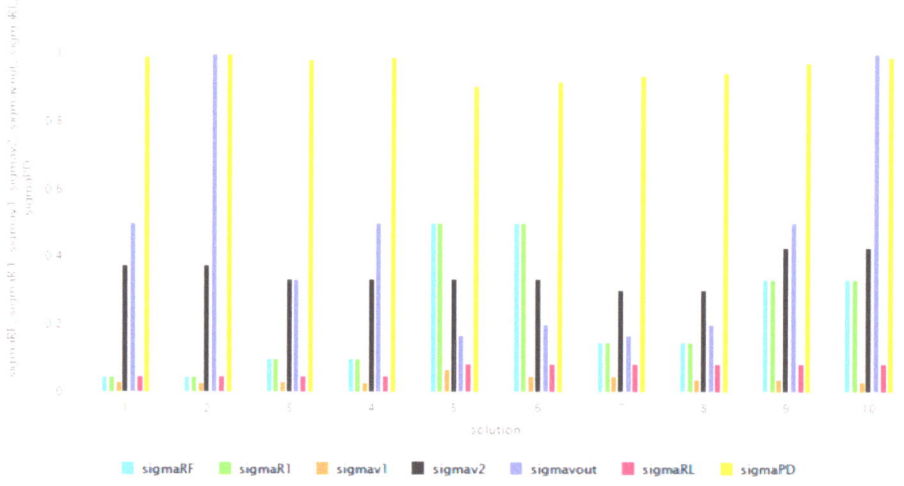

Figure 9. Membership values of randomly chosen solutions at design of subtracting amplifier.

At the second stage, out of all possible 5710 designs, eight are selected—these with the maximal membership values of power dissipation P_D.

For fuzzy graph construction, in addition to the membership values of vertices, it is necessary to know the membership values of the edges, which are calculated similarly to the previous demonstration and are presented in Table 3. The m-polar fuzzy graph itself is shown in Figure 10. Its structure points out the most suitable designs found, which are eight $SS_1 \div SS_8$. They are connected because they share common attributes.

Table 3. Membership values of the edges.

$SS_i \otimes SS_k$	A_1	A_2	A_3	A_4	A_5	A_6	A_7	A_8
$E_{12} = SS_1 \otimes SS_2$	1	1	0.222	0.3	1	1	1	1
$E_{23} = SS_2 \otimes SS_3$	1	1	0.25	0.333	1	1	1	1
$E_{34} = SS_3 \otimes SS_4$	1	1	0.285	0.375	1	1	1	1
$E_{45} = SS_4 \otimes SS_5$	1	1	0.333	0.428	1	1	1	1
$E_{56} = SS_5 \otimes SS_6$	1	1	0.4	0.5	1	1	1	1
$E_{67} = SS_6 \otimes SS_7$	1	1	0.5	0.6	1	1	1	1
$E_{78} = SS_7 \otimes SS_8$	1	1	0.666	0.75	1	1	1	1
...	...	...	...	...	...			

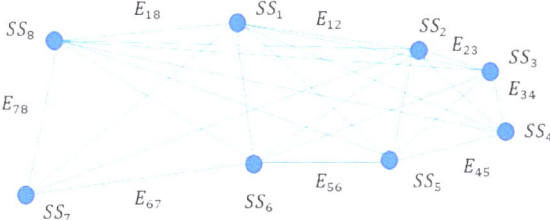

Figure 10. M-polar fuzzy graph at design of subtracting amplifier.

In the third stage, the best solution found is SS_8, after calculating the values of the weighting function (ordinate axes) as it is shown in Figure 11. It seems that SS8 is the design, which satisfies the user requirements in the best way. Moreover, the remaining seven also could be applied because of the closest weighting values to the most suitable design.

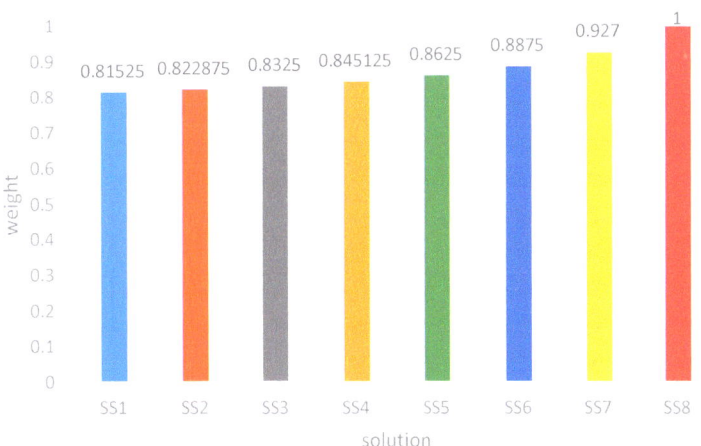

Figure 11. Prioritizing the suitable designs of the subtracting amplifier.

5.3. Summing and Subtracting Amplifier

The electronic circuit of the summing and subtracting amplifier is presented in Figure 12. The output voltage v_{out} is calculated as the sum of the input voltages applied to the non-inverting input $v_{11}, v_{12}, \ldots, v_{1n}$, and the subtraction of the input voltages at the inverting input $v_{21}, v_{22}, \ldots, v_{2n}$. So, the implemented function of this amplifier can be presented in the following form [44]:

$$v_{out} = (k_{11}v_{11} + k_{12}v_{12} + \cdots + k_{1n}v_{1n}) - (k_{21}v_{21} + k_{22}v_{22} + \cdots + k_{2n}v_{2n}) \qquad (7)$$

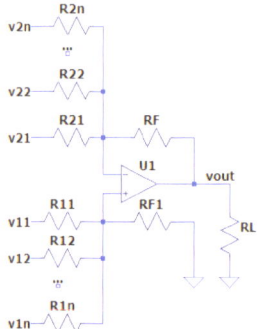

Figure 12. Summing and subtracting amplifier [44].

The sizing coefficients are defined as the ratio between the feedback resistor and the respective input resistor $k_{1j} = \frac{R_{F1}}{R_{1j}}$ and $k_{2j} = \frac{R_F}{R_{2j}}$.

If the specification says that the amplifier must satisfy the following equation:

$$v_{out} = (9v_{11} + 2v_{12}) - (6v_{21} + 3v_{22}) \qquad (8)$$

then the first stage includes the dataset preparation, which is similar to the one shown in Table 1 and the aim is the design with minimal power dissipation P_D to be found considering a given load R_L. The examined design solutions are 144,414: $S_1 \div S_{144414}$ as they possess 11 common attributes: $A_1 = R_F$, $A_2 = R_1$, $A_3 = R_2$, $A_4 = R_3$, $A_5 = R_4$, $A_6 = v_1$, $A_7 = v_2$, $A_8 = v_3$, $A_9 = v_4$, $A_{10} = v_{out}$, $A_{11} = P_D$. All obtained solutions are assigned predicted membership values for each attribute $\sigma^{S_i}(A_j)$ through the usage of a deep learning algorithm. The prediction chart of membership values of power dissipation P_D is presented in Figure 13. The predictive model is evaluated, and it is characterized by very small errors: RMSE = 0.0187, AE = 0.0098, RE = 1.57%.

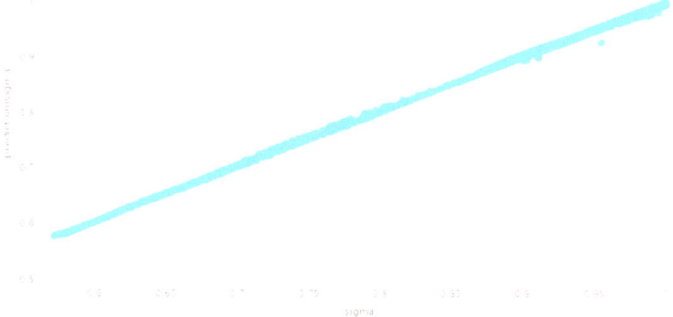

Figure 13. The prediction chart of the membership values of power dissipation (sigma) at design of summing and subtracting amplifier.

The membership values of ten random solutions out of 144,414 found solutions are presented in Figure 14.

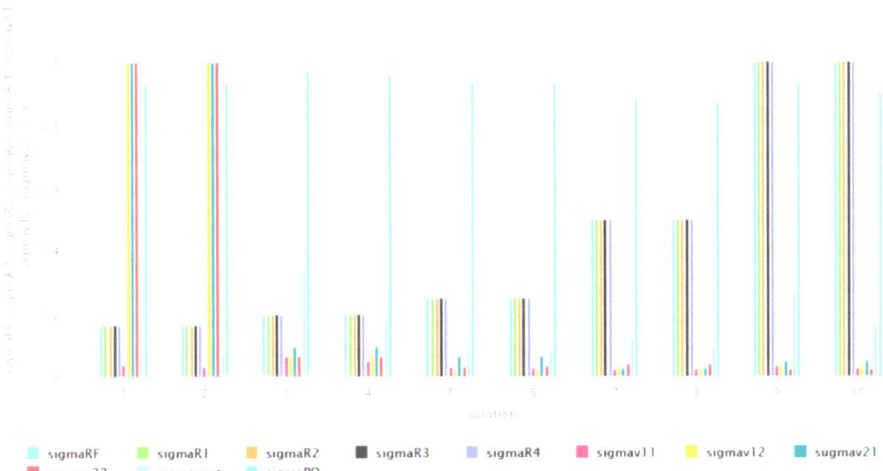

Figure 14. Sigma membership values of ten random designs of summing and subtracting amplifier.

Stage two includes selecting the most suitable solutions from 144,414 designs. The membership values of the most appropriate 39 designs $SS_1 \div SS_{39}$ are calculated, as well as the membership functions of the edges (SS_i, SS_k) taking into account the attributes $A_1 \div A_{11}$. Then, the m-polar fuzzy graph is constructed.

In the third stage, for every attribute, the membership values of the edges between the vertices are calculated and the weights of the solutions are presented in Figure 15. The best solution found is SS_1 according to the predefined user specification.

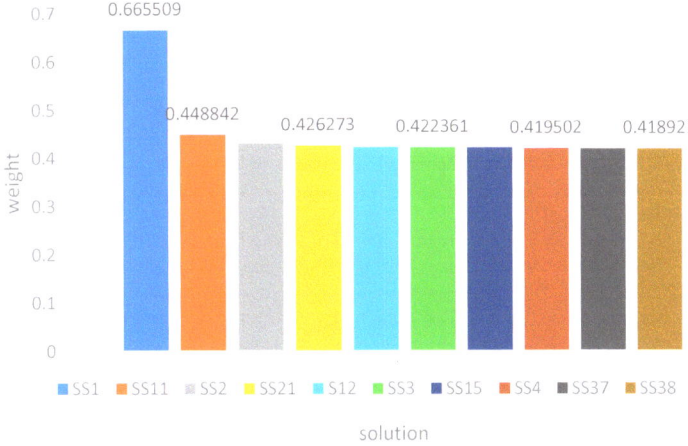

Figure 15. Prioritizing the suitable designs of the summing and subtracting amplifier.

6. Conclusions

The paper presents a novel method for supporting the design process of analog amplifiers based on the concepts of m-polar fuzzy graphs and deep learning techniques. For the verification of the proposed method, three electronic circuits are designed: an inverting summing amplifier, a subtracting amplifier (differential amplifier), and a summing/subtracting amplifier with an operational amplifier as the solved problem is related to finding the designs with the smallest power dissipation and performing a certain mathematical function. The findings point out that:

- The synergetic combination of m-polar fuzzy graphs theory and DL leads to obtaining the most suitable solutions only in three stages, extremely reducing the number of repetitive tasks concerning the calculation of the values of designs' attributes, their comparison, and design selection.
- DL is a suitable approach when expert opinion could be predicted and used for further analysis. In this work, the membership functions of attributes are predicted instead of the expert votes to be gathered. The created predictive models are evaluated, and it is proved that they are characterized with high precision since the obtained errors are very small: RMSE is from 0.0032 to 0.0187, AE is from 0.022 to 0.098, and RE is between 0.27% and 1.57%.
- Fuzzy graph construction gives a possibility for very fast finding the eligible designs, proposes apparatus for their prioritization, and an opportunity for reaching the best design according to a given predefined user specification.

The method can be applied to the design of any electronic circuit to assist a designer in decision-making when the task is multivariate and the environment is complex and uncertain. A promising future work includes further exploration of the advantages and applicability of the m-polar fuzzy graphs theory and deep learning in support of circuit designers and analysts.

Author Contributions: Conceptualization, M.D. and M.I.; methodology, M.I.; software, M.I.; validation M.I. and M.D.; formal analysis M.D. and M.I.; investigation M.I. and M.D.; resources, M.D. and M.I.; writing—original draft preparation, M.D. and M.I.; writing—review and editing, M.D. and M.I. All authors have read and agreed to the published version of the manuscript.

Funding: This research is supported by Bulgarian National Science Fund in the scope of the project "Exploration of the application of statistics and machine learning in electronics" under contract number КП-06-Н42/1.

Data Availability Statement: Not applicable.

Conflicts of Interest: The authors declare no conflict of interest.

References

1. Palomera-Garcia, R. Revisiting Matrix Theory and Electric Circuit Analysis. 2007. Available online: https://www.ineer.org/Events/ICEE2007/papers/140.pdf (accessed on 1 December 2022).
2. Mina, R.; Chadi, J.; Sakr, G.E. A Review of Machine Learning Techniques in Analog Integrated Circuit Design Automation. *Electronics* **2022**, *11*, 435. [CrossRef]
3. Gao, Y.; He, J. An approach to reducing the time complexity of analog active circuit evolutionary design. In Proceedings of the 2015 11th International Conference on Natural Computation (ICNC), Zhangjiajie, China, 15–17 August 2015; pp. 1098–1102. [CrossRef]
4. Zhang, X.; Xia, P.; He, J. Distributed Computation Framework for Circuit Evolutionary Design Under CS Network Architecture. In Proceedings of the 2018 IEEE 18th International Conference on Communication Technology (ICCT), Chongqing, China, 8–11 October 2018; pp. 232–236. [CrossRef]
5. Chen, C.-H.; Yang, Y.-S.; Chen, C.-Y.; Hsieh, Y.-C.; Tsai, Z.-M.; Li, Y. Circuit-Simulation-Based Design Optimization of 3.5 GHz Doherty Power Amplifier via Multi-Objective Evolutionary Algorithm and Unified Optimization Framework. In Proceedings of the 2020 IEEE International Symposium on Radio-Frequency Integration Technology (RFIT), Hiroshima, Japan, 2–4 September 2020; pp. 76–78. [CrossRef]
6. Fang, Y.; Pong, M.H. A Bayesian Optimization and Partial Element Equivalent Circuit Approach to Coil Design in Inductive Power Transfer Systems. In Proceedings of the 2018 IEEE PELS Workshop on Emerging Technologies: Wireless Power Transfer (Wow), Montreal, QC, Canada, 3–7 June 2018; pp. 1–5. [CrossRef]
7. Lyu, W.; Yang, F.; Yan, C.; Zhou, D.; Zeng, X. Batch Bayesian Optimization via Multi-Objective Acquisition Ensemble for Automated Analog Circuit Design. In Proceedings of the 35th International Conference on Machine Learning, Stockholm, Sweden, 10–15 July 2018.

8. Li, Q.; Shih, T.-Y. Non-Foster Matching Circuit Synthesis Using Artificial Neural Networks. In Proceedings of the 2021 IEEE Radio and Wireless Symposium (RWS), San Diego, CA, USA, 17–22 January 2021; pp. 11–13. [CrossRef]
9. Xuefang, X.; Qinghao, Z.; Yun, L. The Fault Analysis of Analog Circuit Based on BP Neural Network. In Proceedings of the 2021 4th International Conference on Advanced Electronic Materials, Computers and Software Engineering (AEMCSE), Changsha, China, 26–28 March 2021; pp. 128–131. [CrossRef]
10. Kaufman, A. *Introduction à la Théorie des Sous-Ensembles flous à l'usage des Ingénieurs: Applications à la Linguistique, à la Logique et à la Sémantique*; Masson et cie 1: Échandens, Switzerland, 1973.
11. Rosenfield, A. Fuzzy graphs. In *Fuzzy Sets and Their Application*; Zadeh, L., Fu, K., Shimura, M., Eds.; Academic Press: New York, NY, USA, 1975; pp. 77–95.
12. Yeh, R.; Bang, S. Fuzzy graphs, fuzzy relations, and their applications to cluster analysis. In *Fuzzy Sets and Their Applications*; Zadeh, L.A., Fu, K.S., Shimura, M., Eds.; Academic Press: New York, NY, USA, 1975; pp. 125–149.
13. Samanta, S.; Pal, M. Fuzzy tolerance graphs. *Int. J. Latest Trends Math.* **2011**, *1*, 57–67.
14. Samanta, S.; Pal, M. Fuzzy threshold graphs. *CIIT Int. J. Fuzzy Syst.* **2011**, *3*, 360–364.
15. Pal, M.; Samanta, S.; Rashmanlou, H. Some results on interval-valued fuzzy graphs. *Int. J. Comput. Sci. Electron. Eng.* **2015**, *3*, 205–211.
16. Pramanik, T.; Samanta, S.; Pal, M. Interval-valued fuzzy planar graphs. *Int. J. Mach. Learn. Cybern.* **2016**, *7*, 653–664. [CrossRef]
17. Rashmanlou, H.; Pal, M. Balanced interval-valued fuzzy graphs. *J. Phys. Sci.* **2013**, *17*, 43–57.
18. Samanta, S.; Pal, M. Fuzzy k-competition graphs and p-competition fuzzy graphs. *Fuzzy Inform. Eng.* **2013**, *5*, 191–204. [CrossRef]
19. Samanta, S.; Akram, M.; Pal, M. M-step fuzzy competition graphs. *J. Appl. Math Comput.* **2015**, *47*, 461–472. [CrossRef]
20. Javaid, M.; Kashif, A.; Rashid, T. Hesitant Fuzzy Graphs and Their Products. *Fuzzy Inf. Eng.* **2020**, *12*, 238–252. [CrossRef]
21. Zhang, W. Bipolar fuzzy sets and relations: A computational framework for cognitive modeling and multiagent decision analysis. In Proceedings of the First International Joint Conference of The North American Fuzzy Information Processing Society Biannual Conference. The Industrial Fuzzy Control and Intellige, San Antonio, TX, USA, 18–21 December 1994; pp. 305–309.
22. Zhang, W. Bipolar fuzzy sets. In Proceedings of the 1998 IEEE International Conference on Fuzzy Systems Proceedings, IEEE World Congress on Computational Intelligence, Anchorage, AK, USA, 4–9 May 1998; pp. 835–840.
23. Akram, M. Bipolar fuzzy graphs. *Inf. Sci.* **2011**, *181*, 5548–5564. [CrossRef]
24. Chen, J.; Li, S.; Ma, S.; Wang, X. M-polar fuzzy sets: An extension of bipolar fuzzy sets. *Sci. World J.* **2014**, *2014*, 416530. [CrossRef] [PubMed]
25. Ghorai, G.; Pal, M. A Study on m-polar Fuzzy Planar Graphs. *Int. J. Comput. Sci. Math.* **2016**, *7*, 283–292. [CrossRef]
26. Ghorai, G.; Pal, M. On some operations and density of m-polar fuzzy graphs. *Pac. Sci. Rev. A Nat. Sci. Eng.* **2015**, *17*, 14–22. [CrossRef]
27. Ghorai, G.; Pal, M. Some isomorphic properties of m-polar fuzzy graphs with applications. *Springer Plus* **2016**, *5*, 2104. [CrossRef]
28. Akram, M.; Akmal, R.; Alshehri, N. On m polar fuzzy graph structures. *Springer Plus* **2016**, *5*, 1448. [CrossRef]
29. Mahapatra, T.; Pal, M. An investigation on m-polar fuzzy tolerance graph and its application. *Neural Comput. Appl.* **2022**, *34*, 3007–3017. [CrossRef]
30. Akram, M. *M-Polar Fuzzy Graphs, Theory, Methods & Applications*; Springer Nature Switzerland AG: Cham, Switzerland, 2019.
31. Pal, M.; Samanta, S.; Ghorai, G. *Modern Trends in Fuzzy Graph Theory*; Springer Nature Singapore Pte Ltd.: Singapore, 2020.
32. Mathew, S.; Mordeson, J.; Malik, D. *Fuzzy Graph Theory*; Springer International Publishing: Cham, Switzerland, 2018.
33. AL-Hawary, T. Complete fuzzy graphs. *Int. J. Math. Comb.* **2011**, *4*, 26–34.
34. Bera, S.; Pal, M. On m-Polar Interval-valued Fuzzy Graph and its Application. *Fuzzy Inf. Eng.* **2020**, *12*, 71–96. [CrossRef]
35. Akram, M.; Sarwar, M. Novel applications of m-polar fuzzy competition graphs in decision support system. *Neural Comput. Applic* **2018**, *30*, 3145–3165. [CrossRef]
36. Nosratabadi, S.; Mosavi, A.; Keivani, R.; Ardabili, S.; Aram, F. State of the Art Survey of Deep Learning and Machine Learning Models for Smart Cities and Urban Sustainability. In *Engineering for Sustainable Future*; Várkonyi-Kóczy, A., Ed.; Inter-Academia 2019; Lecture Notes in Networks and Systems; Springer: Cham, Switzerland, 2020; Volume 101.
37. Alom, M.Z.; Taha, T.M.; Yakopcic, C.; Westberg, S.; Sidike, P.; Nasrin, M.S.; Asari, V.K. State-of-the-Art Survey on Deep Learning Theory and Architectures. *Electronics* **2019**, *8*, 292. [CrossRef]
38. Dieste-Velasco, M.I.; Diez-Mediavilla, M.; Alonso-Tristán, C. Regression and ANN Models for Electronic Circuit Design. *Complexity* **2018**, *2018*, 7379512. [CrossRef]
39. Devi, G.; Tilwankar, G.; Zele, R. Automated Design of An[alog Circuits using Machine Learning Techniques. In Proceedings of the 2021 25th International Symposium on VLSI Design and Test (VDAT), Surat, India, 16–18 September 2021; pp. 1–6. [CrossRef]
40. Wang, Z.; Luo, X.; Gong, Z. Application of Deep Learning in Analog Circuit Sizing. In Proceedings of the 2018 2nd International Conference on Computer Science and Artificial Intelligence, Shenzhen, China, 8–10 December 2018; pp. 571–575. [CrossRef]
41. Hasani, R.M.; Haerle, D.; Baumgartner, C.F.; Lomuscio, A.R.; Grosu, R. Compositional neural-network modeling of complex analog circuits. In Proceedings of the 2017 International Joint Conference on Neural Networks (IJCNN), Anchorage, AK, USA, 14–19 May 2017; pp. 2235–2242. [CrossRef]
42. Dutta, R.; James, A.; Raju, S.; Jeon, Y.J.; Foo, C.S.; Chai, K.T.C. Automated Deep Learning Platform for Accelerated Analog Circuit Design. In Proceedings of the 2022 IEEE 35th International System-on-Chip Conference (SOCC), Belfast, UK, 16 May 2022; pp. 1–5. [CrossRef]

43. Budak, A.F.; Gandara, M.; Shi, W.; Pan, D.Z.; Sun, N.; Liu, B. An Efficient Analog Circuit Sizing Method Based on Machine Learning Assisted Global Optimization. *IEEE Trans. Comput.-Aided Des. Integr. Circuits Syst.* **2022**, *41*, 1209–1221. [CrossRef]
44. Ivanova, M. *Analog Electronics*; Technical University of Sofia: Sofia, Bulgaria, 2020; ISBN 978-619-167-423-7. (In Bulgarian)
45. Kuehl, T. *Top Questions on Op Amp Power Dissipation—Part 2*; Texas Instruments: Dallas, TX, USA, 2014; Available online: https://e2e.ti.com/blogs_/archives/b/precisionhub/posts/top-2-questions-on-op-amp-power-dissipation-part-2 (accessed on 1 December 2022).
46. Texas Instruments. *OPAx322x 20-MHz, Low-Noise, 1.8-V, RRI/O, CMOS Operational Amplifier with Shutdown*; Texas Instruments: Dallas, TX, USA, 2016; Available online: https://www.ti.com/lit/ds/symlink/opa322.pdf?ts=1673723341954&ref_url=https%253A%252F%252Fwww.ti.com%252Fproduct%252FOPA322 (accessed on 1 December 2022).

Disclaimer/Publisher's Note: The statements, opinions and data contained in all publications are solely those of the individual author(s) and contributor(s) and not of MDPI and/or the editor(s). MDPI and/or the editor(s) disclaim responsibility for any injury to people or property resulting from any ideas, methods, instructions or products referred to in the content.

Article

FADS: An Intelligent Fatigue and Age Detection System

Mohammad Hijji [1,*,†], Hikmat Yar [2,†], Fath U Min Ullah [3], Mohammed M. Alwakeel [1], Rafika Harrabi [1], Fahad Aradah [1], Faouzi Alaya Cheikh [4], Khan Muhammad [5,*] and Muhammad Sajjad [2,4,*]

[1] Faculty of Computers and Information Technology, University of Tabuk, Tabuk 47711, Saudi Arabia
[2] Digital Image Processing Laboratory, Islamia College Peshawar, Peshawar 25000, Pakistan
[3] Department of Software Convergence, Sejong University, Seoul 143-747, Republic of Korea
[4] The Software, Data and Digital Ecosystems (SDDE) Research Group, Department of Computer Science, Norwegian University of Science and Technology (NTNU), 2815 Gjøvik, Norway
[5] Visual Analytics for Knowledge Laboratory (VIS2KNOW Lab), Department of Applied Artificial Intelligence, School of Convergence, College of Computing and Informatics, Sungkyunkwan University, Seoul 03063, Republic of Korea
* Correspondence: m.hijji@ut.edu.sa (M.H.); khanmuhammad@g.skku.edu (K.M.); muhammad.sajjad@ntnu.no (M.S.)
† These authors contributed equally to this work.

Abstract: Nowadays, the use of public transportation is reducing and people prefer to use private transport because of its low cost, comfortable ride, and personal preferences. However, personal transport causes numerous real-world road accidents due to the conditions of the drivers' state such as drowsiness, stress, tiredness, and age during driving. In such cases, driver fatigue detection is mandatory to avoid road accidents and ensure a comfortable journey. To date, several complex systems have been proposed that have problems due to practicing hand feature engineering tools, causing lower performance and high computation. To tackle these issues, we propose an efficient deep learning-assisted intelligent fatigue and age detection system (FADS) to detect and identify different states of the driver. For this purpose, we investigated several neural computing-based methods and selected the most appropriate model considering its feasibility over edge devices for smart surveillance. Next, we developed a custom convolutional neural network-based system that is efficient for drowsiness detection where the drowsiness information is fused with age information to reach the desired output. The conducted experiments on the custom and publicly available datasets confirm the superiority of the proposed system over state-of-the-art techniques.

Keywords: artificial intelligence; age prediction; deep learning (DL); drowsiness detection; neural computing; smart surveillance

MSC: 68T07

1. Introduction

Modern cities are linked with crossroads and mass communication channels for rapid transportation that facilitate the daily commutes of millions of people [1]. Despite this, road accidents happen, which is one of the highest causes of people's injuries and deaths. The victims of road accidents often have a permanent disability that stays throughout their life. Accidents are roadside injuries that cause an average of 3242 deaths on a daily basis, which is higher than any other single source in the world [2]. Road crashes are very generic worldwide and are annually estimated by the Association for Safe International Road Travel, 2013, showing that the ratio of deaths in road accidents each year is approximately 1.3 million, where 20–50 million people are injured or permanently disabled. Unless urgent actions are taken, roadside injuries are anticipated to become the fifth leading reason for death by 2030 [3]. Every year, around 328,000 crashes occur in the U.S., which has an annual cost to society of millions of dollars [4]. One of the main reasons for road-related accidents is the inability of drivers due to age. In most of these accidents, the driver is either under- or overage to drive the vehicle. Another reason is that the drivers risk their own life or the lives of the others around them either due to stress, sleepiness, fatigue, drowsiness, or under the influence of alcohol.

Among the above-mentioned reasons, drowsiness is the most common factor. Driver fatigue or drowsiness is a human state where the victim is unaware of its surroundings. Due to the sleep-deprived state, the driver does not know what may be happening in their surroundings, which reduces their attentiveness and leads to road accidents. Millions of people are killed and injured every year due to the driver's state such as sleeping while driving [5,6]. Drowsiness decreases the attentiveness, head, and gaze of the drivers due to which the ratio of road accidents is also increasing. Some studies have revealed that driver drowsiness causes 20% of road accidents, resulting in 50% of serious injuries or death [7–10]. Drivers are usually aware of their drowsiness and can decide to continue or stop driving to rest as most fatal accidents are caused by tired drivers. According to the National Highway Traffic Safety Administration [7], 56,000 crashes occur every year where drowsiness or fatigue was cited by the police as a causal factor that leads to 1550 fatalities and 40,000 nonfatal injuries on an average basis. Similarly, 15% to 44% of crashes take place in the U.S. and Australia [8–10], where 18.6%–30% of heavy vehicle crashes involve fatigue [11,12]. About 30,000 vehicle crashes have caused injuries that were also due to fatigue [13]. Moreover, fatigued or sleepy commercial vehicle drivers have a 21 times greater risk of causing fatal accidents, and safety-related drivers had higher drowsiness levels than other drivers [7]. Along with other issues such as fatigue and drowsiness, most of the accidents involved either underage or overage drivers. According to the report, it is estimated that over 2000 drivers between the ages of 13–19 died in the U.S in 2009 [14].

Several approaches have been suggested by the research community to overcome road accidents. Most of them are working on scalar sensors to detect the driver's heartbeat and temperature while several rely on vision sensors. Most of the existing studies have used complex networks that are costly and difficult to deploy over edge devices. Similarly, the existing methods are limited to detecting the different states of the drivers including age. Thus, to tackle the problems and challenges, we proposed "FADS: An Intelligent Fatigue and Age Detection System" by using several realistic approaches for drowsiness detection and age classification based on facial feature analysis to keep underage and overage people from driving when an alarm is generated when the driver's state is detected as drowsy, angry, or sad. We used a lightweight CNN that is easily deployable over an edge device (e.g., Jetson Nano) to perform real-time processing [7], making it suitable for smart surveillance and the Internet of vehicles. The major contributions of the proposed FADS are summarized as follows:

- We developed a DL-assisted FADS for driver mood detection from an easy-to-deploy resource-constrained vision sensor. Addressing this issue of complex systems, it can overcome high computational costs and ensure the real-time detection of the driver's mood.
- Age is an important factor in avoiding most of the accidents, and for this purpose, the proposed FADS extracts facial features to classify the driver's age. If the classified age is beyond the defined threshold (age <18 and age >60), then an alert is generated to notify the nearby vehicles and the authorized department. Another influencing factor that causes road accidents is drowsiness or driver moods such as anger or sadness. Therefore, their prediction is also performed by the facial features using a lightweight CNN. These factors can avoid most accidents and ensure safe vehicle driving.
- Due to data unavailability, we created a new dataset for FADS as a step toward the smart system, which includes five classes (i.e., active, angry, sad, sleepy, and yawning). Furthermore, a UTKFace dataset was categorized into three classes (i.e., underage (age <18), middle-age (age ≥ 18 or ≤ 60), and overage (>60)) for detailed analysis. This categorization further enhances FADS by fusing dual features to reach an optimum outcome, which is needed for smart surveillance.

- Extensive experiments were conducted from different aspects and the results over the baseline CNNs confirm that the proposed FADS achieved state-of-the-art performance on the standard and the new dataset in terms of lower model complexity and good accuracy.

The remainder of the paper is structured as follows. A compact literature review is presented in Section 2. We cover the proposed FADS in Section 3. The experimental results of FADS and its comparisons are given in Section 4. Section 5 concludes our work with some future research directions.

2. Literature Review

Facial images can be used for the analysis of a driver's behavior based on drowsiness or driver mood detection (i.e., anger or sadness). For instance, the technique named "DriCare" [15] used face landmarks and their key points to detect faces and track them for driver fatigue detection. This includes eye blinking, eye closure, or yawning. Next, Verma et al. [16] enhanced the strategy using two VGG16 CNN parallels to detect the driver expressions. First, the region of interest was detected, which was fed to the VGG16 model as the input while the face landmarks and key points were used as the input in the second VGG16 model. Their combined results were used for fatigue detection. In another approach [17], a dataset called DROZY (ULG Multi-modality Drowsiness dataset) was developed for drowsiness detection. Tsaur et al. [18] proposed a real-time system for driver abnormality detection using edge-fog computing and achieved a promising performance. Furthermore, Xing et al. [19] attempted to detect seven different tasks performed by the drivers such as normal driving, using a mobile phone, checking left and right mirrors, and setting up video devices in a vehicle. They extracted 42 different features using a Kinect camera and used random forests for classification. Yu et al. [20] employed a condition-adaptive representation method for driver drowsiness detection. Their system was evaluated on the NTHU driver drowsiness detection video dataset, which outperformed the state-of-the-art methods based on visual analysis. In the next approach, Dua et al. [21] used four different DL models such as ResNet, FlowImageNet, VGG-FaceNet, and AlextNet for drowsiness detection. However, the time complexity and limited accuracy restricted their system from real-world deployment. Abdelmalik et al. [22] proposed a four-tier approach for driver drowsiness detection consisting of face detection and alignment, pyramid multi-level face representation, face description using multi-level features extraction, and features subset selection. Likewise, in [23], the authors used a DL approach for eye state classification in static facial images, where they fused two deep neural networks for a better decision. Recurrent convolutional neural networks (R-CNN) have also played an important role in detecting the driver's state such as normal blinking or a falling asleep situation from the sequences of the frames [15,24]. Ghoddoosian et al. [25] presented a technique for eye blink detection based on hierarchical multi-scale long short-term memory.

Aside from drowsiness detection, age classification based on facial feature analysis is a trending area due to its wide range of applications such as human–computer interaction, security, and age-oriented commercial advertisement. Several traditional methods [26,27] and DL methods [28–31] have been presented for age classification. For instance, [32] presented a technique in which they used VGG16 CNN architecture for age classification by creating the dataset "IMDB-WIKI". Similarly, Shen et al. [33] presented deep regression forests for end-to-end feature learning for age estimation. The authors in [34] predicted age using a directed acyclic graph CNN. Furthermore, Lou et al. [35] presented an expression invariant age classification method by concurrently learning the age and expression. They studied the correlation between age and expression by deploying a graphical model that adopted a hidden layer. In [36], the researchers presented an ordinal DL mechanism by learning features for both age estimation and face representation.

In the aforementioned techniques, several researchers have individually contributed to the area of driver fatigue detection and age classification. However, none of them could detect the different driver states along with their age. Therefore, we proposed FADS for both driver fatigue detection and age classification, restricting underage and overage people from driving, and generating an alarm in the case of detecting drivers in a fatigued state.

3. Fatigue and Age Detection System

In this section, the proposed system is explained in detail. First, the face was detected using an improved Faster R-CNN algorithm. Next, different CNN models were used to examine various facial features for age classification and the driver mood or state detection. Finally, we fused both the driver's age and mood information to provide an effective solution. Furthermore, the proposed system was deployable over edge devices where a vision sensor captures the live stream images/videos, and an edge device was mounted on top of the dashboard in a car. The video stream is processed, and the age and overall state of the driver are predicted in real-time. The proposed system was divided into the following steps: face detection, drowsiness detection, age classification, and fusion strategy, as demonstrated in Figure 1.

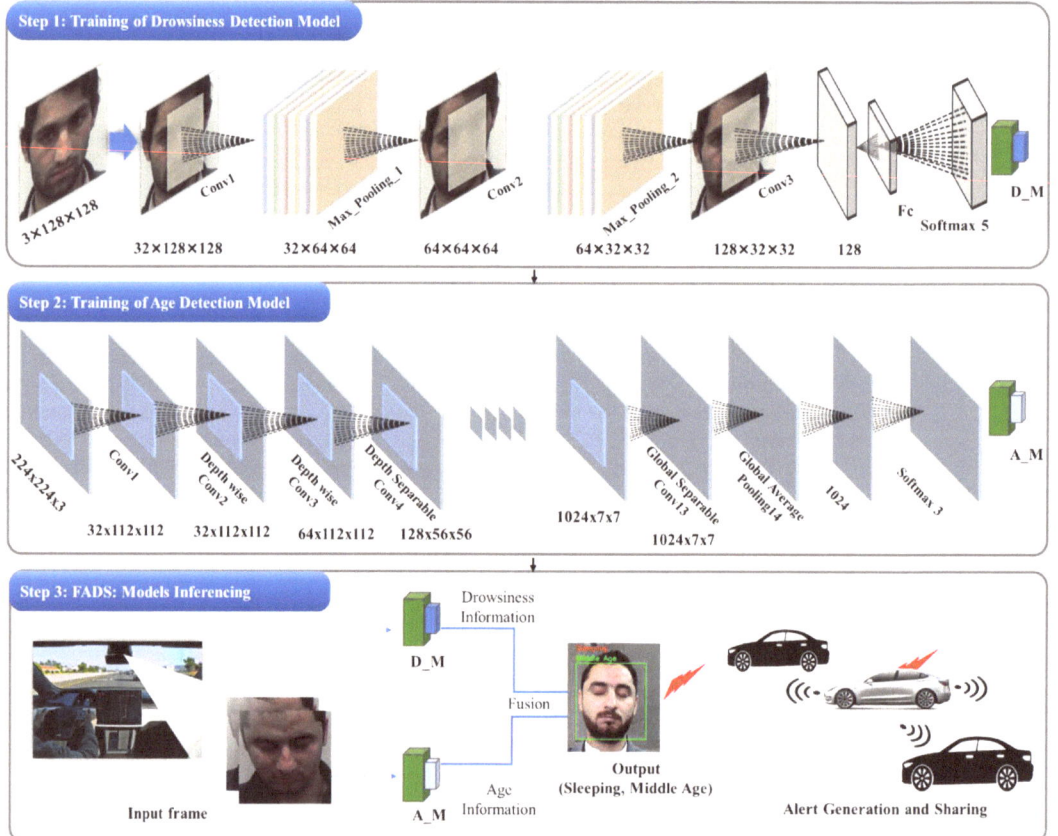

Figure 1. The proposed FADS consists of four different stages: (1) receives an input frame from the edge device, (2) input frames are received where face detection is performed via an efficient algorithm. (3) Two CNN networks are employed for feature extraction and classification purposes and (4) an output label is obtained from the fusion information, which is further sent to the nearest vehicle and authorized authorities for safety concerns.

3.1. Face Detection

Face detection is the most imperative problem that has been intensively surveyed in the last few decades. Early researchers were mainly concerned with hand-crafted feature extraction methods [37,38]. However, there are some limitations in these techniques. They often require experts in the field of image processing to extract effective and useful features where each component is optimized individually, making the entire pipeline of the detection process often sub-optimal. Therefore, Sun et al. [39] proposed a technique to extend the state-of-the-art Faster R-CNN method [40]. Their approach increased the existing Faster R-CNN approach by fusing multiple important schemes, consisting of feature fusion, multi-scale training, and hard negative mining. We employed a similar strategy to Sun et al. [39] to capture the face images. This strategy contained two main steps: a region proposal network to capture the regions of interest and a Fast R-CNN network to classify the region into its corresponding category. Sun et al. [39] trained a Faster R-CNN with the WIDER Face dataset [31]. Furthermore, the targeted dataset was used to test the model to generate hard negatives, which were then fed into the network during training as a second step. By training these hard negative samples, the trained model was capable of generating a lower false positive rate. Moreover, their model was fine-tuned on the FDDB dataset. In the final phase, they employed the multi-scale training process and adopted a feature-fusing strategy to improve the model performance. For the entire training procedure, an end-to-end model training strategy such as Faster R-CNN was used due to its effective performance. Finally, the resulting detection bounding boxes were converted into ellipses as the regions of human faces. Therefore, we employed an improved Faster R-CNN approach for efficient and accurate face detection in the FADS, whose sample results are visualized in Figure 2.

Figure 2. Visual demonstration of the face detection algorithm used in the proposed system.

3.2. Driver Drowsiness Detection

Inspired by the performance of the MobileNet [41–43], GoogleNet [44], SqueezeNet with deep autoencoder [45], Inception [46,47], Darknet [48,49], and ConvLSTM [50] models, we implemented a new custom network consisting of three convolution layers, where a max-pooling layer was used after each convolutional layer and two dense layers, as demonstrated in Figure 3. In the first convolutional layer, the input image size was $128 \times 128 \times 3$ with 32 different kernels. Each kernel's size was 3×3 with a one-pixel stride, on which the pooling operation was applied. The output of the first convolutional layer is the input of the second convolutional layer, and there was a total of 64 kernels having a size of 3×3. The third convolution layer had 128 kernels with the size of 3×3 connected to the output of the previous layer. The fully connected layer had 64 neurons, which fed the output to the softmax classifier to classify the input source image into their corresponding classes. The custom architecture learns 2.2 million parameters compared to all of the above-the mentioned models such as the AlexNet [51], Vgg16 [52], MobileNet [41], GoogleNet [44], Inception [46], and MobileNet [41] models. Custom architecture computes an extremely smaller number of parameters due to the size of the input image and several filters selected during the convolution.

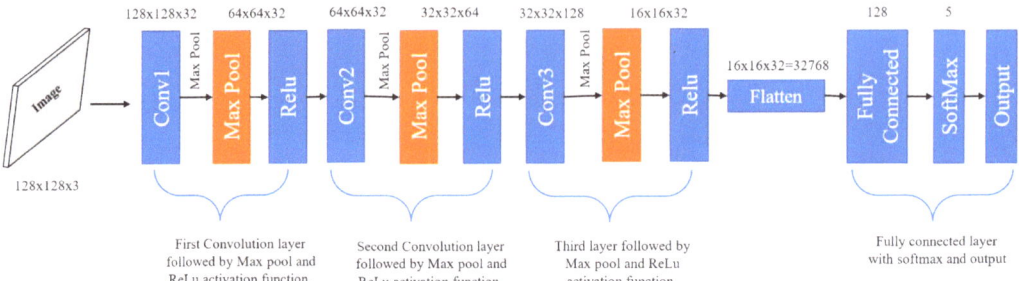

Figure 3. Layer-by-layer architecture of the proposed system.

3.3. Driver Age Classification

For driver age classification in the FADS, we used a fine-tuned lightweight model for prominent facial feature extraction and classification. In the proposed system, we employed MobileNet, a depth-wise separable convolutional neural network, which is a lightweight DCNN and provides an efficient system for embedded vision applications. In this model, the depth-wise separable convolutions are composed of point convolution filters (PCF) and depth-wise convolution filters (DCF). The DCF performs a single convolution on each channel, and PCF combines the output of the DCF linearly with 1×1 convolutions, as shown in Figure 4. The output of the depth-wise separable convolution using RGB images with a 3×3 kernel size and a movement interval of 1 is given in Equations (1) and (2) [53].

$$\hat{O} = \sum_{i=1}^{3}\sum_{j=1}^{3} K_{j,i,c} \cdot F_{x+i, y+j-1, c} \quad (1)$$

$$O = \sum_{c=1}^{3} \check{K}_{c,n} + \hat{O}_{x,y,v} \quad (2)$$

where $\hat{O}$ shows the output result of the depth-wise convolution; K is the kernel; and F is the input. In Equation (2), O represents the output of the pointwise convolution and $\check{K}$ is the kernel of the 1×1 convolution. We employed the above-mentioned strategy of PCF and DCF and modified the MobileNet architecture consisting of 28 convolution layers including deep convolutional layers (point convolution layer) 1×1, batch-normalization, ReLu activation, average pooling, and a softmax layer. In this architecture, the ReLU activation is employed in each convolutional layer to perform a thresholding operation, where each input value less than 0 is set to 0, and positive values remain the same. MobileNet consists of a pooling layer strategy, which summarizes the outputs of neighboring groups of neurons. The pooling layer is used for dimensionality reduction, which influences the duration of the network training, and the output neurons are equal to the number of classes in the dataset recognized by the network. Finally, softmax is used for probabilities to classify a driver's age. This probability is the basis for making the final decision about the classification result. In summary, we employed an efficient MobileNet model for driver age classification into three different categories such as underage, middle age, and overage.

3.4. Fusion Strategy in FADS

This subsection explains the fusion strategy of the proposed system to achieve the desired output. The CNN-based architecture was used for face detection, drowsiness detection, and driver age classification. Our system consisted of three steps. (1) The input frames were acquired from the vision sensor mounted with Jetson Nano, where the face is detected through an improved Faster R-CNN. The detected face was cropped from the entire image and fed into CNN for drowsiness detection and driver age classification. (2) The detected face was processed by our new custom CNN architecture that performed the drowsiness detection of the individual driver present in the entire frame. (3) The driver age was computed and classified using a customized version of the MobileNet architecture. Finally, we fused these models during the inference time to achieve the desired output, as shown in Figure 1.

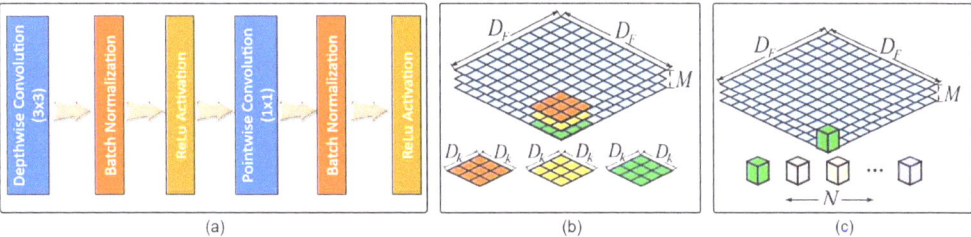

Figure 4. MobileNetV2 architecture where (**a**) represents the depth-wise and pointwise layers followed by batch-normalization and the ReLU activation function, (**b**) depth-wise convolutional layer, and (**c**) pointwise convolutional layer [54].

4. Results and Discussion

This section provides a detailed explanation of the hardware configurations, the datasets used for the driver age classification and drowsiness detection, and the training and testing process in the evaluation. Furthermore, quantitative, and qualitative assessments were performed with the state-of-the-art for both driver age and drowsiness detection. For the training process, we categorized both datasets into three subclasses (i.e., training, testing, and validation) with the proportion of 70%, 10%, and 20%, respectively.

4.1. System Configuration and Evaluation

The proposed system was trained using NVidia GPU GTX 1070 GPU, which has 8 GB of RAM and a 2.9 GHz processor. The operating system, programming language, and libraries used in our work are listed in Table 1.

Table 1. Software specification and libraries used for the proposed system.

Name	Configuration
OS	Window 10
Programming language and IDE	Jupyter Notebook, Python 3.7.2
Libraries	TensorFlow, PyLab, Numpy, Keras, Matplotlib
Imaging libraries	OpenCV 4.0, Scikit-Image, Scikit-Learn

In the computer vision domain, the trained CNN is mostly assessed by conducting quantitative analysis via commonly used different evaluation parameters including accuracy, F1-measure, precision, and recall (sensitivity). These evaluation metrics can be easily calculated from the confusion matrix by forwarding the predicted and actual labels. The mathematical expression of accuracy, precision, recall, and F1-measure are given in Equations (3)–(6), respectively. The accuracy is considered on the major evaluation matrix to evaluate the overall performance of the system.

$$Accuracy = \left(\frac{TPV + TNV}{TPV + TNV + FPV + FNV} \right) \quad (3)$$

$$Precision = \left(\frac{TPV}{TPV + FPV} \right) \quad (4)$$

$$Recall = \left(\frac{TPV}{TPV + FNV} \right) \quad (5)$$

$$F1-measure = 2*\left(\frac{P*R}{P+R}\right) \tag{6}$$

4.2. Dataset Explanation

In the proposed system, we used two datasets: a custom dataset for drowsiness detection and a UTKFace [55] dataset for age classification. The custom dataset had images that were collected from different sources. We used UTKFace [55] for he acquisition of better results. Each dataset was properly cleaned and labeled. These datasets are described as follows.

The UTKFace is a large-scale publicly available dataset for facial feature analysis to predict age, ranging from 0 to 116 years. It consists of 23,708 RGB facial images, having a resolution of 200 × 200 pixels with .jpg extensions, and annotations of age, ethnicity, and gender. The images cover large variations in pose, occlusion, illumination, facial expression, and resolution. This dataset can be used in a variety of vision-related tasks such as age regression, age estimation, face detection, and landmark localization. In this research, we converted all images to one standard JPG format and made three classes from the dataset, as demonstrated in Figure 5. Their corresponding age factors are given in Table 2 such as underaged (6–18), middle-aged (18–60), and overaged (60+).

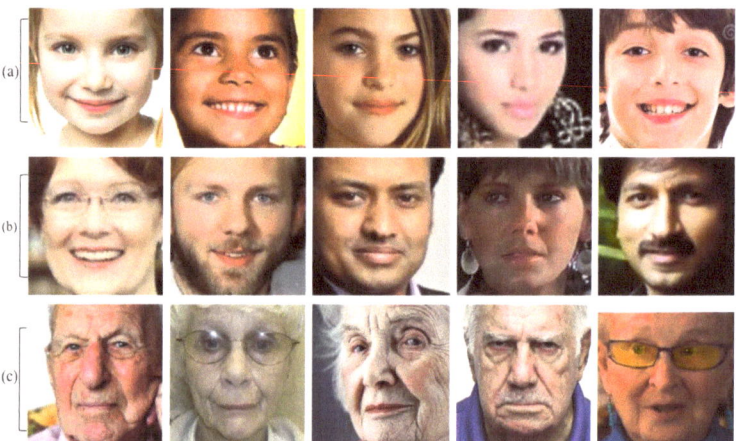

Figure 5. Sample images of the modified datasets for age classification (**a**) underage, (**b**) middle age, and (**c**) overage.

Table 2. The information about the ages (in years) of different people.

Class	Age Group
Underage	6–16
Middle age	18–60
Overage	60+

Furthermore, we collected a custom dataset for drowsiness detection. It contains three classes (i.e., active, sleeping, and yawning). Each class had two thousand images, having a resolution of 124 * 124 with three channels Red (G), Green (G), and Blue (B). In the creation of this dataset, a total of 40 university students participated, whose ages were in the range of 16–35 years. Furthermore, the angry and sad classes were taken from the publicly available KDEF [56] dataset. Figure 6 shows sample images of our dataset.

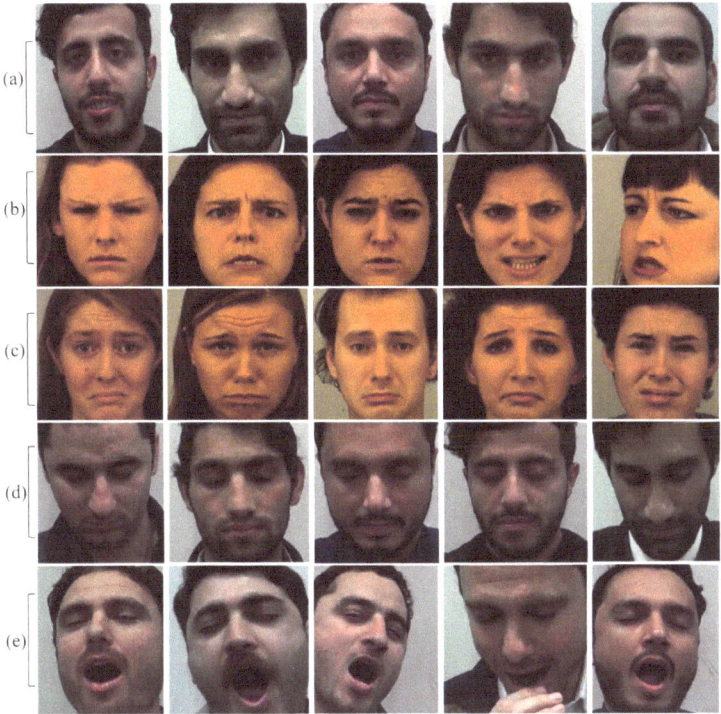

Figure 6. Sample images of the custom dataset. (**a**) Sample images of the active class, (**b**) angry, (**c**) sad, (**d**) sleeping class, (**e**) yawning class.

4.3. Performance Comparison of Different Edge Devices

A single-board-computer or System-on-Chip (SoC) is becoming popular among the research community because of its versatility in different video streaming and machine learning-based applications [57,58]. SoC consists of input and output ports along with enough memory and disk space to run certain applications smoothly. However, the major issue with these devices is that they are usually incapable of using DL models as they do not have enough neural computation capabilities needed for DL model implementation in real-time. For this purpose, we conducted a survey and came up with several different options, among which we selected Nvidia's Jetson Nano as a prime candidate for this application. A list of available options is given in Table 3.

Table 3. Comparison of the different prototyping platforms along with their specifications.

Board	Chip	RAM	OS
Udoo [59]	ARM Cortex A9	1 GB	Debian, Android
Phidgets [60]	SBC	64 MB	Linux
Beagle Bone [61]	ARM AM335 @ 1 Ghz	512 MB	Linux Angstrom
Raspberry Pi 4 [62,63]	Broadcom BCM2711 Processor	2 GB, 4 GB, 8 GB	Raspbian
Jetson Nano [64]	1.43 Ghz Quad Core Cortex A57	4 GB	All Linux Distro

The Jetson Nano has most of the required capabilities compared to other platforms. It is a small-size high-performance computer that can run modern AI applications at low cost and low power. Recently, the AI community is leaning more toward using Jetson Nano as a computational platform for real-time applications because it can run different

AI-based systems for applications (i.e., image segmentation, object detection, and image classification). The Jetson Nano can be powered by micro-USB and it comes with wide-ranging I/O interfaces including general purpose input/output (GPIO) pins that provide an ease of implementation for different sensors to provide an easy-to-use interface for different sensors, as explained in [64].

4.4. Results of Drowsiness Detection

In this subsection, we discuss the experimental results of the drowsiness detection in terms of the confusion matrix, accuracy, and loss graphs, as shown in Figures 7 and 8. During testing, the proposed system was evaluated for each class, and we found that the accuracy of active, sad, and yawning was far better than the angry and sleeping classes, which was 97% and 98%, respectively. We performed various experiments with the help of the above-mentioned dataset. The experiments were performed with different parameters such as a different number of training epochs for the purpose of achieving high accuracy. In Figure 8, we can see that the training accuracy started from 62% and the validation accuracy started from 55% in the first epoch. After each epoch, the accuracy of training and validation improved. In the third epoch, the training accuracy intercepts the line of validation accuracy. Finally, after 30 epochs, the level of training accuracy reached 98% and the validation accuracy reached 97%, as shown in Figure 8.

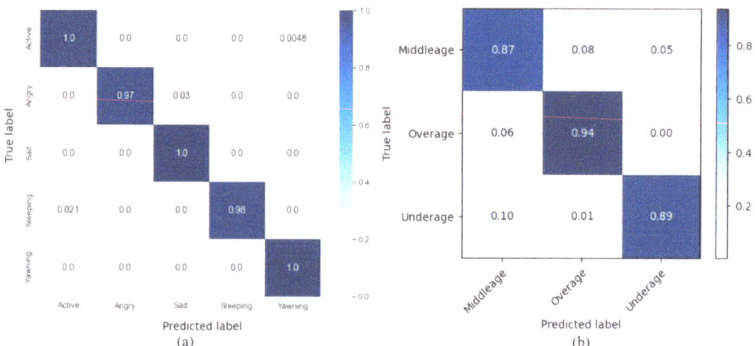

Figure 7. Confusion matrix of our system to validate the class-wise performance. (**a**) Confusion matrix of the drowsiness dataset (**b**) Confusion matrix for the age dataset.

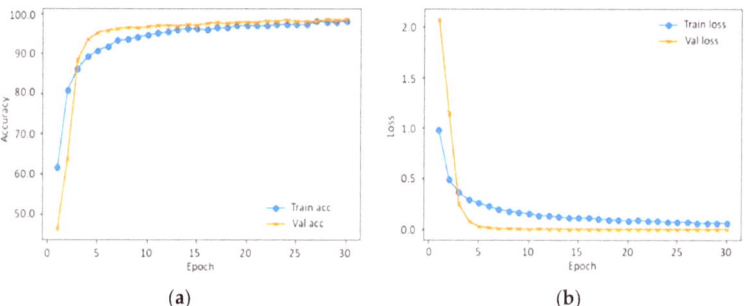

Figure 8. Training/validation accuracy and loss, where (**a**) represents the accuracy and (**b**) represents a loss of drowsiness detection.

We used different evaluation metrics (i.e., recall, precision, and F1-measure) for the performance validation. The results obtained by the proposed system using the drowsiness dataset are given in Table 4. The comparative analysis of the proposed system is given in Table 5, where it was compared with four state-of-the-art systems. The proposed system reached an accuracy of 98%, where the accuracy of AlexNet, VGG16, ResNet50, and MobileNet was 94.0%, 98.3%, 88.0%, and 93.5%, respectively.

Table 4. Results of the drowsiness detection in terms of the precision, recall, and F1-score.

Driver State	Precision	Recall	F1-Measure
Active	0.98	1	0.99
Angry	1	0.97	0.98
Sad	0.97	1	0.98
Sleeping	1	0.98	0.99
Yawning	0.99	1	0.99

Table 5. Comparison of different DL architectures on the custom drowsiness detection dataset in terms of model size, parameters, and accuracy.

Technique	Model Size (MB)	Parameters (Million)	Accuracy (%)
AlexNet [51]	233	60	94.0
VGG16 [52]	528	138	98.3
ResNet50 [65]	98	20	88.0
MobileNet [41]	13	4.2	93.5
The proposed system	15	2.2	98.0

4.5. Results of Age Classification

A detailed explanation of age classification using different DL architectures is given in this section. In Figure 9, we demonstrated the accuracy of MobileNet, which had the highest accuracy in our experiments. Figure 9 represents the training accuracy, which started from 64% in the first epoch, whereas the validation accuracy started from 56%. After each epoch, the accuracy of training and validation showed a certain fluctuation.

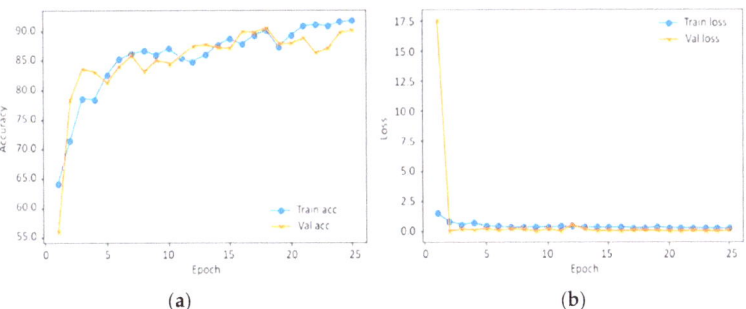

Figure 9. The proposed system's training and validation accuracy and loss of age classification where (**a**) is the accuracy and (**b**) is the loss.

Finally, after the 25th epoch, the level of training accuracy reached 91%, the validation accuracy reached 89%, and the loss of training and validation was nearly 0%, as shown in Figure 9.

The reports generated in Table 6 demonstrate the result of the age estimation using metrics such as the F1-measure, recall, and precision. The experimental evaluation based on accuracy is given in Table 7, where the proposed system obtained an average accuracy of 90%, which surpassed AlexNet, VGG16, and ResNet50 by achieving the higher value of 13%, 9%, and 6%, respectively.

Table 6. Performance of the proposed system over the modified UTKFace dataset in terms of the precision, recall, and f1-score.

Age Classes	Precision	Recall	F1-Measure
Middle age	0.88	0.84	0.86
Overage	0.90	0.97	0.93
Underage	0.92	0.88	0.90

Table 7. Comparison of different DL architectures on the modified UTKFace dataset.

Technique	Accuracy (%)
AlexNet [51]	77.0
VGG16 [52]	81.0
ResNet50 [65]	84.0
The proposed system	90.0

4.6. Time Complexity Analysis

In this section, we discuss the time complexity analysis of the proposed system and compare it with various versions of deep CNNs as given in Table 8. We show the frame per second (FPS) of four different fused methods such as (AlexNet and MobileNet), (VGG16 and MobileNet), (ResNet50 and MobileNet), and the proposed (MobileNet and custom CNN) over CPU, GPU, and Jetson Nano. The CPU system used for running time analysis was an Intel(R) Core (TM) i3-4010u CPU @ 1.70 GHz with 4 GB RAM. To validate the system performance, we calculated the FPS using Equation (7). We verified that the proposed system was significantly faster than other CNN architectures over CPU, GPU, and Jetson Nano. In Table 8, it can be seen that a lower FPS is associated with VGG16 + MobileNet (i.e., 5.73 FPS on CPU and 33.07 FPS on GPU). The FPS of AlexNet + MobileNet using CPU, GPU, and Jetson Nano was 6.37, 39.87, and 8.01, respectively. The ResNet50 + MobileNet achieved a higher FPS compared with AlexNet + MobileNet and VGG16 + MobileNet. However, our model outperformed the ResNet50 + MobileNet by achieving a higher FPS of 4.98, 12.53, and 5.31, respectively. The better FPS shows that our system is easily deployable over resource-constrained devices.

Table 8. Comparison of different DL architectures on the custom drowsiness detection dataset.

Method Fusion	CPU	GPU	Jetson Nano
AlexNet + MobileNet	6.37	39.87	8.01
VGG16 + MobileNet	5.73	33.07	6.78
ResNet50 + MobileNet	8.90	42.50	13.12
The proposed system	13.88	55.03	18.43

4.7. Qualitative Analysis of the Proposed System

In this section, we demonstrate the visual results of the proposed system, as shown in Figure 10. First, the proposed system was able to accurately detect the face in the entire image. Next, the system could classify the age and state of the driver. In Figure 10a, a set of sample images are shown from the Internet and Figure 10b shows a few sample images taken from a real-time scenario of a camera. In the last row of Figure 10a,b, there was a wrong classification due to the visual similarity of each class. These results show the efficiency and effectiveness of the proposed model and can be deployed for age and various states of driver detection, as evidenced by the quantitative and qualitative analysis.

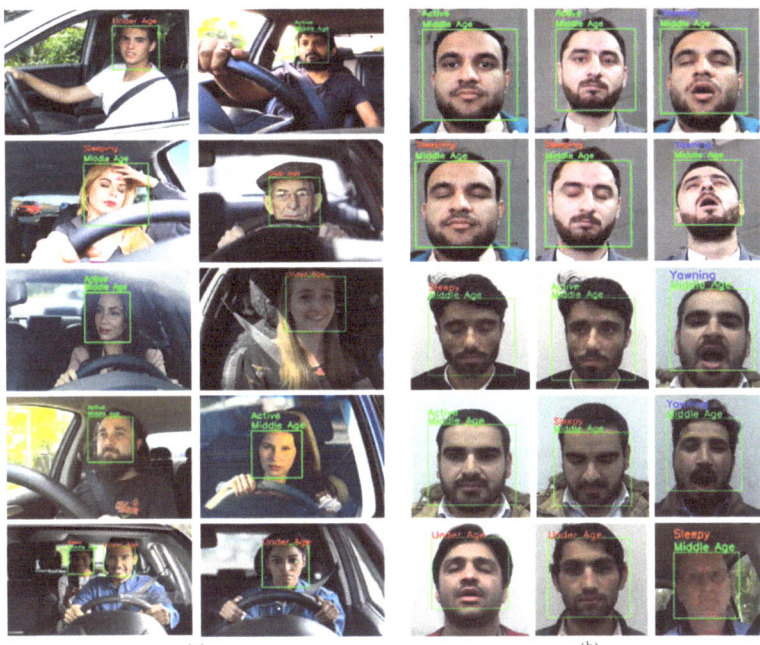

Figure 10. The visual results analysis of the proposed system over images taken from the Internet and from a real scenario. (**a**) Images were taken from the Internet to check the different states and ages of the drivers. (**b**) Images were taken from a real-time scenario for drowsiness detection and age classification.

5. Conclusions and Future Work

In this study, an intelligent fatigue and age detection system (FADS) was proposed for the safety of drivers, helping to prevent plenty of human losses and increasing the intelligence level of vehicles for smart surveillance. The proposed FADS was tested on different platforms for comparison and real-time applicability. A custom CNN model was suitable for low-power hardware, which was deployable over Nvidia's Jetson Nano to achieve portability and a relatively good inference time. After extensive experimentation, we chose two different CNN architectures for driver drowsiness detection and driver age classification based on face feature analysis. The driver's drowsiness detection was achieved using a custom dataset and for age classification, we modified the UTKFace dataset. For experimental evaluation, we evaluated different DL architectures such as AlexNet, VGG16, ResNet50, MobileNetV2, and a 3-layer custom CNN for driver drowsiness detection. The Custom CNN model provided better results and reached an accuracy of 98% for drowsiness detection, whereas MobileNetV2 provided good results in terms of 90% accuracy using the UTKFace dataset. Finally, the results of both models were fused during inference time to ease the deployment for real-time assistance. The developed system is helpful for aged people to prevent them from vehicle accidents. In addition, the results shown in Section 4.6 exhibit the deployability of the proposed method over resource-constrained devices to reduce heavy computation and consumption.

Future work of this study can consider different scenarios such as optimizing a single end-to-end network for its usage in an embedded system to reduce the computational and financial costs without affecting the performance. Next, a federated learning mechanism can be designed for the development of an online model to improve the edge learning capability of FADS. The current dataset can be extended by adding the fatigue levels of people of different ages.

Author Contributions: Conceptualization, M.M.A.; Methodology, M.H., H.Y., F.U.M.U., M.M.A., R.H., F.A.C. and K.M.; Software, H.Y. and F.A.; Validation, F.A., F.A.C., K.M. and M.S.; Formal analysis, F.U.M.U., R.H., F.A.C., K.M. and M.S.; Investigation, M.H. and K.M.; Resources, R.H.; Writing—original draft, H.Y.; Writing—review & editing, M.H., H.Y., F.A.C. and M.S.; Supervision, K.M.; Project administration, K.M.; Funding acquisition, M.H. and K.M. All authors have read and agreed to the published version of the manuscript.

Funding: This work was supported by the Deanship of Scientific Research at the University of Tabuk through Research No. 0254-1443-S.

Data Availability Statement: Not applicable.

Acknowledgments: The authors extend their appreciation to the Deanship of Scientific Research at the University of Tabuk for funding this work through Research No. 0254-1443-S.

Conflicts of Interest: The authors report that there are no competing interests to declare.

References

1. Lin, C.; Han, G.; Du, J.; Xu, T.; Peng, Y. Adaptive traffic engineering based on active network measurement towards software defined internet of vehicles. *IEEE Trans. Intell. Transp. Syst.* **2020**, *22*, 3697–3706. [CrossRef]
2. Peden, M.; Scurfield, R.; Sleet, D.; Mohan, D.; Hyder, A.A.; Jarawan, E.; Mathers, C. *World Report on Road Traffic Injury Prevention*; World Health Organization: Geneva, Switzerland, 2004.
3. World Health Organization. *Association for Safe International Road Travel. Faces behind Igures: Voices of Road Trafic Crash Victims and Their Families*; OMS: Genebra, Switzerland, 2007.
4. National Safety Council. Drivers are Falling Asleep Behind the Wheel. 2020. Available online: https://www.nsc.org/road/safety-topics/fatigued-driver (accessed on 1 January 2023).
5. Vennelle, M.; Engleman, H.M.; Douglas, N.J. Sleepiness and sleep-related accidents in commercial bus drivers. *Sleep Breath.* **2010**, *14*, 39–42. [CrossRef] [PubMed]
6. de Castro, J.R.; Gallo, J.; Loureiro, H. Tiredness and sleepiness in bus drivers and road accidents in Peru: A quantitative study. *Rev. Panam. Salud Publica (Pan Am. J. Public Health)* **2004**, *16*, 11–18.
7. Lenné, M.G.; Jacobs, E.E. Predicting drowsiness-related driving events: A review of recent research methods and future opportunities. *Theor. Issues Ergon. Sci.* **2016**, *17*, 533–553. [CrossRef]
8. Tefft, B.C. Prevalence of motor vehicle crashes involving drowsy drivers, United States, 1999–2008. *Accid. Anal. Prev.* **2012**, *45*, 180–186. [CrossRef]
9. Armstrong, K.; Filtness, A.J.; Watling, C.N.; Barraclough, P.; Haworth, N. Efficacy of proxy definitions for identification of fatigue/sleep-related crashes: An Australian evaluation. *Transp. Res. Part F Traffic Psychol. Behav.* **2013**, *21*, 242–252. [CrossRef]
10. Centers for Disease Control and Prevention Drowsy driving-19 states and the District of Columbia, 2009–2010. *MMWR Morb. Mortal. Wkly. Rep.* **2013**, *61*, 1033–1037.
11. Williamson, A.; Friswell, R. The effect of external non-driving factors, payment type and waiting and queuing on fatigue in long distance trucking. *Accid. Anal. Prev.* **2013**, *58*, 26–34. [CrossRef]
12. Hassall, K. Do 'safe rates'actually produce safety outcomes? A decade of experience from Australia. In *HVTT14: International Symposium on Heavy Vehicle Transport Technology*, 14th ed.; HVTT Forum: Rotorua, New Zealand, 2016.
13. Kalra, N. *Challenges and Approaches to Realizing Autonomous Vehicle Safety*; RAND: Santa Monica, CA, USA, 2017.
14. Ballesteros, M.F.; Webb, K.; McClure, R.J. A review of CDC's Web-based Injury Statistics Query and Reporting System (WISQARS™): Planning for the future of injury surveillance. *J. Saf. Res.* **2017**, *61*, 211–215. [CrossRef]
15. Deng, W.; Wu, R. Real-time driver-drowsiness detection system using facial features. *IEEE Access* **2019**, *7*, 118727–118738. [CrossRef]
16. Zhao, L.; Wang, Z.; Wang, X.; Liu, Q. Driver drowsiness detection using facial dynamic fusion information and a DBN. *IET Intell. Transp. Syst.* **2017**, *12*, 127–133. [CrossRef]
17. Massoz, Q.; Langohr, T.; François, C.; Verly, J.G. The ULg multimodality drowsiness database (called DROZY) and examples of use. In Proceedings of the 2016 IEEE Winter Conference on Applications of Computer Vision (WACV), Lake Placid, NY, USA, 7–10 March 2016; IEEE: Piscataway, NJ, USA, 2016; pp. 1–7.
18. Tsaur, W.J.; Yeh, L.Y. DANS: A Secure and Efficient Driver-Abnormal Notification Scheme with I oT Devices Over I o V. *IEEE Syst. J.* **2018**, *13*, 1628–1639. [CrossRef]
19. Xing, Y.; Lv, C.; Zhang, Z.; Wang, H.; Na, X.; Cao, D.; Velenis, E.; Wang, F.-Y. Identification and analysis of driver postures for in-vehicle driving activities and secondary tasks recognition. *IEEE Trans. Comput. Soc. Syst.* **2017**, *5*, 95–108. [CrossRef]
20. Yu, J.; Park, S.; Lee, S.; Jeon, M. Driver drowsiness detection using condition-adaptive representation learning framework. *IEEE Trans. Intell. Transp. Syst.* **2018**, *20*, 4206–4218. [CrossRef]
21. Dua, M.; Singla, R.; Raj, S.; Jangra, A. Deep CNN models-based ensemble approach to driver drowsiness detection. *Neural Comput. Appl.* **2021**, *33*, 3155–3168. [CrossRef]

22. Moujahid, A.; Dornaika, F.; Arganda-Carreras, I.; Reta, J. Efficient and compact face descriptor for driver drowsiness detection. *Expert Syst. Appl.* **2021**, *168*, 114334. [CrossRef]
23. Karuna, Y.; Reddy, G.R. Broadband subspace decomposition of convoluted speech data using polynomial EVD algorithms. *Multimed. Tools Appl.* **2020**, *79*, 5281–5299. [CrossRef]
24. Ji, Y.; Wang, S.; Zhao, Y.; Wei, J.; Lu, Y. Fatigue state detection based on multi-index fusion and state recognition network. *IEEE Access* **2019**, *7*, 64136–64147. [CrossRef]
25. Ghoddoosian, R.; Galib, M.; Athitsos, V. A realistic dataset and baseline temporal model for early drowsiness detection. In Proceedings of the IEEE/CVF Conference on Computer Vision and Pattern Recognition Workshops, Long Beach, CA, USA, 16–17 June 2019; IEEE: Piscataway, NJ, USA, 2019.
26. Sai, P.-K.; Wang, J.-G.; Teoh, E.-K. Facial age range estimation with extreme learning machines. *Neurocomputing* **2015**, *149*, 364–372. [CrossRef]
27. Lu, J.; Liong, V.E.; Zhou, J. Cost-sensitive local binary feature learning for facial age estimation. *IEEE Trans. Image Process.* **2015**, *24*, 5356–5368. [CrossRef]
28. Huerta, I.; Fernández, C.; Segura, C.; Hernando, J.; Prati, A. A deep analysis on age estimation. *Pattern Recognit. Lett.* **2015**, *68*, 239–249. [CrossRef]
29. Ranjan, R.; Zhou, S.; Chen, J.C.; Kumar, A.; Alavi, A.; Patel, V.M.; Chellappa, R. Unconstrained age estimation with deep convolutional neural networks. In Proceedings of the IEEE International Conference on Computer Vision Workshops, Santiago, Chile, 7 13 December 2015; pp. 109–117.
30. Han, H.; Jain, A.K.; Wang, F.; Shan, S.; Chen, X. Heterogeneous face attribute estimation: A deep multi-task learning approach. *IEEE Trans. Pattern Anal. Mach. Intell.* **2017**, *40*, 2597–2609. [CrossRef] [PubMed]
31. Dornaika, F.; Arganda-Carreras, I.; Belver, C. Age estimation in facial images through transfer learning. *Mach. Vis. Appl.* **2019**, *30*, 177–187. [CrossRef]
32. Rothe, R.; Timofte, R.; van Gool, L. Deep expectation of real and apparent age from a single image without facial landmarks. *Int. J. Comput. Vis.* **2018**, *126*, 144–157. [CrossRef]
33. Shen, W.; Guo, Y.; Wang, Y.; Zhao, K.; Wang, B.; Yuille, A.L. Deep regression forests for age estimation. In Proceedings of the IEEE Conference on Computer Vision and Pattern Recognition, Salt Lake City, UT, USA, 18–23 June 2018; pp. 2304–2313.
34. Taheri, S.; Toygar, Ö. On the use of DAG-CNN architecture for age estimation with multi-stage features fusion. *Neurocomputing* **2019**, *329*, 300–310. [CrossRef]
35. Lou, Z.; Alnajar, F.; Alvarez, J.M.; Hu, N.; Gevers, T. Expression-invariant age estimation using structured learning. *IEEE Trans. Pattern Anal. Mach. Intell.* **2017**, *40*, 365–375. [CrossRef]
36. Liu, H.; Lu, J.; Feng, J.; Zhou, J. Group-aware deep feature learning for facial age estimation. *Pattern Recognit.* **2017**, *66*, 82–94. [CrossRef]
37. Ullah, F.U.M.; Obaidat, M.S.; Ullah, A.; Muhammad, K.; Hijji, M.; Baik, S.W. A Comprehensive Review on Vision-based Violence Detection in Surveillance Videos. *ACM Comput. Surv.* **2022**, *55*, 1–44. [CrossRef]
38. Sajjad, M.; Nasir, M.; Ullah, F.U.M.; Muhammad, K.; Sangaiah, A.K.; Baik, S.W. Raspberry Pi assisted facial expression recognition framework for smart security in law-enforcement services. *Inf. Sci.* **2019**, *479*, 416–431. [CrossRef]
39. Sun, X.; Wu, P.; Hoi, S.C. Face detection using deep learning: An improved faster RCNN approach. *Neurocomputing* **2018**, *299*, 42–50. [CrossRef]
40. Ren, S.; He, K.; Girshick, R.; Sun, J. Faster r-cnn: Towards real-time object detection with region proposal networks. In *Advances in Neural Information Processing Systems*; MIT Press: Cambridge, MA, USA, 2015; pp. 91–99.
41. Howard, A.G.; Zhu, M.; Chen, B.; Kalenichenko, D.; Wang, W.; Weyand, T.; Andreetto, M.; Adam, H. Mobilenets: Efficient convolutional neural networks for mobile vision applications. *arXiv* **2017**, arXiv:1704.04861.
42. Ullah, W.; Ullah, A.; Hussain, T.; Khan, Z.A.; Baik, S.W. An Efficient Anomaly Recognition Framework Using an Attention Residual LSTM in Surveillance Videos. *Sensors* **2021**, *21*, 2811. [CrossRef]
43. Yar, H.; Hussain, T.; Khan, Z.A.; Koundal, D.; Lee, M.Y.; Baik, S.W. Vision Sensor-Based Real-Time Fire Detection in Resource-Constrained IoT Environments. *Comput. Intell. Neurosci.* **2021**, *2021*, 5195508. [CrossRef]
44. Szegedy, C.; Liu, W.; Jia, Y.; Sermanet, P.; Reed, S.; Anguelov, D.; Erhan, D.; Vanhoucke, V.; Rabinovich, A. Going deeper with convolutions. In Proceedings of the IEEE Conference on Computer Vision and Pattern Recognition, Boston, MA, USA, 7–12 June 2015; pp. 1–9.
45. Khan, S.U.; Hussain, T.; Ullah, A.; Baik, S.W. Deep-ReID: Deep features and autoencoder assisted image patching strategy for person re-identification in smart cities surveillance. *Multimed. Tools Appl.* **2021**, 1–22. [CrossRef]
46. Szegedy, C.; Vanhoucke, V.; Ioffe, S.; Shlens, J.; Wojna, Z. Rethinking the inception architecture for computer vision. In Proceedings of the IEEE Conference on Computer Vision and Pattern Recognition, Las Vegas, NV, USA, 27–30 June 2016; pp. 2818–2826.
47. Yar, H.; Hussain, T.; Agarwal, M.; Khan, Z.A.; Gupta, S.K.; Baik, S.W. Optimized Dual Fire Attention Network and Medium-Scale Fire Classification Benchmark. *IEEE Trans. Image Process.* **2022**, *31*, 6331–6343. [CrossRef]
48. Redmon, J.; Farhadi, A. Darknet: Open Source Neural Networks in C. 2013. Available online: https://pjreddie.com/darknet/ (accessed on 1 January 2023).
49. Ullah, F.U.M.; Obaidat, M.S.; Muhammad, K.; Ullah, A.; Baik, S.W.; Cuzzolin, F.; Rodrigues, J.J.P.C.; de Albuquerque, V.H.C. An intelligent system for complex violence pattern analysis and detection. *Int. J. Intell. Syst.* **2021**, *37*, 10400–10422. [CrossRef]

50. Ullah, F.U.M.; Muhammad, K.; Haq, I.U.; Khan, N.; Heidari, A.A.; Baik, S.W.; de Albuquerque, V.H.C. AI assisted Edge Vision for Violence Detection in IoT based Industrial Surveillance Networks. *IEEE Trans. Ind. Inform.* **2021**, *18*, 5359–5370. [CrossRef]
51. Krizhevsky, A.; Sutskever, I.; Hinton, G.E. Imagenet classification with deep convolutional neural networks. *Commun. ACM* **2017**, *60*, 84–90. [CrossRef]
52. Simonyan, K.; Zisserman, A. Very deep convolutional networks for large-scale image recognition. *arXiv* **2014**, arXiv:1409.1556.
53. Wang, H.; Lu, F.; Tong, X.; Gao, X.; Wang, L.; Liao, Z.J.E.R. A model for detecting safety hazards in key electrical sites based on hybrid attention mechanisms and lightweight Mobilenet. *Energy Rep.* **2021**, *7*, 716–724. [CrossRef]
54. Bi, C.; Wang, J.; Duan, Y.; Fu, B.; Kang, J.-R.; Shi, Y. MobileNet based apple leaf diseases identification. *Mob. Netw. Appl.* **2020**, *27*, 172–180. [CrossRef]
55. Rothe, R.; Timofte, R.; van Gool, L. Dex: Deep expectation of apparent age from a single image. In Proceedings of the IEEE International Conference on Computer Vision Workshops, Santiago, Chile, 7–13 December 2015; pp. 10–15.
56. Sajjad, M.; Zahir, S.; Ullah, A.; Akhtar, Z.; Muhammad, K. Human behavior understanding in big multimedia data using CNN based facial expression recognition. *Mob. Netw. Appl.* **2020**, *25*, 1611–1621. [CrossRef]
57. Zhang, T.; Han, G.; Yan, L.; Peng, Y. Low-Complexity Effective Sound Velocity Algorithm for Acoustic Ranging of Small Underwater Mobile Vehicles in Deep-Sea Internet of Underwater Things. *IEEE Internet Things J.* **2022**, *10*, 563–574. [CrossRef]
58. Sun, F.; Zhang, Z.; Zeadally, S.; Han, G.; Tong, S. Edge Computing-Enabled Internet of Vehicles: Towards Federated Learning Empowered Scheduling. *IEEE Trans. Veh. Technol.* **2022**, *71*, 10088–10103. [CrossRef]
59. Rizzo, A.; Burresi, G.; Montefoschi, F.; Caporali, M.; Giorgi, R. Making IoT with UDOO. *IxD&A* **2016**, *30*, 95–112.
60. Nasir, M.; Muhammad, K.; Ullah, A.; Ahmad, J.; Baik, S.W.; Sajjad, M. Enabling automation and edge intelligence over resource constraint IoT devices for smart home. *Neurocomputing* **2022**, *491*, 494–506. [CrossRef]
61. Nayyar, A.; Puri, V.A.; Puri, V. A review of Beaglebone Smart Board's-A Linux/Android powered low cost development platform based on ARM technology. 9th International Conference on Future Generation Communication and Networking (FGCN), Jeju, Republic of Korea, 25–28 November 2015; IEEE: Genebra, Switzerland; pp. 55–63.
62. Yar, H.; Imran, A.S.; Khan, Z.A.; Sajjad, M.; Kastrati, Z. Towards smart home automation using IoT-enabled edge-computing paradigm. *Sensors* **2021**, *21*, 4932. [CrossRef]
63. Jan, H.; Yar, H.; Iqbal, J.; Farman, H.; Khan, Z.; Koubaa, A. Raspberry pi assisted safety system for elderly people: An application of smart home. 2020 First International Conference of Smart Systems and Emerging Technologies (SMARTTECH), Riyadh, Saudi Arabia, 3–5 November 2020; IEEE: Genebra, Switzerland, 2020; pp. 155–160.
64. Cass, S. Nvidia makes it easy to embed AI: The Jetson nano packs a lot of machine-learning power into DIY projects-[Hands on]. *IEEE Spectr.* **2020**, *57*, 14–16. [CrossRef]
65. He, K.; Zhang, X.; Ren, S.; Sun, J. Deep residual learning for image recognition. In Proceedings of the IEEE Conference on Computer Vision and Pattern Recognition, Las Vegas, NV, USA, 27–30 June 2016; pp. 770–778.

Disclaimer/Publisher's Note: The statements, opinions and data contained in all publications are solely those of the individual author(s) and contributor(s) and not of MDPI and/or the editor(s). MDPI and/or the editor(s) disclaim responsibility for any injury to people or property resulting from any ideas, methods, instructions or products referred to in the content.

Article

Applying Neural Networks on Biometric Datasets for Screening Speech and Language Deficiencies in Child Communication

Eugenia I. Toki [1,2], Giorgos Tatsis [1,3], Vasileios A. Tatsis [1,4], Konstantinos Plachouras [1], Jenny Pange [2] and Ioannis G. Tsoulos [5,*]

1. Department of Speech and Language Therapy, School of Health Sciences, University of Ioannina, Panepistimioupoli B′, 45500 Ioannina, Greece
2. Laboratory of New Technologies and Distance Learning, Department of Early Childhood Education, School of Education, University of Ioannina, 45110 Ioannina, Greece
3. Physics Department, University of Ioannina, 45110 Ioannina, Greece
4. Department of Computer Science & Engineering, University of Ioannina, 45110 Ioannina, Greece
5. Department of Informatics and Telecommunications, University of Ioannina, 47150 Kostaki Artas, Greece
* Correspondence: itsoulos@uoi.gr

Abstract: Screening and evaluation of developmental disorders include complex and challenging procedures, exhibit uncertainties in the diagnostic fit, and require high clinical expertise. Although typically, clinicians' evaluations rely on diagnostic instrumentation, child observations, and parents' reports, these may occasionally result in subjective evaluation outcomes. Current advances in artificial intelligence offer new opportunities for decision making, classification, and clinical assessment. This study explores the performance of different neural network optimizers in biometric datasets for screening typically and non-typically developed children for speech and language communication deficiencies. The primary motivation was to give clinicians a robust tool to help them identify speech disorders automatically using artificial intelligence methodologies. For this reason, in this study, we use a new dataset from an innovative, recently developed serious game collecting various data on children's speech and language responses. Specifically, we employed different neural network approaches such as Artificial Neural Networks (ANNs), K-Nearest Neighbor (KNN), Support Vector Machines (SVM), along with state-of-the-art Optimizers, namely the Adam, the Broyden–Fletcher–Goldfarb–Shanno (BFGS), Genetic algorithm (GAs), and Particle Swarm Optimization algorithm (PSO). The results were promising, while Integer-bounded Neural Network proved to be the best competitor, opening new inquiries for future work towards automated classification supporting clinicians' decisions on neurodevelopmental disorders.

Keywords: SmartSpeech; neural networks; optimization; genetic algorithms; biometrical data

MSC: 92B20

1. Introduction

Neurodevelopmental disorders (NDs) are complex conditions affecting brain functions, altering neurological development, and causing difficulties in social, cognitive, learning, communication, behavior, and emotional functioning [1–3]. DSM-5 provides a framework for diagnosis and describes that Neurodevelopmental Disorders (NDs), among others, mainly include [1–4]:

- Autism Spectrum Disorders (ASD): are characterized by deficits in (i) social communication and social interaction and (ii) restricted repetitive patterns of behavior, interests, and activities.
- Attention Deficit Hyperactivity Disorder (ADHD): is characterized by inattention, impulsiveness, and hyperactivity, interfering with daily activities and functioning.

- Intellectual Disability (ID): comprises impairments of general mental abilities that impact adaptive functioning (determine how well an individual copes with everyday tasks) in the conceptual, social, and practical domains. [4].
- Specific Learning Disorder (SLD): is characterized by difficulties in learning and processing specific academic skills, such as reading, writing, or mathematics, despite normal intelligence and adequate educational opportunities. These symptoms can affect academic and daily functioning.
- Communication Disorders (CD): involve language disorder, speech sound disorder, childhood-onset fluency disorder, and social (pragmatic) communication disorder (difficulties in the social uses of verbal and nonverbal communication).

NDs commonly onset throughout development stages from young infancy to adolescence and persist into adulthood or may go undiagnosed until one is an adult [1]. The deficits' severity in NDs varies and may co-occur with other disorders. These deficits can affect the quality of life for individuals and their families, causing significant care needs and extensive community assets [5,6].

Speech and language deficiencies can be early indicators of many neurodevelopmental disorders. In addition, effective communication is critical to human development and social interaction, suggesting developmental continuity from early years to later life [1,7]. To screen and diagnose the NDs' various features, clinicians commonly rely on diagnostic instrumentation, child observations, perceived behaviors, parent interviews, and testing, occasionally resulting in subjective evaluation [5]. However, since clinical evaluations include complex, challenging, non-standardized, multiparametric procedures, and uncertainties in the diagnostic fit, they require high-level clinical expertise and objective measurements [8]. Moreover, early identification and treatment of speech and language deficiencies can help diminish NDs' impact on an individual's overall development and functioning [9]. Thus, there is a highly demanding need to contribute to the need for additional support in eliminating the over- or under-diagnosed child [10].

Recent advancements and innovations in artificial intelligence (AI) spark great interest in their potential benefits in speech and language pathology and special education for individuals with developmental disabilities, learning disabilities, articulation disorders, voice disorders, and more [9,11–16]. Computer science, mathematical algorithms, AI, and other emerging technologies introduce new prospects to support clinical decision-making [10] primarily for an accurate diagnosis, even in rare medical conditions [17–19]. The current literature documents the growing attention in AI algorithms and automated measurement tools for decision-making, classification, and clinical assessment in communication deficiencies and NDs in research [8–10,20]. The results of a pilot study of an integrated technology solution, including a serious game using machine learning models and a mobile app for monitoring ADHD behaviors, indicate ML's potential in ADHD prediction based on gameplay data [8]. The applicability of eye-tracking data to aid the early screening of autism in children reveals that using ML methods strongly suggests that eye-tracking data can help clinicians for a quick and reliable autism screening [10]. In addition, for the classification of developmental delay, the use of AI, serious games, and fine motor movements captured from touching a mobile display have been suggested [9]. Moreover, online gamified testing with a predictive machine learning model for individuals with dyslexia reports results that correctly detect over 80% of the participants with dyslexia, presenting the potential of using a ML approach for dyslexia screening [20].

Hence, this study aims to assist clinicians' decision-making and support evaluation procedures. To screen typically and non-typically developed children for speech and language communication deficiencies, various neural networks adopting different optimizers have been implemented and tested in a new biometric dataset to automatically classify the individuals.

This study is organized into sections as follows: Section 1 explains the significance of clinical evaluation procedures for NDs and speech and language deficiencies in children, the importance of early and objective evaluation procedures, and includes a short description of the research's motivation; Section 2 summarizes the required background knowledge on neural networks and the implemented optimizers; Section 3 presents the methods used in this paper, including the dataset and how the implemented neural networks have been formulated in this research work; followed by Section 4 that presents the experimental results, in which the provided results are discussed. Finally, the paper concludes with Section 5, presenting the conclusions, limitations, and suggestions for future research.

2. Background Information

This section briefly provides the required background information for this study and the corresponding algorithms. Specifically, it is devoted to Artificial Neural Networks (ANNs), K-Nearest Neighbor (KNN), Support Vector Machines (SVM), and the corresponding optimizers used in work, namely the Adam optimizer, the Broyden–Fletcher–Goldfarb–Shanno (BFGS), Genetic algorithm (GAs), and Particle Swarm Optimization algorithm (PSO).

ANNs are parametric machine learning tools [21,22] that utilize a series of parameters commonly called weights or processing units. These tools have found application in a variety of scientific areas, such as physics [23–25], the solution of differential equations [26,27], agriculture [28,29], chemistry [30–32], economics [33–35], and health [36,37]. In addition, recently, neural networks have been used in solar radiation prediction [38], 3D printing [39], and lung cancer research [40].

A neural network typically uses a special function, called the activation function, that decides whether a neuron should be activated or not. A commonly used activation function is the sigmoid function, defined as [21,22]:

$$\sigma(x) = \frac{1}{1+e^x} \tag{1}$$

The neural network has hidden nodes and each one is expressed as,

$$o_i(x) = \sigma\left(w_i^T x + \theta_i\right) \tag{2}$$

where w_i is the weight vector, and θ_i is the bias of the ith node. A neural network can be defined as in the following equation,

$$(x) = \sum_{i=1}^{H} v_i o_i(x) \tag{3}$$

where H is the total number of processing units, and v_i stands for the output weight of the ith node.

The training error of the neural network is defined as:

$$E\left(N\left(\vec{x},\vec{w}\right)\right) = \sum_{i=1}^{M}\left(N\left(\vec{x}_i,\vec{w}\right) - y_i\right)^2 \tag{4}$$

where the set $(\vec{x}_i, y_i)$, $i = 1, \ldots, M$ is the training dataset for the neural network, $\vec{x}_i$ stands for the input vector, and y_i stands for the assigned class. Essentially, the training of the artificial neural network includes the determination of the optimal vector of $\vec{w}$ parameters through the minimization of Equation (4). During recent years, a variety of optimization methods have been proposed to minimize this equation such as the Back Propagation method [41,42], the RPROP method [43–45], Quasi Newton methods [46,47], Simulated Annealing [48,49], GAs [50,51], and PSO [52,53].

BFGS is a widely used iterative optimization in various fields [54], including machine learning algorithms [55]. Specifically, it approximates the inverse of the Hessian matrix (the matrix of second-order partial derivatives) to determine the search direction in which the objective function should be minimized. Furthermore, it updates the approximation in each iteration based on the gradient information. BFGS has good convergence properties, is well-suited for problems with high dimensionality, and is often used in machine learning to optimize the weights of neural networks. Despite its popularity, BFGS can be sensitive to the choice of initial guess and may converge to a suboptimal solution in the case of non-convex objective functions. However, convex optimization problems present fast and reliable results.

GAs are a class of heuristic search algorithms inspired by the mechanics of natural selection and genetics [56–58]. Precisely, GAs initialize a population of candidate solutions by forming the corresponding problem's parameters into chromosomes. The population evolves through the application of genetic operators, such as selection, crossover, and mutation. First, a fitness function is used to evaluate their value, and the best-performing individuals are then selected to create a new population in the next generation. This process is repeated until a satisfactory solution or termination criteria are met. GAs have been applied to a wide range of optimization problems, including scheduling, resource allocation, and neural networks, and have shown to be effective and efficient in many cases.

PSO is a computational optimization method introduced by Eberhart and Kennedy in 1995 [59]. The main inspiration came from the social behavior of birds in a flock. PSO initiates a population of particles representing candidate solutions to probe the search space. Their positions are adjusted based on their own best solution and the overall best solution found by the swarm or a predefined neighborhood. The algorithm iterates continuously, and the best-found solution is reported. PSO proposed parameters such as population number, inertia weight, cognitive and social acceleration, and maximum velocity. The linear decrease in inertia weight determines how much the particles are influenced by their previous velocity over time. The self-adaptation of inertia weight allows the swarm to transition from exploring the solution space to exploiting the best-known solution, effectively guiding the search toward the global optimum [60–62].

The INN is an advanced method of training artificial neural networks which identifies the optimal interval for initializing and training artificial neural networks [63]. The location of the optimal interval is performed using rules evolving from a genetic algorithm. The method has two phases: (i) an attempt is made to locate the optimal interval, and (ii) the artificial neural network is initialized and trained in this interval using a global optimization method, such as a genetic algorithm. The method has been tested on various categorization and function learning data, and the experimental results were incredibly encouraging [63].

The Adam optimizer is an adaptive gradient-based optimization technique frequently used in machine learning algorithms [64]. The technique keeps a different learning rate for the supplied neural network weights and adapts the learning rate as needed throughout training. Adam is a standard optimization method that is well-known for being efficient and for being able to handle sparse gradients.

The K-Nearest Neighbor (KNN) algorithm is a straightforward but effective classification algorithm [65,66]. This algorithm differs as it does not use a training dataset to build a model. It operates by locating the k training samples closest to the new data point in the feature space and assigning it to the majority class or average value of these k neighbors. KNN's simplicity puts it in the top selections, but its performance is sensitive to the choice of the number of nearest neighbors and the distance metric used.

Support Vector Machines (SVM) [67,68] is another popular and effective supervised classification algorithm. The method finds the best decision boundary that maximally separates the classes by maximizing the margin. The margin refers to the distance between the decision boundary and the closest data points from each class. SVMs are known to be effective in handling complex data distributions, and their performance is less sensitive to overfitting than other machine learning algorithms. However, SVMs can be computationally expensive and require careful kernel function and hyperparameter selection to achieve optimal performance.

3. Materials and Methods

We designed a serious game to collect and process players' responses. This serious game contains numerous activities on screening/assessment procedures for NDs [69]. The game data are processed on a dedicated server back-end service to examine early clinical screening/diagnostic patterns on specified domains or skills towards automated indications.

This study is part of the "Smart Computing Models, Sensors, and Early diagnostic speech and language deficiencies indicators in Child Communication" research project with the acronym "SmartSpeech". SmartSpeech is an ongoing research project funded by the Region of Epirus in Greece and the European Regional Development Fund (ERDF).

3.1. Data Description

The sample in our analysis consisted of children, with a total of 435 participants with an average age of 9 years, of which 224 were males and 211 were females. The 339 participants had typical development (TD) (with no NDs), whereas 96 had NDs. We categorized them according to DSM-5. More specifically, 17 had ASD, 18 had ADHD, 8 had ID, 19 had SLD, and 42 had CD. Some of the participants exhibited more than one disorder. The sample recruitment was conducted after various calls through health and educational sectors supporting TD and non-TD children. Parents were informed of the nature and scope of the project, the procedures, and the project approval by the Research Ethics Committee of the University of Ioannina, Greece (Reg. Num.: 18435/15.5.2020), which complies with the General Data Protection Regulation GDPR. Next, they signed the parent consent form.

The participation process included registration in the database and the completion of questionnaires about the child's developmental profile. Then, guided by the clinician, the child played the interactive game explicitly designed for this purpose. Overall, at the end of the process, the variables we used in the analysis came from the game's scores and the bio-signal measurements, i.e., heart rate and eye-tracking measurements.

The SmartSpeech game is designed in the Unity environment [70] and generates several variables regarding scores on the game's activity performances and biometric data. The developed game activities represent the overall performance according to the known developmental skills that children typically acquire. Several activities correspond to these specific speech and language skills [69]. In addition to the scores based on direct responses of the child/player via the touchscreen with clicks and hand movements, other biometric data were also measured, namely voice, heart rate (HR), and gaze.

Voice was recorded in mp3 files when the child needed to answer verbally in a posing question. For this purpose, a speech-to-text program was used [71], for which a Greek model was trained [72] and used. The child was required to give about 40 verbal replies, including but not restricted to naming objects, fruits, vegetables, and characters' names. The SmartSpeech game using this speech-to-text program transcribes the audio files into text and then matches the child's response with the correct answer in a manner of the true–false outcome.

During the gameplay, the child wore a smartwatch with dedicated software developed, which continuously captured the heart rate values in bpm (beats per minute) units. For every game activity, we took the signal for the corresponding period and calculated three metrics: the mean, the standard deviation, and the range of HR. Ideally, we would like to have had the heart rate variability (HRV) in hand, but due to hardware limitations, this was not possible. Hence, we used the dispersion statistics above as an alternative to the mean baseline.

Furthermore, the game presented the child with several visual stimuli to detect the areas on the screen that attracted the player's focus. We conducted this procedure by eye-tracking software [73] executed during the game by capturing the child's gaze via the tablet's camera. When the viewer focused on a specific area, this led to a particular metric called a fixation. The software gave these fixations, and we computed three standard variables in eye-tracking [74]. These were:

1. The number of fixations (fixation count—FC);

2. The time that passed until the first fixation (time to first fixation—TTFF);
3. The total duration of fixations (time spent—TS).

3.2. Data Formulation & Methods Description

As for the eye-tracking variables, the filtering process left only the fixations count and the time spent on areas of interest. The time to first fixation had many missing values and was removed from the dataset.

The dataset is divided into three subsets that correspond to the categories of (i) game scores, (ii) heart rate statistics, and (iii) eye-tracker metrics. Each of these subsets constitutes the set of input variables to the classification process. Several missing and non-valid data were filtered out. Thus, our dataset was forced to reduce the number of cases, although our initial dataset was larger. Ultimately, this also reduced instances of the pathological population. The following tables summarize the input variables. Table 1 shows the variables of the game scores. In total, 30 variables corresponded to the types described in Tables 1–3 that summarize the variables from the heart rate and eye tracking accordingly. A total of 15 HR variables used all statistical means, standard deviations, and range. As for the eye-tracking variables, the filtering process left only the fixations' count and time spent on areas of interest, 16 variables in total. The time to first fixation had many missing values and was removed from the dataset.

Table 1. Variables from scores of the game activities.

Variable Description	Count
Object recognition	6
Click on objects	7
Vocal intensity	1
Verbal response	6
Memory task	2
Emotion recognition	3
Hearing test	1
Puzzle solving	2
Moving objects	2
Total	30

Table 2. Variables from the heart rate statistics.

Variable Description	Count
Mean HR	5
HR standard deviation	5
HR total range	5
Total	15

Table 3. Variables from the eye-tracking metrics.

Variable Description	Count
Fixation counts	10
Time spent	6
Total	16

Table 4 shows the target variables defining the classes that were used. These variables are binary, meaning either they had the condition or not. The Disorder variable denotes TD and non-TD children. ASD, ADHD, ID, SLD, and CD variables suggest more specifically the disorder as described above, according to DSM-5.

Descriptive statistics for the variables (means, St Ds) are summarized in Appendix A.

Table 4. Target variables defining the classes.

Target Variable Name	Description
Disorder	Any disorder of the following
ASD	Autism spectrum disorder
ADHD	Attention deficit hyperactivity disorder
ID	Intellectual disability
SLD	Specific learning disorder
CD	Communication disorders

4. Application Details and Experimental Results

In this section, the application details of the applied classifiers and their corresponding parameterization are described in detail, followed by the experimental results.

Seven different classifiers were considered to assess their performance on the provided dataset. Specifically, each neural network employed one layer with ten neurons, and four different optimizers were adopted accordingly, namely BFGS [75], genetic algorithm [58,74,76], PSO [77], and Adam [64]. The same population and chromosome number for PSO and genetic algorithms was used, accordingly, $N = 200$. At the same time, the parameters of the rest of the optimizers remained the same as in the original papers. Furthermore, an INN rule construction method, a KNN method [65] with five neighborhoods, and an SVM method [68] (using the freely available library libsvm [78]) were also considered for the comparisons. Finally, the maximum number of iterations was set at 200 for fair comparisons.

The datasets were split into ten subsets using the 10-fold cross-validation technique to estimate their performance reasonably. Nine of the produced subsets were used for training, and the remaining one was used for the test. Thirty independent experiments were conducted for each instance, calculating each algorithm and the average classification error. For this purpose, different seed numbers were also used for each experiment using the drand48() random number generator of the C programming language. The experiments were performed using the freely available in-house software from https://github.com/itsoulos/IntervalGenetic (accessed on 15 February 2023). The cells in the experimental tables describe average results on the corresponding test set.

Additionally, neural network neurons were changed to examine the performance disturbances, and, specifically, they ranged from 4 to 14. This experiment was conducted for the three disorder datasets, and the results are graphically demonstrated in Figures 1–3. Observing the related graphs shows that 8–10 processing nodes usually achieve the lowest values in the control dataset in almost all techniques.

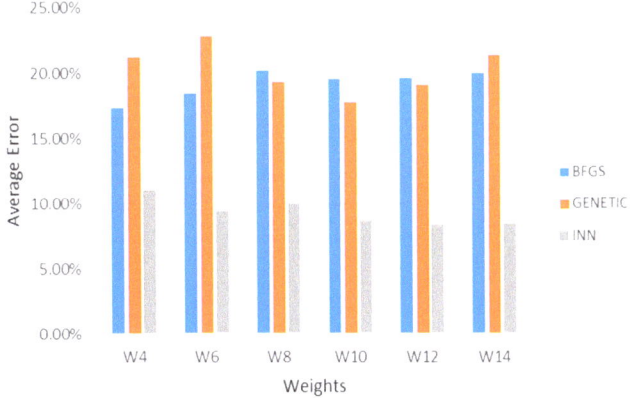

Figure 1. Eye-tracking results.

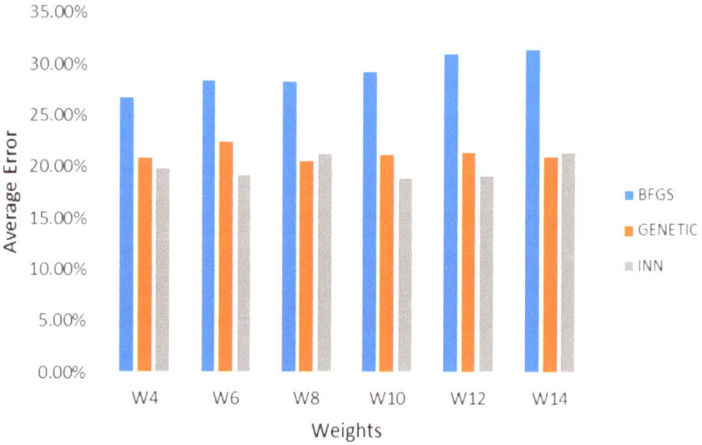

Figure 2. Heart rate results.

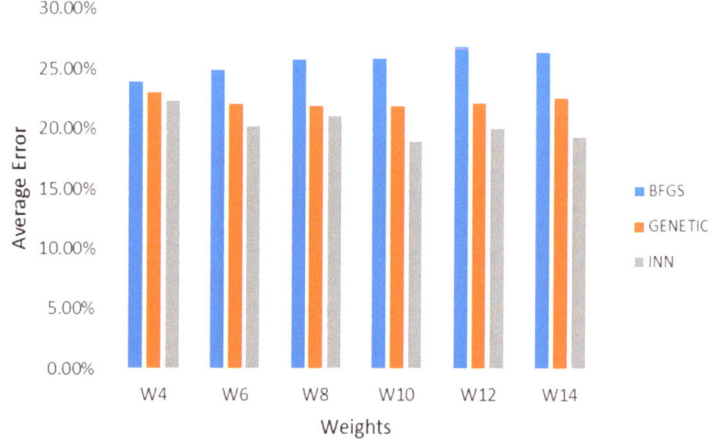

Figure 3. Game play results.

Next, Tables 5–7 compare the utilized classification methods using error rate (%) for the eye-tracking, heart rate, and game responses datasets.

Table 5. Comparison of classification methods using error rate (%) for the eye-tracking dataset.

Class	BFGS	Genetic	PSO	INN	ADAM	KNN	SVM
Disorder	19.53	17.76	14.14	8.67	15.46	28.78	24.53
ASD	27.07	24.32	25.59	19.94	25.66	27.00	24.97
ADHD	27.30	23.42	23.42	19.69	25.45	29.00	30.00
ID	40.30	31.50	37.33	26.63	37.10	42.00	32.00
SLD	29.44	25.68	25.88	22.49	26.17	25.00	27.33
CD	29.46	25.49	27.96	23.78	26.97	26.33	24.67
Average	28.85	24.70	25.72	20.20	26.14	29.69	27.25

Table 6. Comparison of classification methods using error rate (%) for the heart rate dataset.

Class	BFGS	Genetic	PSO	INN	ADAM	KNN	SVM
Disorder	29.26	21.26	22.09	18.95	20.54	22.22	24.33
ASD	32.31	20.61	23.19	19.83	20.85	22.22	20.12
ADHD	34.11	21.11	26.69	19.89	21.43	25.56	23.33
ID	31.13	20.89	24.98	19.61	20.71	23.89	20.00
SLD	32.98	19.17	22.02	19.06	20.65	23.89	23.18
CD	31.61	20.17	20.48	20.02	21.22	23.89	24.63
Average	31.90	20.54	23.25	19.56	20.90	23.61	22.60

Table 7. Comparison of classification methods using error rate (%) for the game scores dataset.

Class	BFGS	Genetic	PSO	INN	ADAM	KNN	SVM
Disorder	25.92	21.92	24.37	18.95	23.45	24.42	23.49
ASD	28.84	20.96	25.42	20.39	23.05	23.49	26.66
ADHD	29.64	21.54	29.62	21.52	23.47	26.51	24.80
ID	29.62	22.64	26.67	21.25	23.47	26.51	24.80
SLD	30.22	22.59	27.15	21.95	21.56	26.28	29.33
CD	30.61	22.25	25.92	22.57	23.28	23.25	23.91
Average	29.14	21.98	26.53	21.11	23.02	24.81	25.86

Table 8 shows the precision and recall metrics indicatively for Genetic and INN when applied to this study's datasets.

Table 8. Precision and recall metrics for Genetic and INN for corresponding datasets.

Dataset	Genetic Precision	Genetic Recall	INN Precision	INN Recall
Eye Tracking	0.76	0.73	0.88	0.84
Heart Rate	0.66	0.63	0.70	0.66
Game Responses	0.69	0.58	0.75	0.62

As can be observed from the tables presenting the experimental results, BFGS achieved lower results for all instances and all datasets since it is a local optimization algorithm. Specifically, BFGS achieved an average error rate of 28.85% for the eye-tracking dataset, while the genetic and PSO have marginally better results. INN achieved the best average error rate, namely 20.02%, indicating that it can detect the best areas that the weights can range. INN is conducted in two phases. During the first phase, a branch and bound algorithm locates the most promising intervals for the neural network parameters. In the second phase, a genetic algorithm optimizes the neural network inside the interval located in the first phase. Additionally, the experimental results indicated that Adam slightly overcomes SVM and KNN in most cases.

Moreover, the average error rate concerning whether an individual has a disorder is more profitable since the corresponding data are more extensive than those from other instances, such as ID and ADHD. Furthermore, as mentioned in Section 3.2, gathering data on child populations is challenging, with missing data reported. For instance, regarding eye-track activities, there is in-line evidence of difficulty obtaining continuous and valid measurements due to the child's spontaneous movements [79]. The same patterns were also applied in the heart rate and the game score datasets. INN proved better than the rest, achieving a classification error of around 20% aligning with the results of precision and recall rates. It is clear that the classification for "Disorder" for the eye-tracking dataset, thus screening between TD and non-TD children, reports the best results for all the optimizers. The highest performance is clearly achieved using the INN optimizer (8.67% error rate).

In the same way as this study, others have looked into the potential of drag-and-drop data as a digital biomarker and proposed a classification model to categorize kids with developmental disorders [9]. They created an algorithm for a deep-learning convolutional neural network model with promising findings suggesting diagnoses of developmental disorders. In a different study, the potential for the early detection of developmental impairments in children was explored, using diagnostic information from the International Classification of Diseases (ICD) and supplementary information, including prescription history, treatment duration, and frequency records [80]. By combining four algorithms, namely k-nearest neighbor, random forest, logistic regression, and gradient boosting, they created the best model for the early diagnosis of impairments. Their classification model for detecting disorders yielded high accuracy outcomes, just as in our study. It also specified delivering diagnoses around a year earlier than the usual diagnostic age.

5. Conclusions

Screening and evaluating speech and language deficiencies and NDs is a challenging, rigorous, and complex procedure that may occasionally result in misleading outcomes due to uncertainties in the diagnostic fit, subjective evaluation, and clinical expertise. Delayed or inaccurate evaluation eliminates chances for early identification and treatment, while if detected in time, it can help diminish NDs' impact on an individual's overall development and functioning. This highlights the significance of this study, using artificial intelligence for automatic classification.

For this reason, in this study, a first attempt to enhance the clinician's decision-making assessment was conducted using machine learning methodologies. Specifically, the collected data provided by a novel, recently developed serious game were used as a test bed to estimate the classification performance of the proposed neural network algorithms. The provided dataset includes a variety of variables stemming from the game, along with biometrical data from a total of 435 participants. The experiments were conducted in a series of different neural networks adopting a variety of optimizers, and the average classification error was collected.

The results were promising, opening new inquiries for future research. INN proved to be the most competitive algorithm, achieving an average classification error of 20%. This performance may be further improved by using different optimization and machine learning methodologies, and/or by increasing the number of participants, which we will thoroughly examine in future work. The results of this study are expected to contribute towards developing an innovative digital approach to support health care. They may be valuable tools for the early identification of NDs, delivering objective metrics complementary to the clinician's diagnosis, reducing screening and diagnostic costs, and enriching clinician efficiency.

Author Contributions: Conceptualization, E.I.T. and I.G.T.; methodology, E.I.T. and I.G.T.; software, I.G.T.; validation, K.P., J.P. and V.A.T.; formal analysis, E.I.T., G.T., V.A.T and I.G.T.; investigation, V.A.T. and K.P.; resources, G.T. and K.P.; data curation, G.T., K.P. and J.P.; writing—original draft preparation, E.I.T., G.T, V.A.T. and I.G.T.; writing—review and editing, E.I.T., J.P. and I.G.T.; visualization, G.T. and I.G.T.; supervision, E.I.T. and I.G.T.; project administration, E.I.T.; funding acquisition, E.I.T. All authors have read and agreed to the published version of the manuscript.

Funding: This research was funded by the Region of Epirus, project titled "Smart Computing Models, Sensors, and Early diagnostic speech and language deficiencies indicators in Child Communication", with code HP1AB-28185, supported from the European Regional Development Fund (ERDF).

Institutional Review Board Statement: The study was conducted in accordance with the Declaration of Helsinki, and approved by the Research Ethics Committee of the UNIVERSITY OF IOANNINA, Greece (protocol code 18435 and date of approval 15 May 2020).

Informed Consent Statement: Informed consent was obtained from all subjects involved in the study.

Data Availability Statement: The participants of this study did not give written consent for their data to be shared publicly, so due to privacy restrictions and the sensitive nature of this research data sharing is not applicable to this article.

Acknowledgments: We wish to thank all the participants for their valuable contribution in this study.

Conflicts of Interest: The authors declare no conflict of interest. The funders had no role in the design of the study; in the collection, analyses, or interpretation of data; in the writing of the manuscript; or in the decision to publish the results.

Appendix A

Table A1. Means and standard deviations of game score variables for all classes.

Variables		TD	ASD	ADHD	ID	SLD	CD
Object recognition	Var1	88.44 ± 28.07	90.00 ± 15.97	88.89 ± 32.34	79.00 ± 35.42	91.16 ± 24.53	91.19 ± 23.44
	Var2	73.40 ± 31.61	62.29 ± 26.05	70.11 ± 32.23	62.25 ± 37.59	85.84 ± 28.04	69.52 ± 30.27
	Var3	90.67 ± 23.03	88.00 ± 16.75	90.67 ± 25.14	79.00 ± 35.42	91.21 ± 26.90	87.19 ± 26.61
	Var4	87.16 ± 22.36	82.71 ± 18.14	88.00 ± 22.93	87.63 ± 16.04	82.16 ± 18.66	77.86 ± 22.81
	Var5	76.40 ± 29.45	55.18 ± 29.64	75.72 ± 30.71	62.25 ± 37.59	78.16 ± 29.21	70.07 ± 30.81
	Var6	86.43 ± 26.27	76.12 ± 28.64	89.33 ± 20.53	94.38 ± 15.91	91.68 ± 18.65	80.19 ± 29.02
Click on objects	Var1	6.20 ± 15.94	12.41 ± 20.18	3.72 ± 4.73	9.50 ± 23.10	4.32 ± 10.98	5.90 ± 10.59
	Var2	67.90 ± 22.28	61.47 ± 18.27	64.78 ± 22.92	74.50 ± 19.65	67.21 ± 16.05	65.02 ± 22.09
	Var3	4.94 ± 11.23	13.53 ± 24.99	7.22 ± 16.16	6.13 ± 12.22	6.00 ± 11.21	6.21 ± 14.07
	Var4	68.42 ± 37.35	54.06 ± 40.26	58.44 ± 38.88	35.88 ± 35.41	67.00 ± 34.52	48.14 ± 34.59
	Var5	79.35 ± 29.41	75.29 ± 29.61	87.78 ± 20.74	42.50 ± 49.50	86.32 ± 28.33	74.76 ± 33.66
	Var6	43.42 ± 39.10	22.00 ± 30.43	34.17 ± 32.30	19.75 ± 38.25	32.84 ± 34.72	22.93 ± 27.20
	Var7	90.86 ± 28.87	88.24 ± 33.21	94.44 ± 23.57	75.00 ± 46.29	94.74 ± 22.94	88.10 ± 32.78
Vocal intensity	Var1	37.73 ± 27.11	51.18 ± 23.12	47.67 ± 26.14	37.13 ± 26.25	56.21 ± 25.99	45.05 ± 29.03
Verbal response	Var1	17.11 ± 37.72	5.88 ± 24.25	11.11 ± 32.34	12.50 ± 35.36	21.05 ± 41.89	7.14 ± 26.07
	Var2	18.08 ± 23.68	26.06 ± 23.04	20.17 ± 20.91	5.25 ± 5.01	23.42 ± 25.62	12.38 ± 13.06
	Var3	21.32 ± 22.10	18.47 ± 16.30	18.44 ± 15.81	7.38 ± 9.02	11.58 ± 13.03	13.00 ± 14.18
	Var4	11.49 ± 15.74	19.41 ± 16.74	18.33 ± 16.87	12.38 ± 17.08	19.11 ± 16.74	10.21 ± 15.44
	Var5	24.48 ± 43.06	11.76 ± 33.21	22.22 ± 42.78	0.00 ± 0.00	10.53 ± 31.53	9.52 ± 29.71
	Var6	9.94 ± 20.78	11.65 ± 16.26	11.00 ± 19.61	0.00 ± 0.00	6.95 ± 17.67	5.50 ± 14.42
Memory task	Var1	18.06 ± 28.66	12.29 ± 20.16	17.17 ± 28.46	19.38 ± 37.84	14.53 ± 26.12	16.81 ± 28.71
	Var2	44.18 ± 40.66	16.24 ± 27.72	48.50 ± 35.76	17.88 ± 27.54	56.53 ± 40.23	33.12 ± 34.76
Emotion recognition	Var1	87.61 ± 33.00	64.71 ± 49.26	83.33 ± 38.35	87.50 ± 35.36	78.95 ± 41.89	83.33 ± 37.72
	Var2	82.89 ± 37.72	76.47 ± 43.72	77.78 ± 42.78	75.00 ± 46.29	78.95 ± 41.89	85.71 ± 35.42
	Var3	8.85 ± 28.44	5.88 ± 24.25	11.11 ± 32.34	0.00 ± 0.00	10.53 ± 31.53	4.76 ± 21.55
Hearing test	Var1	27.08 ± 27.08	28.24 ± 31.67	32.22 ± 27.56	7.50 ± 14.88	22.11 ± 20.97	20.00 ± 20.24
Puzzle solving	Var1	107.90 ± 63.30	108.00 ± 58.62	113.00 ± 43.56	71.00 ± 46.62	109.58 ± 54.94	130.74 ± 61.90
	Var2	68.84 ± 31.74	70.12 ± 32.42	74.78 ± 31.04	68.25 ± 42.13	74.37 ± 29.53	70.12 ± 30.29
Moving objects	Var1	72.98 ± 23.15	66.65 ± 28.17	69.06 ± 22.26	56.38 ± 33.49	72.95 ± 23.54	69.14 ± 19.48
	Var2	29.37 ± 18.39	29.71 ± 14.19	34.67 ± 24.18	27.88 ± 19.93	33.47 ± 18.33	31.90 ± 17.75

Table A2. Means and standard deviations of heart-rate variables for all classes.

Variables		TD	ASD	ADHD	ID	SLD	CD
HR Mean	Var1	84.22 ± 15.53	84.20 ± 10.20	85.54 ± 13.32	84.80 ± 8.72	80.59 ± 11.74	85.25 ± 12.13
	Var2	84.98 ± 15.98	85.23 ± 13.27	89.34 ± 15.48	82.70 ± 6.02	81.03 ± 10.01	86.01 ± 15.07
	Var3	84.88 ± 14.44	85.80 ± 13.99	86.34 ± 14.20	84.53 ± 8.65	80.54 ± 9.45	84.22 ± 16.18
	Var4	86.14 ± 16.66	86.90 ± 16.03	88.10 ± 12.81	80.88 ± 7.57	80.26 ± 14.67	87.11 ± 14.89
	Var5	85.57 ± 16.76	86.53 ± 11.80	85.37 ± 13.81	83.15 ± 5.74	80.04 ± 14.82	84.16 ± 18.43
HR Std	Var1	2.96 ± 2.78	4.60 ± 2.38	4.30 ± 3.36	3.03 ± 0.88	2.23 ± 1.33	4.43 ± 3.43
	Var2	3.76 ± 2.64	3.08 ± 1.80	5.44 ± 4.64	2.88 ± 1.97	2.33 ± 0.59	4.00 ± 3.35
	Var3	4.20 ± 2.49	3.23 ± 1.20	5.14 ± 2.02	4.05 ± 1.30	4.79 ± 2.93	5.69 ± 3.15
	Var4	2.64 ± 2.14	1.48 ± 0.36	3.69 ± 3.12	4.05 ± 2.76	3.71 ± 2.10	3.39 ± 2.61
	Var5	0.66 ± 0.86	1.38 ± 2.25	1.06 ± 1.38	1.85 ± 2.01	0.60 ± 0.48	1.14 ± 1.51

Table A2. Cont.

Variables		TD	ASD	ADHD	ID	SLD	CD
HR Range	Var1	10.62 ± 8.72	18.40 ± 12.27	14.27 ± 11.75	9.98 ± 2.94	7.43 ± 4.04	15.78 ± 11.14
	Var2	13.88 ± 8.19	11.20 ± 4.94	19.67 ± 14.20	12.40 ± 8.75	10.59 ± 3.87	14.91 ± 10.87
	Var3	17.67 ± 10.13	14.33 ± 2.72	21.47 ± 6.82	16.28 ± 5.82	17.77 ± 8.69	22.83 ± 10.30
	Var4	9.17 ± 6.90	5.30 ± 1.55	12.40 ± 9.96	13.38 ± 8.00	13.99 ± 7.92	11.61 ± 8.35
	Var5	1.61 ± 2.16	4.08 ± 6.81	2.66 ± 3.15	5.08 ± 6.24	1.40 ± 1.16	3.18 ± 4.30

Table A3. Means and standard deviations of eye-tracking variables for all classes.

Variables		TD	ASD	ADHD	ID	SLD	CD
Fixation counts	Var1	8.32 ± 5.98	7.00 ± 8.04	7.27 ± 6.15	5.57 ± 5.26	5.12 ± 4.76	4.91 ± 5.43
	Var2	11.38 ± 6.77	9.06 ± 7.37	12.87 ± 8.48	12.29 ± 5.19	12.71 ± 6.76	11.70 ± 7.39
	Var3	3.36 ± 2.38	2.06 ± 1.91	4.07 ± 3.26	2.86 ± 2.27	3.29 ± 2.80	2.64 ± 2.04
	Var4	3.36 ± 2.38	0.76 ± 1.02	1.58 ± 1.31	0.86 ± 0.79	1.60 ± 1.37	1.40 ± 1.19
	Var5	2.10 ± 2.13	1.94 ± 2.44	2.40 ± 3.14	2.14 ± 3.98	2.53 ± 2.15	2.27 ± 2.83
	Var6	2.10 ± 2.13	1.15 ± 1.54	1.29 ± 1.61	0.71 ± 1.00	1.48 ± 1.64	0.97 ± 0.90
	Var7	1.47 ± 1.64	1.31 ± 1.35	1.20 ± 1.57	1.43 ± 0.98	1.18 ± 1.43	1.30 ± 1.19
	Var8	1.99 ± 1.98	1.62 ± 1.78	2.27 ± 1.62	2.57 ± 1.99	2.00 ± 2.21	1.48 ± 1.96
	Var9	1.28 ± 1.45	1.25 ± 1.44	1.80 ± 1.90	2.00 ± 2.00	1.59 ± 1.81	1.52 ± 1.64
	Var10	2.32 ± 2.39	1.50 ± 1.59	2.13 ± 2.03	2.00 ± 1.29	1.82 ± 2.04	1.70 ± 1.93
Time spent	Var1	4.83 ± 4.25	3.36 ± 3.91	2.72 ± 2.52	2.38 ± 2.64	2.10 ± 2.34	2.61 ± 3.43
	Var2	5.55 ± 3.96	3.71 ± 3.67	6.02 ± 4.41	4.13 ± 2.09	6.13 ± 4.13	5.78 ± 3.86
	Var3	0.58 ± 0.77	0.87 ± 1.35	0.27 ± 0.31	0.92 ± 0.87	0.36 ± 0.40	0.55 ± 0.73
	Var4	0.66 ± 0.79	0.73 ± 0.93	0.97 ± 0.97	0.84 ± 0.74	0.58 ± 0.64	0.51 ± 0.79
	Var5	0.57 ± 0.80	0.38 ± 0.59	0.76 ± 0.92	0.97 ± 0.95	0.69 ± 0.83	0.75 ± 0.95
	Var6	0.68 ± 0.82	0.46 ± 0.64	0.57 ± 0.76	0.56 ± 0.35	0.53 ± 0.76	0.52 ± 0.71

References

1. Thapar, A.; Cooper, M.; Rutter, M. Neurodevelopmental Disorders. *Lancet Psychiatry* **2017**, *4*, 339–346. [CrossRef]
2. Harris, J.C. New Classification for Neurodevelopmental Disorders in DSM-5. *Curr. Opin. Psychiatry* **2014**, *27*, 95–97. [CrossRef]
3. American Psychiatric Association (APA). *Diagnostic and Statistical Manual of Mental Disorders: DSM-5*, 5th ed.; American Psychiatric Association, Ed.; American Psychiatric Association: Washington, DC, USA, 2013; ISBN 978-0-89042-554-1.
4. DSM-5 Intellectual Disability Fact Sheet. Available online: https://www.psychiatry.org/File%20Library/Psychiatrists/Practice/DSM/APA_DSM-5-Intellectual-Disability.pdf (accessed on 28 January 2023).
5. Hyman, S.L.; Levy, S.E.; Myers, S.M.; Council on children with disabilities, Section on Developmental and Behavioral Pediatrics; Kuo, D.Z.; Apkon, S.; Davidson, L.F.; Ellerbeck, K.A.; Foster, J.E.A.; Noritz, G.H.; et al. Identification, Evaluation, and Management of Children With Autism Spectrum Disorder. *Pediatrics* **2020**, *145*, e20193447. [CrossRef] [PubMed]
6. Bishop, D.V.M.; Snowling, M.J.; Thompson, P.A.; Greenhalgh, T. CATALISE consortium CATALISE: A Multinational and Multidisciplinary Delphi Consensus Study. Identifying Language Impairments in Children. *PLoS ONE* **2016**, *11*, e0158753. [CrossRef] [PubMed]
7. Hobson, H.; Kalsi, M.; Cotton, L.; Forster, M.; Toseeb, U. Supporting the Mental Health of Children with Speech, Language and Communication Needs: The Views and Experiences of Parents. *Autism Dev. Lang. Impair.* **2022**, *7*, 239694152211011. [CrossRef]
8. Pandria, N.; Petronikolou, V.; Lazaridis, A.; Karapiperis, C.; Kouloumpris, E.; Spachos, D.; Fachantidis, A.; Vasiliou, D.; Vlahavas, I.; Bamidis, P. Information System for Symptom Diagnosis and Improvement of Attention Deficit Hyperactivity Disorder: Protocol for a Nonrandomized Controlled Pilot Study. *JMIR Res. Protoc.* **2022**, *11*, e40189. [CrossRef]
9. Kim, H.H.; An, J.I.; Park, Y.R. A Prediction Model for Detecting Developmental Disabilities in Preschool-Age Children Through Digital Biomarker-Driven Deep Learning in Serious Games: Development Study. *JMIR Serious Games* **2021**, *9*, e23130. [CrossRef] [PubMed]
10. Kanhirakadavath, M.R.; Chandran, M.S.M. Investigation of Eye-Tracking Scan Path as a Biomarker for Autism Screening Using Machine Learning Algorithms. *Diagnostics* **2022**, *12*, 518. [CrossRef]
11. Wang, X.; Yang, S.; Tang, M.; Yin, H.; Huang, H.; He, L. HypernasalityNet: Deep Recurrent Neural Network for Automatic Hypernasality Detection. *Int. J. Med. Inf.* **2019**, *129*, 1–12. [CrossRef]
12. Muppidi, A.; Radfar, M. Speech Emotion Recognition Using Quaternion Convolutional Neural Networks. In Proceedings of the ICASSP 2021–2021 IEEE International Conference on Acoustics, Speech and Signal Processing (ICASSP), Toronto, ON, Canada, 6–11 June 2021; IEEE: Toronto, ON, Canada, 2021; pp. 6309–6313.

13. Kadiri, S.R.; Javanmardi, F.; Alku, P. Convolutional Neural Networks for Classification of Voice Qualities from Speech and Neck Surface Accelerometer Signals. In Proceedings of the Interspeech 2022, Incheon, Republic of Korea, 18–22 September 2022; pp. 5253–5257.
14. Georgoulas, G.; Georgopoulos, V.C.; Stylios, C.D. Speech Sound Classification and Detection of Articulation Disorders with Support Vector Machines and Wavelets. In Proceedings of the 2006 International Conference of the IEEE Engineering in Medicine and Biology Society, New York, NY, USA, 30 August 2006; pp. 2199–2202.
15. Georgopoulos, V.C. Advanced Time-Frequency Analysis and Machine Learning for Pathological Voice Detection. In Proceedings of the 2020 12th International Symposium on Communication Systems, Networks and Digital Signal Processing (CSNDSP), Oline, 20–22 July 2020; pp. 1–5.
16. Georgopoulos, V.C.; Chouliara, S.; Stylios, C.D. Fuzzy Cognitive Map Scenario-Based Medical Decision Support Systems for Education. In Proceedings of the 2014 36th Annual International Conference of the IEEE Engineering in Medicine and Biology Society, Chicago, IL, USA, 26–30 August 2014; IEEE: Chicago, IL, USA, 2014; pp. 1813–1816.
17. Brasil, S.; Pascoal, C.; Francisco, R.; dos Reis Ferreira, V.; Videira, P.A.; Valadão, G. Artificial Intelligence (AI) in Rare Diseases: Is the Future Brighter? *Genes* **2019**, *10*, 978. [CrossRef]
18. Hirsch, M.C.; Ronicke, S.; Krusche, M.; Wagner, A.D. Rare Diseases 2030: How Augmented AI Will Support Diagnosis and Treatment of Rare Diseases in the Future. *Ann. Rheum. Dis.* **2020**, *79*, 740–743. [CrossRef]
19. Chen, Y.; Li, Y.; Wu, M.; Lu, F.; Hou, M.; Yin, Y. Differentiating Crohn's Disease from Intestinal Tuberculosis Using a Fusion Correlation Neural Network. *Knowl.-Based Syst.* **2022**, *244*, 108570. [CrossRef]
20. Rello, L.; Baeza-Yates, R.; Ali, A.; Bigham, J.P.; Serra, M. Predicting Risk of Dyslexia with an Online Gamified Test. *PLoS ONE* **2020**, *15*, e0241687. [CrossRef]
21. Bishop, C. *Neural Networks for Pattern Recognition*; Oxford University Press: Oxford, UK, 1995.
22. Cybenko, G. Approximation by Superpositions of a Sigmoidal Function. *Math. Control. Signals Syst.* **1989**, *2*, 303–314. [CrossRef]
23. Baldi, P.; Cranmer, K.; Faucett, T.; Sadowski, P.; Whiteson, D. Parameterized Neural Networks for High-Energy Physics. *Eur. Phys. J. C* **2016**, *76*, 1–7. [CrossRef]
24. Valdas, J.J.; Bonham-Carter, G. Time Dependent Neural Network Models for Detecting Changes of State in Complex Processes: Applications in Earth Sciences and Astronomy. *Neural Netw.* **2006**, *19*, 196–207. [CrossRef]
25. Carleo, M.T.G. Solving the Quantum Many-Body Problem with Artificial Neural Networks. *Science* **2017**, *355*, 602–606. [CrossRef]
26. Shirvany, Y.; Hayati, M.; Moradian, R. Multilayer Perceptron Neural Networks with Novel Unsupervised Training Method for Numerical Solution of the Partial Differential Equations. *Appl. Soft Comput.* **2009**, *9*, 20–29. [CrossRef]
27. Malek, A.; Beidokhti, R.S. Numerical Solution for High Order Differential Equations Using a Hybrid Neural Network—Optimization Method. *Appl. Math. Comput.* **2006**, *183*, 260–271. [CrossRef]
28. Topuz, A. Predicting Moisture Content of Agricultural Products Using Artificial Neural Networks. *Adv. Eng. Softw.* **2010**, *41*, 464–470. [CrossRef]
29. Escamilla-García, A.; Soto-Zarazúa, G.M.; Toledano-Ayala, M.; Rivas-Araiza, E.; Gastélum-Barrios, A. Abraham, Applications of Artificial Neural Networks in Greenhouse Technology and Overview for Smart Agriculture Development. *Appl. Sci.* **2020**, *10*, 3835. [CrossRef]
30. Shen, L.; Wu, J.; Yang, W. Multiscale Quantum Mechanics/Molecular Mechanics Simulations with Neural Networks. *J. Chem. Theory Comput.* **2016**, *12*, 4934–4946. [CrossRef] [PubMed]
31. Manzhos, S.; Dawes, R.; Carrington, T. Neural Network-based Approaches for Building High Dimensional and Quantum Dynamics-friendly Potential Energy Surfaces, Int. *J. Quantum Chem.* **2015**, *115*, 1012–1020. [CrossRef]
32. Wei, J.N.; Duvenaud, D.; Aspuru-Guzik, A. Neural Networks for the Prediction of Organic Chemistry Reactions. *ACS Cent. Sci.* **2016**, *2*, 2–725. [CrossRef] [PubMed]
33. Falat, L.; Pancikova, L. Quantitative Modelling in Economics with Advanced Artificial Neural Networks. *Procedia Econ. Financ.* **2015**, *34*, 194–201. [CrossRef]
34. Namazi, M.; Shokrolahi, A.; Maharluie, M.S. Detecting and Ranking Cash Flow Risk Factors via Artificial Neural Networks Technique. *J. Bus. Res.* **2016**, *69*, 1801–1806. [CrossRef]
35. Tkacz, G. Neural Network Forecasting of Canadian GDP Growth. *Int. J. Forecast.* **2001**, *17*, 57–69. [CrossRef]
36. Baskin, I.I.; Winkler, D.; Igor, V.; Tetko, A. Renaissance of Neural Networks in Drug Discovery. *Expert Opin. Drug Discov.* **2016**, *11*, 785–795. [CrossRef] [PubMed]
37. Bartzatt, R. Prediction of Novel Anti-Ebola Virus Compounds Utilizing Artificial Neural Network (ANN). *Chem. Fac. Publ.* **2018**, *49*, 16–34.
38. Yadav, A.K.; Chandel, S.S. Solar Radiation Prediction Using Artificial Neural Network Techniques: A Review. *Renew. Sustain. Energy Rev.* **2014**, *33*, 772–781. [CrossRef]
39. Mahmood, M.A.; Visan, A.I.; Ristoscu, C.; Mihailescu, I.N. Artificial Neural Network Algorithms for 3D Printing. *Materials* **2021**, *14*, 163. [CrossRef] [PubMed]
40. Prisciandaro, E.; Sedda, G.; Cara, A.; Diotti, C.; Spaggiari, L.; Bertolaccini, L. Artificial Neural Networks in Lung Cancer Research: A Narrative Review. *J. Clin. Med.* **2023**, *12*, 880. [CrossRef]
41. Rumelhart, D.E.; Hinton, G.E.; Williams, R.J. Learning Representations by Back-Propagating Errors. *Nature* **1986**, *323*, 533–536. [CrossRef]

42. Chen, T.; Zhong, S. Privacy-Preserving Backpropagation Neural Network Learning. *IEEE Trans. Neural Netw.* **2009**, *20*, 1554–1564. [CrossRef] [PubMed]
43. Riedmiller, M.; Braun, A.H. Direct Adaptive Method for Faster Backpropagation Learning: The RPROP Algorithm. In Proceedings of the IEEE International Conference on Neural Networks, San Francisco, CA, USA, 28 March–1 April 1993; pp. 586–591.
44. Pajchrowski, T.; Zawirski, K.; Nowopolski, K. Neural Speed Controller Trained Online by Means of Modified RPROP Algorithm. *IEEE Trans. Ind. Inform.* **2015**, *11*, 560–568. [CrossRef]
45. Hermanto, R.P.S.; Suharjito, D.; Nugroho, A. Waiting-Time Estimation in Bank Customer Queues Using RPROP Neural Networks. *Procedia Comput. Sci.* **2018**, *135*, 35–42. [CrossRef]
46. Robitaille, B.; Marcos, B.; Veillette, M.; Payre, G. Modified Quasi-Newton Methods for Training Neural Networks. *Comput. Chem. Eng.* **1996**, *20*, 1133–1140. [CrossRef]
47. Liu, Q.; Liu, J.; Sang, R.; Li, J.; Zhang, T.; Zhang, Q. Fast Neural Network Training on FPGA Using Quasi-Newton Optimization Method. *IEEE Trans. Very Large Scale Integr. (VLSI) Syst.* **2018**, *26*, 1575–1579. [CrossRef]
48. Yamazaki, A.; De Souto, M.C.P.; Ludermir, T.B. Optimization of neural network weights and architectures for odor recognition using simulated annealing. In Proceedings of the 2002 International Joint Conference on Neural Networks, Honolulu, HI, USA, 12–17 May 2002; pp. 527–533.
49. Da, Y.; Xiurun, G. An Improved PSO-Based ANN with Simulated Annealing Technique. *Neurocomputing* **2005**, *63*, 527–533. [CrossRef]
50. Leung, F.H.F.; Lam, H.K.; Ling, S.H.; Tam, P.K.S. Tuning of the Structure and Parameters of a Neural Network Using an Improved Genetic Algorithm. *IEEE Trans. Neural Netw.* **2003**, *14*, 79–88. [CrossRef]
51. Yao, X. Evolving Artificial Neural Networks. *Proc. IEEE* **1999**, *87*, 1423–1447.
52. Zhang, C.; Shao, H.; Li, Y. Particle Swarm Optimisation for Evolving Artificial Neural Network. In Proceedings of the Smc 2000 conference proceedings, Nashville, TN, USA, 8–11 October 2000; pp. 2487–2490.
53. Yu, Q.; Liu, Z.-H.; Lei, T.; Tang, Z. Subjective Evaluation of the Frequency of Coffee Intake and Relationship to Osteoporosis in Chinese Men. *J. Health Popul. Nutr.* **2016**, *35*, 1–7. [CrossRef] [PubMed]
54. Hery, M.A.; Ibrahim, M.; June, L.W. BFGS Method: A New Search Direction. *Sains Malays.* **2014**, *43*, 1591–1597.
55. Christou, V.; Miltiadous, A.; Tsoulos, I.; Karvounis, E.; Tzimourta, K.D.; Tsipouras, M.G.; Anastasopoulos, N.; Tzallas, A.T.; Giannakeas, N. Evaluating the Window Size's Role in Automatic EEG Epilepsy Detection. *Sensors* **2022**, *22*, 9233. [CrossRef] [PubMed]
56. Holland, J.H. *Adaptation in Natural and Artificial Systems: An Introductory Analysis with Applications to Biology, Control, and Artificial Intelligence*; MIT Press: Cambridge, MA, USA, 1992.
57. Goldberg, D.E. *Genetic Algorithms*; Pearson Education India: Noida, India, 2013.
58. Michalewicz, Z. Genetic Algorithms+ Data Structures= Evolution Programs. *Comput. Stat.* **1996**, 372–373.
59. Kennedy, J.; Eberhart, R. Particle Swarm Optimization. In Proceedings of the ICNN'95—International Conference on Neural Networks, Perth, WA, Australia, 27 November–1 December 1995; IEEE: Perth, WA, Australia, 1995; Volume 4, pp. 1942–1948.
60. Epitropakis, M.G.; Plagianakos, V.P.; Vrahatis, M.N. Evolving Cognitive and Social Experience in Particle Swarm Optimization through Differential Evolution: A Hybrid Approach. *Inf. Sci.* **2012**, *216*, 50–92. [CrossRef]
61. Wang, W.; Wu, J.-M.; Liu, J.-H. A Particle Swarm Optimization Based on Chaotic Neighborhood Search to Avoid Premature Convergence. In Proceedings of the 2009 Third International Conference on Genetic and Evolutionary Computing, Guilin, China, 14–17 October 2009; IEEE: Guilin, China, 2009; pp. 633–636.
62. Eberhart, R.C.; Shi, Y. Tracking and Optimizing Dynamic Systems with Particle Swarms. In Proceedings of the 2001 Congress on Evolutionary Computation (IEEE Cat. No.01TH8546), Seoul, Republic of Korea, 27–30 May 2001; IEEE: Seoul, Republic of Korea, 2001; Volume 1, pp. 94–100.
63. Tsoulos, I.G.; Tzallas, A.; Karvounis, E. A Rule-Based Method to Locate the Bounds of Neural Networks. *Knowledge* **2022**, *2*, 412–428. [CrossRef]
64. Kingma, D.P.; Ba, J. Adam: A Method for Stochastic Optimization. *arXiv* **2014**, arXiv:1412.6980. [CrossRef]
65. Cunningham, P.; Delany, S.J. K-Nearest Neighbour Classifiers—A Tutorial. *ACM Comput. Surv.* **2022**, *54*, 1–25. [CrossRef]
66. Fix, E.; Hodges, J. *Discriminatory Analysis, Nonparametric Discrimination*; USAF School of Aviation Medicine: Randolph Field, OH, USA, 1951.
67. Cortes, C.; Vapnik, V. Support-Vector Networks. *Mach. Learn.* **1995**, *20*, 273–297. [CrossRef]
68. Burges, C.J. A Tutorial on Support Vector Machines for Pattern Recognition. *Data Min. Knowl. Discov.* **1998**, *2*, 121–167. [CrossRef]
69. Toki, E.I.; Zakopoulou, V.; Tatsis, G.; Plachouras, K.; Siafaka, V.; Kosma, E.I.; Chronopoulos, S.K.; Filippidis, D.E.; Nikopoulos, G.; Pange, J.; et al. A Game-Based Smart System Identifying Developmental Speech and Language Disorders in Child Communication: A Protocol Towards Digital Clinical Diagnostic Procedures. In *New Realities, Mobile Systems and Applications*; Auer, M.E., Tsiatsos, T., Eds.; Lecture Notes in Networks and Systems; Springer: Cham, Switzerland, 2022; Volume 411, pp. 559–568. ISBN 978-3-030-96295-1.
70. Unity®2022. Available online: https://unity.com/ (accessed on 15 February 2022).
71. CMUSphinx 2016. Available online: https://cmusphinx.github.io/ (accessed on 20 October 2022).
72. Pantazoglou, F.K.; Papadakis, N.K.; Kladis, G.P. Implementation of the Generic Greek Model for CMU Sphinx Speech Recognition Toolkit. In Proceedings of the eRA-12 International Scientific, Athens, Greece, 24–26 October 2017.

73. SeeSo: Eye Tracking Software 2022. Available online: https://manage.seeso.io/#/console/sdk (accessed on 10 November 2022).
74. Borys, M.; Plechawska-Wójcik, M. Eye-Tracking Metrics in Perception and Visual Attention Research. *EJMT* **2017**, *3*, 11–23.
75. Powell, M.J.D. A Tolerant Algorithm for Linearly Constrained Optimization Calculations. *Math. Program.* **1989**, *45*, 547–566. [CrossRef]
76. Grady, S.A.; Hussaini, M.Y.; Abdullah, M.M. Placement of Wind Turbines Using Genetic Algorithms. *Renew. Energy* **2005**, *30*, 259–270. [CrossRef]
77. Charilogis, V.; Tsoulos, I.G. Toward an Ideal Particle Swarm Optimizer for Multidimensional Functions. *Information* **2022**, *13*, 217. [CrossRef]
78. Chang, C.-C.; Lin, C.-J. LIBSVM: A Library for Support Vector Machines. *ACM Trans. Intell. Syst. Technol. TIST* **2011**, *2*, 1–27. [CrossRef]
79. Sim, G.; Bond, R. Eye Tracking in Child Computer Interaction: Challenges and Opportunities. *Int. J. Child-Comput. Interact.* **2021**, *30*, 100345. [CrossRef]
80. Jeong, S.-H.; Lee, T.R.; Kang, J.B.; Choi, M.-T. Analysis of Health Insurance Big Data for Early Detection of Disabilities: Algorithm Development and Validation. *JMIR Med. Inform.* **2020**, *8*, e19679. [CrossRef]

Disclaimer/Publisher's Note: The statements, opinions and data contained in all publications are solely those of the individual author(s) and contributor(s) and not of MDPI and/or the editor(s). MDPI and/or the editor(s) disclaim responsibility for any injury to people or property resulting from any ideas, methods, instructions or products referred to in the content.

Review

An Overview of Applications of Hesitant Fuzzy Linguistic Term Sets in Supply Chain Management: The State of the Art and Future Directions

Francisco Rodrigues Lima-Junior [1,*], **Mery Ellen Brandt de Oliveira** [2] **and Carlos Henrique Lopes Resende** [3]

[1] Postgraduate Program in Administration, Federal Technological University of Paraná, Curitiba 80230-901, Brazil
[2] Postgraduate Program in Information Management, Federal University of Paraná, Curitiba 80060-000, Brazil; mesbrandt@ufpr.br
[3] Faculty of Economics, University of Coimbra, 3004-512 Coimbra, Portugal; carlos.resende@student.fe.uc.pt
* Correspondence: frjunior@utfpr.edu.br

Citation: Lima-Junior, F.R.; Oliveira, M.E.B.d.; Resende, C.H.L. An Overview of Applications of Hesitant Fuzzy Linguistic Term Sets in Supply Chain Management: The State of the Art and Future Directions. *Mathematics* **2023**, *11*, 2814. https://doi.org/10.3390/math11132814

Academic Editor: Óscar Valero Sierra

Received: 1 June 2023
Revised: 20 June 2023
Accepted: 21 June 2023
Published: 23 June 2023

Copyright: © 2023 by the authors. Licensee MDPI, Basel, Switzerland. This article is an open access article distributed under the terms and conditions of the Creative Commons Attribution (CC BY) license (https://creativecommons.org/licenses/by/4.0/).

Abstract: Supply chain management (SCM) encompasses a wide variety of decision-making problems that affect business and supply chain performance. Since most of these problems involve uncertainty and hesitation on the part of decision makers (DMs), various studies have emerged recently that present SCM applications of techniques based on Hesitant Fuzzy Linguistic Term Sets (HFLTSs) and HFLTS extensions. Given the relevance of this subject and the lack of literature review studies, this study presents a systematic review of HFLTS and HFLTS extension applications to SCM decision-making problems. In order to answer a set of research questions, the selected papers were classified in accordance with a group of factors that are pertinent to the origins of these studies, SCM, HFLTSs, and decision making. The results demonstrated that the Source and Enable processes have been studied with greater frequency, while the most common problems have to do with supplier selection, failure evaluation, and performance evaluation. The companies of the automotive sector predominated in the analyzed studies. Even though most of the studies used techniques based on HFLTSs, we identified applications of seven distinct HFLTS extensions. The main contribution of this study consists of presenting an overview of the use of HFLTSs and their extensions in practical examples of SCM, highlighting trends and research opportunities. It is the first study to analyze applications of decision-making techniques that deal with hesitation in SCM. Therefore, the results can help researchers and practitioners develop new studies that involve the use of HFLTSs and HFLTS extensions in decision-making problems, given that this study systematizes elements that should be considered in the modeling, application, and validation of these methods.

Keywords: multicriteria decision making; systematic literature review; operations management; fuzzy logic; group decision making

MSC: 90B50

1. Introduction

The importance of SCM is recognized by researchers and practitioners as a way to ensure operational efficiency and seek global growth, increased profitability, and stakeholder satisfaction [1,2]. Various studies have corroborated the fact that company performance for supply chain members can be improved by better strategic management of the flow of goods, services, finance, and information throughout the supply chain as a whole [1,3]. In addition to seeking a reduction in costs and improved goods and services, current SCM practices frequently seek to help firms comply with socio-environmental requirements [4] and develop resilient capacities to prevent and/or overcome operational disruptions [2].

Given that SCM requires the integration and alignment of the activities of factories, suppliers, and distributors as well as other chain components, various challenges emerge and contribute to increasing complexity [3]. One of the main difficulties is related to making assertive decisions in the face of the uncertainties that frequently affect the business environment [2]. These uncertainties are due to a wide range of factors, including fluctuations in demand, changes in stakeholder requirements and competition, political conflicts, infectious diseases, and catastrophic events such as earthquakes and hurricanes [5]. As a consequence, the difficulty of obtaining reliable, complete, and updated information leads to many SCM decisions being made based on the knowledge of specialists (or decision makers (DMs)) [6]. In this scenario, quantitative decision-making techniques that use DM linguistic evaluations have been adopted increasingly often in making decisions that are inherent to SCM, such as supplier selection [7] supplier development [8], the selection of emergency logistic plans [9], and risk evaluations [10], among other issues.

The literature features a wide variety of approaches to modeling linguistic information that have been employed to support SCM, with those based on Fuzzy Set Theory (as well as the most recent extensions of this theory) playing a prominent role [11–13]. Among these extensions, Hesitant Fuzzy Linguistic Term Sets (HFLTSs) have been attracting increasing attention. In contrast to traditional decision-making methods, the use of HFLTSs supports complex linguistic expressions when DMs are hesitant to choose the linguistic terms that best represent their preferences [12]. Ever since HFLTSs were proposed by Rodríguez et al. [14], there have been various advances by other researchers that have led to the appearance of extensions such as Extended HFLTSs [15], Proportional HFLTSs [16], and Interval-Valued 2-Tuple HFLTSs [17]. In parallel, a wide array of decision-making processes based on a combination of HFLTSs and MCDM methods have appeared [18,19]. In addition to their being suitable in helping DMs deal with uncertainty and hesitation, the use of HFLTS approaches frequently is justified by their ability to support group decision-making (GDM) processes that are recurrent in SCM [14].

Given the relevance of SCM, there have been a variety of studies devoted to reviews of this subject's literature, including those focused on approaches to managing supply chain risks [2], SCM artificial intelligence techniques [13], SCM machine learning methods [5], supply chain performance evaluation models [6,13], and the application of fuzzy logic in supply chains [20,21], among other areas. However, to the best of our knowledge, there are no reviews of the literature focused on the application of HFLTS techniques to problems inherent to SCM.

Using searches of the ACM Digital Library, EBSCO, El Compendex, Emerald Insight, Google Scholar, IEEE Digital, Science Direct, Scopus, Springer, Taylor & Francis, Web of Science, and the Wiley Online Library databases, we found five literature review studies that covered HFLTSs. Based on Table 1, we may observe that most of the literature review studies that involve HFLTSs are devoted to compiling a comprehensive review of existing theoretical developments to compare linguistic information modeling approaches [11,12,22,23]. There is also a bibliometric review based on metadata from 1080 studies of Hesitant Fuzzy Sets (HFSs) and their extensions [24]. Even though these studies are of great importance in summarizing theoretical knowledge about linguistic information modeling, they do not discuss HFLTS applications to practical problems. Moreover, the conducting of a systematic review of the literature focusing on HFLTS applications for SCM is justified for the following reasons:

1. There is a need to map the contexts in which HFLTS techniques have been applied in order to identify which SCM processes have received the most attention, which are the most studied types of SCM strategies, and which economic sectors are represented by the participating companies in these studies. The realization of a systematic review of this subject has the potential to make a contribution for researchers and practitioners by indicating trends and opportunities for future study;
2. It is important to investigate issues regarding SCM decision-making processes and methods; for example, the types of decision-making problems in which HFLTS techniques are applied, the most often used techniques, and the way the application results are validated. Furthermore, since the use of criteria weights influences the results of

the decision-making problem, it is important to analyze how criteria weighting has been addressed in HFLTS applications and its extensions in SCM;

3. As decision-making processes in SCM usually involve a variety of actors inside and outside of companies, it is relevant to verify whether the adopted methods support GDM or not. In addition, it is important to analyze how they deal with weighting the opinion of decision-makers and whether they provide support for achieving consensus among DMs. Despite the importance of these issues, the factors related to GDM have been omitted in most literature reviews regarding decision making and SCM [6,11–13,22–24];

4. Furthermore, the literature lacks studies that map recent HFLTS extensions and discuss their applications. The present study made it possible to indicate which HFLTS extensions have been most studied in SCM and point out potential areas for the application of HFLTS extensions in future studies. In addition to contributing to the generation of knowledge related to the interface between SCM areas and decision making, the mapping of the state of the art of this subject will also indicate paths for the development of new computational tools that can support managers in SCM decision-making processes.

Table 1. Literature review studies that encompass HFLTSs.

	Liao et al. [22]	Morente-Molinera et al. [23]	Wang et al. [11]	Wang et al. [12]	Yu et al. [24]
Focus	Survey of decision-making theory and HFLTS methodologies	Review of approaches to multi-granular fuzzy linguistic modeling	Review of HFLTS developments and their classification according to computational strategies	Mapping of complex modeling techniques that employ linguistic expressions	Bibliometric analysis of 1080 articles about HFSs and their extensions
Approaches analyzed	HFLTSs, fusion theory, and Hesitant Fuzzy Linguistic preference relationship theory, among others	Fuzzy membership functions, HFLTSs, 2-tuples, and discrete fuzzy numbers, among others		HFLTSs, EHFLTSs, linguistic HFSs, and 2-dimensional linguistic terms, among others	Aggregation operators, information measures, preference relationships, and HFS extensions
Analysis of applications	No	No	Identifies 20 applications in various areas	No	No
Comparison between approaches	Yes	Yes	Yes	Yes	No

Given this information, this study will present a systematic review of the literature about HFLTS and HFLTS extension applications developed to solve SCM decision-making problems in order to answer relevant research questions about this subject (which will be detailed in Section 3). The Preferred Reporting Items for Systematics Reviews and Meta-Analyses (PRISMA) method was adopted to structure this study. The analyzed papers were selected from several databases and classified according to 17 factors. The mapping of these studies made it possible to identify a set of opportunities for the realization of future studies. In terms of the structure of the rest of this article, Section 2 will present a brief review of HFLTSs and HFLTS extensions, Section 3 will describe the methodology for this systematic review of the literature, Section 4 will discuss the results, Section 5 will present recommendations for future studies, and Section 6 will consist of our conclusions and the limitations of this study.

2. HFLTSs and HFLTS Extensions

Ever since it was proposed by Lofti [25], Fuzzy Set Theory has been applied successfully to various problems that involve imprecise, vague, and imperfect information [14,26]. This theory is also the foundation of new approaches to deal with decision-making problems under conditions of uncertainty. These approaches are based on various forms of representing information that are used by DMs to express their preferences [16,27].

Based on Fuzzy Set Theory, Torra [28] proposed Hesitant Fuzzy Set Theory, which provides a framework for supporting DMs in situations in which there are a set of possible values to define the membership of an element. Based on the work of Torra [28], Rodríguez et al. [14] proposed HFLTSs, which differ from previous approaches because they allow DMs to use more than one linguistic term and complex linguistic expressions to represent their preferences. In this manner, HFLTSs offer greater flexibility in eliciting linguistic preferences and at the same time support expressions that are closer to natural language [29].

Considering a set of defined linguistic terms such as $S = \{s_{-\tau}, \cdots, s_0, \cdots, s_\tau\} = \{s_{-2} : \text{very low}, s_{-1} : \text{low}, s_0 : \text{medium}, s_1 : \text{high}, s_2 : \text{very high}\}$, as shown in Figure 1a, an HFLTS defined as $HS(\vartheta) = \{s_i | s_i \in S\}$ consists of a finite ordered subset of linguistic terms of S. Thus, $HS(\vartheta) = \{s_{-2} : \text{very low}, s_{-1} : \text{low}\}$ (Figure 1b) and $HS(\vartheta) = \{s_{-1} : \text{low}, s_0 : \text{medium}, s_1 : \text{high}\}$ are examples of HFLTS. Any other HFLTS consists of at least one of the linguistic terms in S [14]. The HFLTS approach makes it possible for DMs to use linguistic expressions that employ more than one linguistic term simultaneously in each judgment. Examples of linguistic expressions include "at least high" $\{s_1, s_2\}$ (Figure 1c), "between very low and low" $\{s_{-2}, s_{-1}\}$, "at most medium" $\{s_{-2}, s_{-1}, s_0\}$, and "lower than high" $\{s_2\}$ [8,29]. Based on the pioneering work of Rodríguez et al. [14], other authors proposed new operators for HFLTSs [30–32], measures of distance and similarity [33,34], techniques that support GDM [18,19], and consistency and consensus methods for group decision making [35], among other advances. Extensions for HFLTSs also appeared, which opened new possibilities for modeling uncertainty in decision-making processes. Table 2 presents examples of the main HFLTS extensions that are described below:

1. Extended HFLTS (EHFLTS): this makes it possible to create sets of non-consecutive ordered terms based on a combination of HFTLSs given by a group of DMs. As indicated in Equation (1) (Table 1), $S = \{s_{-\tau}, \cdots, s_0, \cdots, s_\tau\}$ represents the set of all possible linguistic terms that express the preferences of the DMs, and s_i indicates each of the linguistic terms chosen by them [15]. An EHFLTS can be especially useful in situations in which the DMs are divided into subgroups and there is no consensus among the DMs, which therefore leads to the need to represent evaluations in non-consecutive terms [44];

2. Hesitant Intuitionistic Fuzzy Linguistic Term Set (HIFLTS): this approach deals with situations in which the DMs evaluate each alternative using a possible linguistic interval and an impossible linguistic interval. As presented in Equation (2), each HIFLTS is composed of the functions h and h', which return finite ordered subsets of consecutive linguistic terms. $h(x)$ and $h'(x)$ indicate the respective possible membership degrees and non-membership degrees of element x and an HIFLTS [18];

3. Interval-Valued Hesitant Fuzzy Linguistic Set (IVHFLS): this is an HFLTS extension based on Interval-Valued HFSs. As shown in Equation (3), IVHFLSs incorporate the possible interval-valued membership degrees that an alternative has in relation to a linguistic term. These membership degrees are quantified by finite numbers of closed intervals that are defined in (0, 1] [36];

4. Proportional HFLTS (PHFLTS): this is formed by the union of the HFLTSs that correspond to the individual assessments of the DMs, which can contain consecutive or non-consecutive linguistic terms. It takes into account proportional information for each generalized linguistic term. As indicated in Equation (4), each linguistic term that makes up an PHFLTS is associated with a p_l value, which denotes the degree of possibility that the alternative carries an assessment value s_l provided by a group of DMs [16]. The values are computed as a function of the terms and values of p_l;

5. Probabilistic Linguistic Term Set (PLTS): this approach adds probability distributions to an HFLTS in order to prevent the loss of any linguistic information provided by the DMs. Thus, PLTSs allow DMs to attribute possible linguistic values to an alternative or criterion at the same time that they reflect the probabilistic information of a group of attributed values. In Equation (5), $L(k)(p(k))$ is made up of the linguistic term

L(k), which is associated with the probability p(k), and #L(p) is the total number of different linguistic terms in L(p) [37];

6. Interval-Valued Dual Hesitant Fuzzy Linguistic Set (IVDHFLTS): this considers a function for the possible membership degrees and another function for the possible degrees that do not belong to the $s_{\theta(x)}$ terms selected by the DMs. In Equation (6), $\tilde{h}(x)$ is a set of closed interval values defined in [0, 1], which denote the possible membership degrees of $s_{\theta(x)}$, and $\tilde{g}(x)$ is a set of closed interval values defined in [0, 1] which represent the possible non-membership degrees [38];

7. Multi-Hesitant Fuzzy Linguistic Term Set (MHFLTS): this is an extension based on HFLTS and HFL elements in which each set can contain repeated and non-consecutive linguistic terms. As shown in Equation (7), considering the sets of terms $\hat{S} = \{s_k | k \in [0,1]\}$ and X as the reference set, an MHFLTS in X is represented by a function H_{MS}, which generates a finite ordered multi-subset of $\hat{S}$. $H_{MS}(x)$ indicates the possible membership degrees of element x in X [39];

8. Double Hierarchy Hesitant Fuzzy Linguistic Term Set (DHHFLTS): this is composed of two independent hierarchies of linguistic terms. As presented in Equation (8), considering $S = \{s_t | t = -\tau, \ldots, -1, 0, 1, \ldots, \tau\}$ as the first hierarchy and $O = \{o_k | k = -\varsigma, \ldots, -1, 0, 1, \ldots, \varsigma\}$ as the second, $s_{t<o_k>}$ is defined as a DHHFLTS in which o_k is the second term of the hierarchy when the first term is s_t. The use of this approach makes it possible for DMs to use expressions such as "very very good", "medium or just right", and "between a little bad and very bad" [40];

9. Hesitant Fuzzy 2-Dimension Linguistic Term Set (HF2DLTS): this was proposed based on the concept of two-dimensional linguistic variables. Each set is made up of possible linguistic terms that represent a DM's assessment of an alternative, with each term having a degree of importance denoted by a linguistic term. As represented in Equation (9), if X is a fixed set and $S^{(1)} = \{\dot{s}_0, \dot{s}_1, \dot{s}_1, \ldots, \dot{s}_{g-1}\}$ and $S^{(2)} = \{\ddot{s}_0, \ddot{s}_1, \ddot{s}_1, \ldots, \ddot{s}_{t-1}\}$ are two sets of linguistic terms, each HF2DLTS has a function $\hat{h}_s(x) = \bigcup_{(\dot{s}_a(x), \ddot{s}_b(x)) \in \hat{h}_s(x)} \{(\dot{s}_a(x), \ddot{s}_b(x))\}$, $\dot{s}_a(x)$ is a set of consecutive terms in $S^{(1)}$, and $\ddot{s}_b(x)$ is a two-dimensional piece of linguistic information that expresses a DM's assessment of the importance of $\dot{s}_a(x)$. The adoption of this approach enables DMs to use linguistic expressions such as "it is certain ($\ddot{s}_4$) that ($\dot{s}_3$) is fair" and "it is uncertain ($\ddot{s}_2$) whether ($\dot{s}_4$) is very good" [41];

10. Probabilistic Hesitant Intuitionistic Linguistic Term Set (PHILTS): this arose from the combination of HIFLTS and PLTS to reflect the probabilities of DM assessments. Thus, as shown in Equation (10), this approach takes into account membership probability data (l(x)p(x)) and non-membership probability data (l'(x)p'(x)), with these probabilities being considered independent [27];

11. Interval-Valued 2-Tuple HFLTS (IV2THFLTS): this is a combination of HFLTS with interval numbers. As presented in Equation (11), each IV2THFLTS has a function $\tilde{I}_A(x)$, defined in a closed subinterval of [0, 1], which denotes the possible interval-valued membership degrees of x in $h_s(x)$. This approach helps DMs avoid a loss in information and improves the accuracy of the decision-making results [17];

12. Dual HFLTS (DHFLTS): this is the result of a combination of HFLTSs and Dual HFSs. As shown in Equation (12), DHFLTSs include the possible membership and non-membership degrees of hs(xi) in the set S. It is useful in very risky decision-making situations in which DMs consider not only the advantages of a decision but also the risks of making this decision [42];

13. Hesitant Picture Fuzzy Linguistic Set (HPFLTS): this is based on Picture 2-Tuple Linguistic Term Sets and arose to avoid the loss of DM information. In Equation (13), each term selected by the DM (s_i^k) is accompanied by the crisp values of the positive membership degrees (μ_i^k), of the undetermined membership degrees (η_i^k), and the negative membership degrees (v_i^k) [43]. Another distinguishing characteristic is that this approach takes into account the refusal information concerning DMs for each assessment.

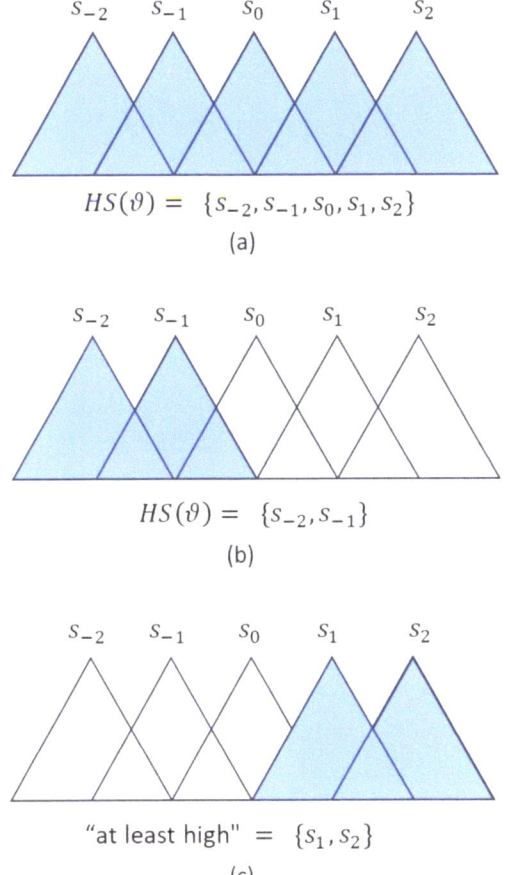

Figure 1. (**a**) Basic Set of Linguistic Terms, (**b**) Examples of Linguistic Terms, and (**c**) Linguistic expressions.

More information about HFLTS extensions can be consulted in the references indicated in Table 2. It is important to emphasize that even though the research presented in Table 2 provides an overview of HFLTS extensions, it is not exhaustive, given that there are other variations that have recently appeared based on the presented approaches.

The following section will present the methodological procedures adopted in this systematic review of the literature.

Table 2. Summary of HFLTS extensions.

ID	Approaches	Author(s)	Mathematical Representation	Example	
1	EHFLTS	Wang [15]	$\{s_i\|s_i \in S\}$	$\{s_2, s_4, s_5\}$	(1)
2	HIFLTS	Beg and Rashid [18]	$\{\langle x, h(x), h'(x)\rangle \| x \in X\}$	$\langle(s_1, s_2, s_3)(s_3, s_4)\rangle$	(2)
3	IVHFLS	Wang et al. [36]	$\{\langle x, s_{\theta(x)}, \Gamma_A(x)\rangle \| x \in X\}$	$\langle s_5, \{[0.4, 0.5], [0.6, 0.7]\}\rangle$	(3)
4	PHFLTS	Chen et al. [16]	$\{(s_i, p_i)\|s_i \in S, i = 0, 1, \ldots, g\}$	$\{(s_4, 0.2), (s_6, 0.6)\}$	(4)
5	PLTS	Pang et al. [37]	$\left\{L^{(k)}\left(p^{(k)}\right)\middle\| L^{(k)} \in S, p^{(k)} \geq 0, k = 1, 2, \ldots, \#L(p), \sum_{k=1}^{\#L}(p)p\right\}$	$\{s_4(0.1), s_5(0.65), s_6(0.2)\}$	(5)
6	IVDHFLTS	Qi et al. [38]	$\left\{\left\langle x, s_{\theta(x)}, \tilde{h}(x), \tilde{g}(x)\right\rangle \middle\| x \in X\right\}$	$\{s_3[0.4, 0.4], [0.5, 0.5]\}, \{[0.3, 0.3], [0.4, 0.4]\}\}$	(6)
7	MHFLTS	Wang et al. [39]	$\{\langle x, h_{MS}(x)\rangle \| x \in X\}$	$\{s_1, s_3, s_{54}\}$	(7)
8	DHHFLTS	Gou et al. [40]	$S_o = \{S_{t<o_k>}\|t = -\tau, \ldots, -1, 0, 1, \ldots, \tau; k - \varsigma, \ldots, -1, 0, 1, \ldots, \varsigma\}$	$\{s_{2<o_1>}, s_{3<o_0>}\}$	(8)
9	HF2DLTS	Liu et al. [41]	$\{\langle x, h_S(x)\rangle \| x \in X\}$	$\{(\ddot{s}_3, \dot{s}_4), (\dot{s}_4, \ddot{s}_4), (\tilde{s}_5, \ddot{s}_3)\}$	(9)
10	PHILTS	Malik et al. [27]	$\{\langle x, l(x)p(x), l'(x)p'(x)\rangle \| s_i \in S\}$	$\{s_2(0.4), s_3(0.1), s_4(0.35)\}, \{s_4(0.3), s_5(0.4)\}$	(10)
11	IV2THFLTS	Si et al. [17]	$\left\{\left\langle x, h_s(x), \tilde{I}_A(x)\right\rangle \middle\| x \in X\right\}$	$\langle\{s_3, s_4\}, [0.7, 0.8]\rangle$	(11)
12	DHFLTS	Zhang et al. [42]	$\{\langle x_i, H_s(x_i), h_s(x_i), g_s(x_i)\rangle \| x_i \in X\}$	$\{s_2\{0.7, 0.5, 0.3\}\{0.3, 0.2\}, s_3\{0.6, 0.4\}\{0.4, 0.2\}\}$	(12)
13	HPFLTS	Yang et al. [43]	$\left\{\left\langle\left(s_i^k\right), \left(\mu_i^k, \eta_i^k, \nu_i^k\right)\right\rangle\middle\| i \in x(0, 1, \ldots, m); k = 1, 2, \ldots, \alpha\right\}$	$\{(s_2, (0.7, 0, 0)\}$	(13)

3. Methodological Procedures

This systematic review of the literature was developed using the PRISMA methodology. According to Page et al. [45], literature review studies "can provide a synthesis of the status of knowledge in a field, from which future priorities can be identified". The PRISMA methodology provides a guide for the preparation of transparent, complete, and concise reviews. It provides an evidence-based minimum set of items that help ensure the robustness and reliability of systematic literature review studies [45].

Figure 2 illustrates the steps of the research method adopted in this study. Stage 1 encompassed the conception and planning of the present study. Stage 2 involved searching, selecting, and ranking studies, while Stage 3 focused on reporting results and identifying opportunities for further research. In Stage 1, based on the analysis of the literature related to SCM and HFLTS (Step 1.1), we identified the need to carry out a systematic review concerning HFLTSs and HFLTS extension applications in SCM (Step 1.2).

Then, we developed the research protocol (Step 1.3) presented in Table 3 based on the PRISMA methodology. For this protocol, the Population, Intervention, Comparison, and Outcome (PICO) were initially specified according to PRISMA [45]. A set of research questions was defined by the authors based on the motivations presented in topics 1 to 4 of the fifth paragraph of Section 1. The research questions RQ1.1 to RQ1.4 dealt with information related to the origin of the selected studies, RQ2.1 to RQ2.4 investigated questions related to SCM, and RQ3.1 to RQ3.8 were focused on questions regarding HFLTSs and decision making. The investigation of these research questions made it possible to provide an overview of the use of HFTLS and HFLTS extension techniques in SCM.

Table 3. Research Protocol Developed for this Study.

Stages	Research Elements	Description
1. Conception and planning	Population	Studies that present decision-making problems in SCM.
	Intervention	HFLTS and HFLTS extension techniques.
	Comparison	Factors related to the origin of these studies, SCM, HFLTSs, and decision making.
	Outcome	- Mapping of the use of HFLTS-based techniques in SCM problems. - Identification of research trends and gaps. - Proposal of directions for future studies.
	Research questions	RQ1.1. What has been the trend in terms of the number of publications since 2012? RQ1.2. Which countries stand out in terms of the production of articles on this subject? RQ1.3. Which journals have published more studies about this subject? RQ1.4. Which are the most cited studies? RQ2.1. What has been the focus of HFLTS techniques in SCM-related problems? RQ2.2. Which SCM processes have received the most attention among the analyzed studies? RQ2.3. With what frequency are these studies devoted to specific types of SCM strategies? RQ2.4. Which economic sectors have received the greatest attention among the identified applications? RQ3.1. Which types of decision-making problems have been addressed by the analyzed studies? RQ3.2. Which HFLTS extension are most frequently used? RQ3.3. How often are techniques integrated in the same problem? What is the frequency of the use of each identified technique? RQ3.4. How are the criteria weights defined? RQ3.5. With what frequency do applications support group decision making? RQ3.6. The weighting of the DMs is considered in which applications? How are these weights defined? RQ3.7. In terms of applications that deal with group decision making, do they use methods to obtain a consensus among the DMs and/or do they perform aggregation operations? RQ3.8. How were the application results validated?
	Keywords	- Hesitant Fuzzy Linguistic Terms Set Synonyms: HFL and Hesitant Fuzzy Linguistic.

Table 3. *Cont.*

Stages	Research Elements	Description
1. Conception and planning	Keywords	- Supply Chain Management — Synonyms: supply chain, customer relationship management, customer service management, demand, distribution, location selection, logistics provider, manufacturing flow, order fulfilment, partner selection, procurement, product development, risk, stock, supplier development, supplier evaluation, supplier selection, and transport.
	Databases	- ACM Digital Library, EBSCO, El Compendex, Emerald Insight, the Google Scholar tool, IEEE Digital, Science Direct, Scopus, Springer, Taylor & Francis, the Web of Science, and the Wiley Online Library.
	Time frame	2012–2023
	Language	- English
	Inclusion criteria	- Studies in English that feature real HFLTS and/or HFLTS extension applications in SCM problems and are approved or published in a peer-reviewed journal.
	Exclusion criteria	- Studies that realize simulated applications of HFLTSs and/or HFLTS extensions in SCM problems. - Studies about SCM decision-making that do not apply HFLTSs or HFLTS extensions; - Studies that apply HFLTSs in problems outside of SCM; - Systematic literature review studies; or - Gray literature.
2. Conducting	Search string	("Hesitant fuzzy linguistic" OR "HFL") AND ("supply chain" OR "customer relationship management" OR "customer service management" OR "demand " OR "logistic provider" OR "manufacturing flow" OR "stock " OR "supplier development" OR "supplier evaluation" OR "supplier selection" OR "location selection" OR "order fulfilment" OR "risk" OR "partner selection" OR "distribution" OR "procurement" OR "product development" OR "transport")
	Filters	Exhibits only journal-published articles in its results; and exhibits articles published in 2012 or later.
	Study selection	- Realized by the two authors in an independent manner through reading titles, keywords, and abstracts.
	Quality assessment	- Realized by the two authors through a complete reading of the article to confirm whether the study performed a real application, was in English, and was peer-reviewed.
	Data extraction	- Performed by the two authors with the help of the Parsifal software through a complete reading of the articles. The factors considered in the data extraction are presented in Table 4. The data generated in this step were exported into MS Excel 2021 (Redmond, WA, USA).
	Classification of the studies	- The classification was performed in accordance with the factors and categories displayed in Table 4.
3. Reporting	Visualization and analysis of results	- Creation of graphs and a publication map using MS Excel 2021; - Creation of tables using MS Word; - Analysis of HFLTS approaches and the problems addressed over the years.
	Summary of the information	- Analysis and summary of the results; - Summary of the results and answers to the research questions; - Comparison with other systematic literature reviews; - Identification of research gaps and proposed recommendations for future studies; and - Conclusion, contributions and limitations.

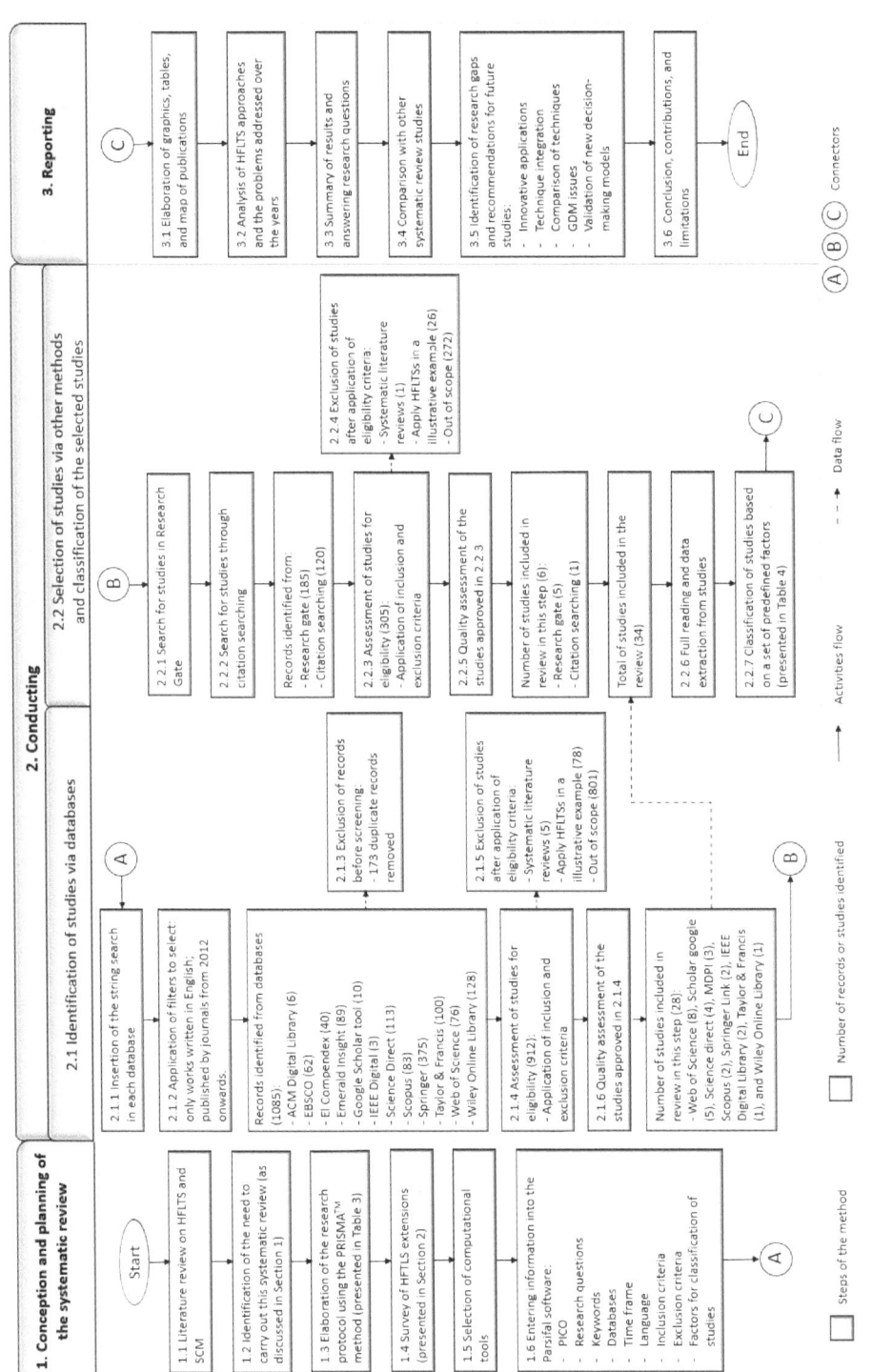

Figure 2. Research Method Adopted for the Systematic Review.

The keywords used in the search string were defined to cover studies that involved the application of HFLTSs in various subareas of SCM. The databases utilized were the ACM Digital Library, EBSCO, El Compendex, Emerald Insight, the Google Scholar tool, IEEE Digital, Science Direct, Scopus, Springer, Taylor & Francis, the Web of Science, and the Wiley Online Library. The time frame considered in our search ranged from 2012 to 2023, given that the first publication concerning HFLTSs dates from 2012 [14]. To generate a reliable and interpretable sample of studies, we only included studies in English that featured real HFLTS and/or HFLTS extension applications in SCM problems that had been approved or published in a peer-reviewed journal.

A survey of HFLTS extensions was carried out in Step 1.4 through searches in the databases indicated in Table 2. The studies shown in Table 1 were also used in this step. The results of this stage provided a theoretical basis for the analysis and classification of studies in the subsequent steps, as discussed in Section 2. The main extensions identified are described in Section 2. In Step 1.5, we chose our computational tools. The Parsifal software was selected to support the selection, assessment, and classification of the studies (in Steps 2.1.3 to 2.1.6 and Steps 2.2.3 to 2.2.7). MS Excel software was chosen to prepare our graphs and a publication map to summarize the findings (in Steps 3.1 to 3.2).

At the end of Stage 1 (Step 1.6), the research protocol information was entered into the Parsifal software to facilitate the selection and classification of studies in Stage 2. The study searches were performed in March 2023. Using the search string in our database search (Step 2.1.1) and after applying the filters (Step 2.1.2) described in Figure 1, we obtained a total of 1085 results. Metadata from these studies were imported into the Parsifal software. Thus, in Step 2.1.3, 173 duplicates were excluded.

When considering the remaining 912 studies, the initial selection was performed in Step 2.1.4 by both authors in an independent manner by reading the titles, keywords, and abstracts. The goal of this step was to only select articles that were within the scope of this review. We excluded books, book chapters, proceedings, literature review articles, and works that did not present real HFLTS applications in SCM, as well as other studies that were not within the scope of this literature review. In total, 884 articles were excluded in the Step 2.1.5 because they did not meet the inclusion criteria. Later, in the assessment of the quality of the articles (2.1.6), both authors read the complete articles to confirm that they performed real applications, were written in English, and had been peer-reviewed. After the quality evaluation, we included 28 studies from these databases.

In Step 2.2, several more studies were collected through a search of the Research Gate (Step 2.2.1) website and a search for the referenced citations (Step 2.2.2) in the studies selected in Step 2.1.6. These studies were also evaluated in terms of their eligibility and quality (Steps 2.2.3 to 2.2.5), which made it possible to identify 6 studies that were included in our review. Thus, a total of 34 studies were included. In Steps 2.2.6 and 2.2.7, all of the selected studies were read in pairs and classified according to the factors presented in Table 4. These factors were defined based on systematic literature reviews focused on SCM [6,13], supplier selection [7], supply management [46], GDM [47], and the modeling of complex linguistic expressions [12]. The categories related to each factor were defined initially based on these studies. Then, the categories were reviewed and updated according to the findings based on a thorough reading of the selected articles.

After the data were extracted and the studies were classified according to Table 4, a set of graphs and tables was prepared to better visualize the results in Step 3.1. The analysis of the frequency of decision-making problems and HFLTSs and the approaches addressed over the years was performed in Step 3.2. In Step 3.3, the obtained information was summarized to answer the research questions presented in Table 3. The obtained results were compared with other systematic review studies regarding this subject in Step 3.4. A framework with recommendations for future research was proposed in Step 3.5 based on the identified research gaps. Finally, the conclusion, contributions and limitations were discussed in Step 3.6.

Table 4. Factors Considered in the Classification of the Selected Studies.

	Factors	Categories	References
Information about the origin of the studies	Year of publication: classified the selected studies by year of publication and analyzes trends over the years.	Frequency of publications per year.	Lima-Junior and Carpinetti [6]
	Country: classified the selected studies according to the country of the first author's affiliated institution.	Frequency of publications per journal.	
	Journal: classified the studies according to the journals in which they were published in order to find the journals that published the most on this subject.	Frequency of publications by country as defined based on the affiliation of the authors of each study.	Resende et al. [7], and Zimmer et al. [46]
	Number of citations: identified the most cited articles based on information from Google Scholar.	Total citations of each study in Google Scholar.	
Factors related to SCM	Application objective: classified the selected studies based on the main objective of the decision-making problem in question.	Failure evaluation, risk evaluation, performance evaluation, supplier selection, and logistics service provider selection, among others.	Riahi et al. [13], Zimmer et al. [46], Lima-Junior and Carpinetti [6]
	SCM processes: based on the application objective, the studies were classified into one of the six main SCM processes.	Source, Plan, Make, Delivery, Return, and Enable.	SCC [48] and Riahi et al. [13]
	Type of SCM strategy: grouped the studies according to the SCM strategy adopted by the participating company in the application.	Agile, digital, flexible, green, lean, resilient, or sustainable.	Lima-Junior and Carpinetti [6] and Resende et al. [7]
	Sector	Automotive, electro-electronics, energy, construction, food, and health, among others.	Riahi et al. [13]
Factors related to decision making and HFLTS	Type of problem: classifies the studies based on the type(s) of decision-making problems in question.	Choice, ordering, or categorization.	Zimmer et al. [46]
	Type of HFLTS approach: identified the frequency of applications of HFLTSs and HFLTS extensions over the years.	HFLTS, EHFLTS, HFLTS, IVHFLS, PHFLTS, PLTS, IVDHFLTS, MHFLTS, DHHFLTS, HF2DLTS, PHLTS, IV2THFLTS, DHFLTS, and HPFLTS.	Wang et al. [12]
	Combination of techniques: identified the frequency with which HFLTSs and HFLTS extensions have been integrated with other techniques.	Frequency of isolated applications and combined applications.	Lambert and Enz [1], Lima-Junior and Carpinetti [6], and Riahi et al. [13]
	Frequency of use of each technique: identified the frequency of use of each the techniques that have been integrated with HFLTSs or HFLTS extensions.	Frequency of the applications for each technique.	

Table 4. *Cont.*

	Factors	Categories	References
	Criteria weights: identified whether the selected studies provided support for criteria weights. When they did, it mapped how the criteria weights were defined.	- Enabled the use or non-use of weights for criteria; - Weights attributed directly by the DMs or calculated weights; - Method(s) adopted for the weight calculations.	Kabak [47]
	Group decision making: investigated whether the selected studies supported group decision-making processes.	- Focused on individual or GDM; - Number of DMs who participated in the application.	Kabak [47]
Factors related to decision making and HFLTS	Weights for DMs: analyzed whether the selected studies consider weights for the DMs. When they did, it analyzed how these weights were defined.	- Considered or ignored the weights for the DMs; - Weights attributed by DMs or calculated by some method.	Kabak [47]
	Consensus among DMs: identified whether the selected studies performed aggregation operations and whether they used methods to obtain a consensus among the DMs.	- Frequency of the use of aggregation methods for the DMs' preferences and the techniques used to obtain consensus in GDM.	Kabak [47]
	Validation: classified the approaches used for validation of the results for each selected study.	- Based on sensitivity analysis, the application of statistical techniques, and comparisons with other methods or real data.	Lima-Junior and Carpinetti [6]

4. Results and Discussion
4.1. Presentation of the Selected Studies
4.1.1. HFLTS-Based Studies

This section presents the selected studies that were based on HFLTS applications. Among them, there were six studies that proposed models focused on supply management. Liao et al. [49] combined HFLTSs with the Best Worst Method (BWM) and Additive Ratio Assessment (ARAS) methods to support supplier selection in a digital supply chain. Dolatabad et al. [50] also proposed a supplier selection model for a digital supply chain; however, they used a fuzzy cognitive map combined with the HFLTS-VIKOR (Vlsekriterijumska Optimizacija I KOmpromisno Resenje in Serbian) method. Liu et al. [51] developed a method that combined HFLTS with Dempster–Shafer evidence theory to achieve consensus in GDM problems. The application was realized in a supplier selection problem for chemical products in a retail supermarket. Lima-Junior and Hsiao [52] developed a model to monitor supplier performance in an automobile factory. In this study, the HFLTS-TOPSIS (Technique for Order of Preference by Similarity to Ideal Solution) method was utilized to classify suppliers in a two-dimensional matrix based on operational performance and supply costs.

Another HFLTS-TOPSIS model based on a bi-dimensional matrix was proposed by Borges et al. [4] to support the segmentation of sustainable suppliers. In this study, the application involved the classification of suppliers in a hydroelectric plant. Finger and Lima-Junior [8] developed an approach based on Quality Function Deployment (QFD) and HFLTS to support decision making during the elaboration of programs to develop sustainable suppliers. The application took place in an automotive firm while considering criteria related to the economic, environmental, and social performance of the suppliers. In addition to the studies by Borges et al. [4] and Finger and Lima-Junior [8], we identified two other studies focused on the promotion of sustainable supply chains. Osiro et al. [19] proposed a method that combined HFLTS with QFD to select evaluation measures for sustainable supply chains. Erol et al. [53] applied the HFLTS-QFD, HFLTS-Delphi, HFLTS-ANP (Analytic Network Process), and HFLTS-TOPSIS methods to analyze barriers to the adoption of circular economics.

There were four studies that involved the application of HFLTSs to performance management. Tüysüz and Şimşek [54] applied an HFLTS-based Analytic Hierarchy Process (AHP) to evaluate the factors that affected the performance of a transport company's affiliates. Pérez-Domínguez et al. [55] evaluated the impact of using lean tools for organizational performance using a combination of AHP and HFLTS-TOPSIS methods. Through a combination of the HFLTS-AHP and HFLTS-MULTIMOORA (Multi-Objective Optimization on the basis of Ratio Analysis Multiplicative Forms) methods, Büyüközkan and Güler [56] created a methodology to support managers in evaluating supply chain analytics tools. Zheng et al. [57] presented a Hesitant Fuzzy Linguistic Decision-Making Trial and Evaluation Laboratory (DEMATEL) model to evaluate the importance of critical success factors in a health company.

In terms of risk management in supply chains, we found three related applications. Wu et al. [58] presented a new approach to risk evaluation in supply chains that integrated HFLTSs, the fuzzy synthetic method, and the eigenvalue method. A pilot application of this approach was realized to evaluate risks in electric vehicle supply chains. Chang et al. [59] combined the Failure Mode and Effect Analysis (FMEA), DEMATEL, and HFLTS methods to analyze risks that failures would occur in an electronic company's production processes. More recently, Qin et al. [9] integrated the swaps method based on the prospect theory with HFLTS. The objective of the application of this proposal was to help select emergency logistics plans during the COVID-19 pandemic.

Finally, we identified two studies related to location selection, one for human resources management and another for packaging design selection. More recently, Wu et al. [60] developed a new method that combined HFLTSs, TODIM (Tomada de Decisão Interativa Multicritério in Portuguese), and DEA. The method was applied for the selection of the most appropriate location a new health management center. Another model for location selection was developed by Ren et al. [35] based on Incomplete Hesitant Fuzzy Linguistic Preference Relations that was applied to selecting the location of hydroelectric plants. Yalçın and Pehlivan [61] proposed a personnel selection model for a manufacturing company. The use of the Fuzzy Combinative Distance-Based Assessment (CODAS) Method based on fuzzy envelopes for HFLTS proved effective in dealing with this problem.

Lima et al. [8] developed a method that combined HFLTS, AHP, QFD, and the Preference Ranking Organization Method for Enriched Evaluation (PROMETHEE) for packaging design selection. This method was applied in an automotive firm, and the results were compared with other multicriteria decision methods.

4.1.2. Studies Based on HFLTS Extensions

Among the 15 studies based on extensions of HFLTS, we identified the use of seven distinct approaches. The use of DHHFLTS predominated, and it was present in seven of these studies. Krishankumar et al. [62] presented a framework for supplier selection based on DHFLTS, which is focused on situations in which the weights of the criteria are unknown. Krishankumar et al. [63] proposed a DHHFLTS-based framework for green supplier selection with partial weight information. Krishankumar et al. [64] applied DHHFLTS to generate a ranking of sustainable suppliers.

In addition to these three studies regarding supply management, we found three applications of DHHFLTS for risk analysis. Shen and Liu [65] evaluated the risk of logistics firms based on a combination of DHHFLTS and FMEA. Dai et al. [66] evaluated the risk of failures in an electronic products firm utilizing DHHFLTS, FMEA, and the K-means algorithm. Similarly, Duan et al. [67] combined DHHFLTS, FMEA, and the K-means algorithm to analyze the risk of failures in a firm in the energy sector. Finally, Wang et al. [68] proposed a study that used DHHFLTS and ORESTE (Organísation, Rangement et Synthèse de Données Relarionnelles in French) to evaluate traffic congestion.

The use of PLTSs has also emerged in SCM problems. Li et al. [69] developed a methodology that integrated PLTS with DEMATEL to evaluate sustainable recycling partners. Zhang et al. [70] proposed the use of PLTSs to deal with sustainable logistics suppliers. Zhang et al. [71] applied PLTSs to supplier selection for a construction company. In this application, the authors also used the BWM and Combined Compromise Solution (CoCoSo) methods based on rough boundary intervals.

The other identified extensions appeared in only one application apiece. Wang et al. [72] proposed an MHFLTS model to support the outsourcing of logistics services, which is especially useful in situations in which the weight information for the criteria is incomplete. Based on the performance indicators of the Supply Chain Operations Reference (SCOR) model, Divsalar et al. [73] created a model to evaluate supply chain performance that integrates the EHFLTS-VIKOR, Fuzzy Delphi, Interval-valued Hesitant Fuzzy, and DANP (DEMATEL-ANP) methods. A distinguishing characteristic of this model was that it combined criteria, paradigms, and Lean, Agile, Resilient, and Green (LARG) practices to improve supply chain performance.

Qu et al. [74] developed a stochastic method based on DHFLTSs and HFLTSs for sustainable supplier selection in a high-tech manufacturing center. Wu et al. [75] combined HPLTSs with the Weighted Cross-Entropy TOPSIS method to support the decision-making process for personnel selection in a firm in the automotive sector. Zolfaghari and Mousavi [75] created a risk evaluation methodology based on FMEA, MULTIMOORA, the Technique of Precise Order Preference (TPOP), and IVHFLTS. A pilot application of this methodology was created to manage failures in a healthcare company.

Based on the characterizations of the studies presented in Sections 4.1.1 and 4.1.2, it was possible to answer the research questions displayed in Table 3. Sections 4.2–4.4 will discuss the obtained results.

4.2. Information about the Origins of These Studies

This section is dedicated to answering the research questions RQ1.1 to RQ1.4 and presenting the classification results for these studies in terms of year of publication, journal, country, and most cited articles. Figure 3 presents the distribution of the number of articles by year of publication (RQ1.1). This figure demonstrates that our subject is quite emergent in the literature, given that the first publication we found occurred in 2017. Roughly 58.8% of the studies were published within the past three years (beginning with 2021). In addition, the trend line of Figure 3 indicates a growth trend in terms of the number of publications about this subject. It is important to mention that the results displayed in the last column of the graph only include studies published in the first few months of 2023, given that our study sample was selected in March 2023.

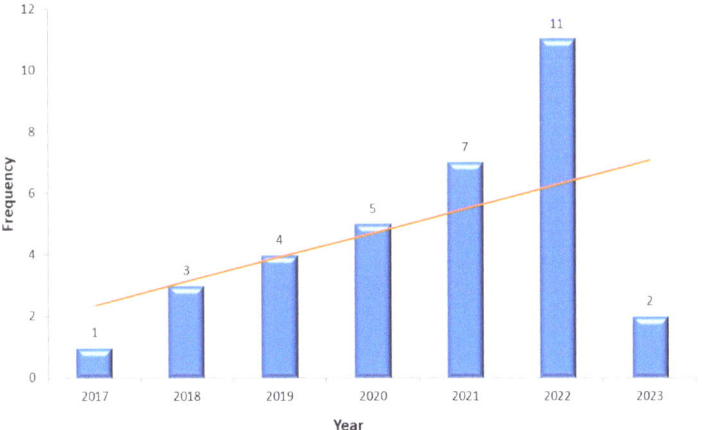

Figure 3. Frequency of Articles Published by Year.

Figure 4a presents the distribution of publications classified according to the country of the first author's affiliated institution for each study (RQ1.2). China had the largest number of publications with 17 studies, which represented 50% of our analyzed sample. It was followed by Brazil (five studies), Turkey (four studies), India (three studies), and Iran (three studies). As shown in Figure 4b, production related to this subject was concentrated in Asia (24 studies), South America (5 studies), and Europe (4 studies).

Table 5 presents the journals in which the selected studies were published (RQ1.3). A wide variety of journals published articles about this subject: there was a total of 27 distinct journals. The journal with the most published articles on this subject was *IEEE Transactions on Engineering Management*. Following it, there was a tie between the journals *Applied Soft Computing, Computers and Industrial Engineering, Environmental Science and Pollution Research, the Journal of Intelligent & Fuzzy Systems*, and *Symmetry*, each with two publications. The other 21 journals had 1 publication apiece.

Table 5. Distribution of Published Articles by Journal.

Journal	2017	2018	2019	2020	2021	2022	2023	Total
IEEE Transactions on Engineering Management						3		3
Applied Soft Computing				1	1			2
Computers and Industrial Engineering						2		2
Environmental Science and Pollution Research						1	1	2
Journal of Intelligent & Fuzzy Systems		1				1		2
Symmetry			1			1		2
Applied Sciences			1					1
Aslib Journal of Information Management							1	1

Table 5. *Cont.*

Journal	2017	2018	2019	2020	2021	2022	2023	Total
Complex & Intelligent Systems	1							1
Complexity				1				1
DYNA					1			1
Energy			1					1
Fuzzy Optimization and Decision Making				1				1
Int. J. of Computational Intelligence Systems				1				1
Int. J. of Environmental Research and Public Health				1				1
Int. J. of Information Technology & Decision Making					1			1
Int. J. of Production Economics						1		1
Int. J. of Strategic Property Management				1				1
Int. Transactions in Operational Research		1						1
J. of Cleaner Production		1						1
J. of Contemporary Administration						1		1
J. of Mathematics					1			1
J. of the Operational Research Society						1		1
Knowledge-Based Systems					1			1
Kybernetes					1			1
Neural Computing & Applications					1			1
Technological and Economic Development of Economy			1					1

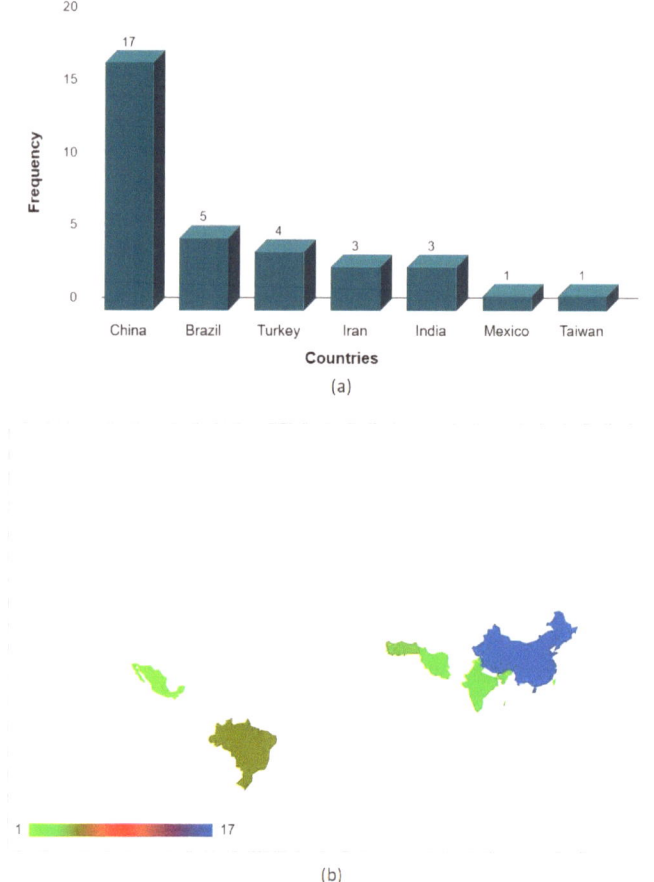

Figure 4. (**a**) Distribution of the Publications by Country and (**b**) a Map of the Publications.

By collecting the number of citations in Google Scholar, it was possible to identify the most influential articles. Table 6 presents the number of citations received by the 11 most cited studies within our analyzed sample (RQ1.4). The article that had the greatest number of citations was Osiro et al. [19] with a total of 86 citations. The works of Wang et al. [72], Yalçın and Pehlivan [72], Tüysüz and Şimşek [54], and Liu et al. [51] followed with 84, 72, 71, and 65 citations, respectively.

Table 6. List of the Most Cited Articles.

Rank	Author(s)	Number of Citations
1st	Osiro et al. [19]	86
2nd	Wang et al. [72]	84
3rd	Yalçın and Pehlivan [61]	72
4th	Tüysüz and Şimşek [54]	71
5th	Liu et al. [51]	67
6th	Wu et al. [58]	56
7th	Wang et al. [68]	56
8th	Liao et al. [49]	41
9th	Duan et al. [67]	32
10th	Erol et al. [53]	28
10th	Chang et al. [59]	28

4.3. Aspects Related to SCM

To answer the research questions RQ2.1 to RQ2.4, this section presents the results regarding the objectives of HFLTS applications in SCM processes, their industrial sector, and the company's type of SCM strategy. In order to identify the most common decision-making problems in SCM that have been addressed using HFLTSs and HFLTS extensions, the selected studies were classified based on the application objective. Based on this, to elucidate which SCM processes received the greatest attention in these applications, each study was classified as dealing with one of the following SCM processes: Source, Plan, Make, Deliver, Return, and Enable. These are the six main SCM processes according to the Supply Chain Operations Reference (SCOR) model, which is an SCM reference that has been widely adopted by practitioners and researchers [48]. Table 7 displays the results of the classification of our studies in terms of the objectives of their applications and the associated SCM processes (RQ2.1). Studies dealing with supplier selection were the most frequent, totaling 23.5% of the sample. They were followed by failure evaluations (11.8%) and performance evaluations (8.8%). Figure 5 exhibits the frequency of application objectives over the years. This figure indicates the dominant relevance of the supplier selection problem over the years.

According to the results displayed in Figure 6a, the Source process had the most associated applications and represented 38.2% of the studies. This process is dedicated to acquiring goods and hiring external services that are necessary to meet actual or planned demand. It was followed by the Enable process with 29.4% of the studies. The high frequency of applications related to the Source process seemed to be related to the fact that this process requires decision-making processes that affect several areas of the business. The main one is supplier selection, which determines part of the supply chain structure and also influences production costs, product quality, and customer satisfaction. Similarly, the Enable process encompasses other decision-making processes that are essential for supply chain structuring and operation, such as personnel selection, location selection, performance evaluation, and risk assessment [48].

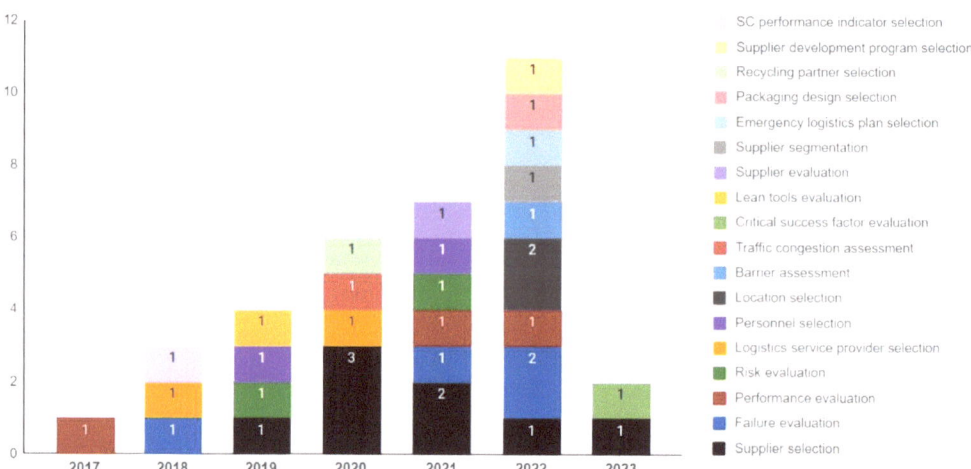

Figure 5. Frequency of Application Objectives over the Years.

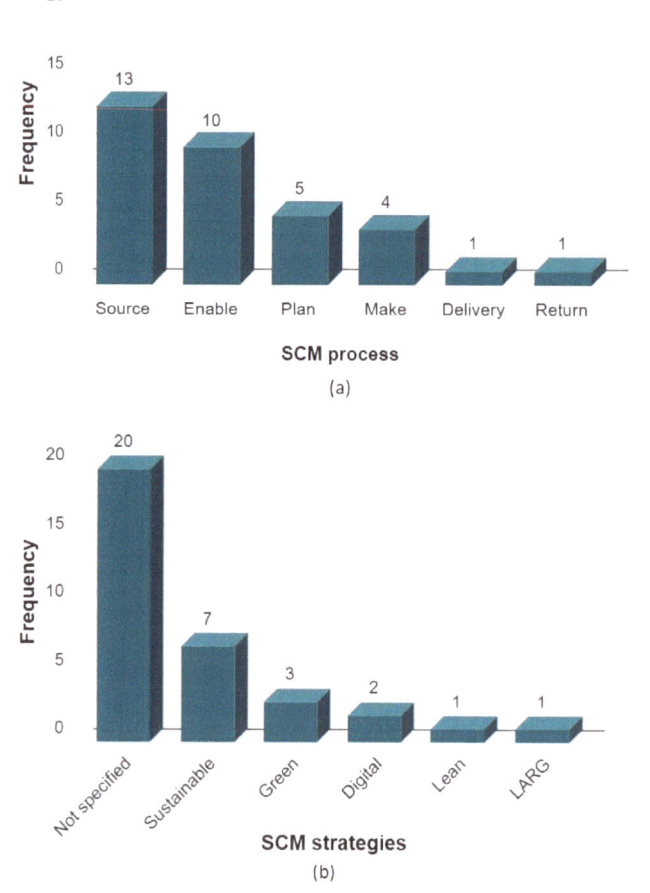

Figure 6. (**a**) SCM Process Results and (**b**) Types of SCM Strategies.

The Plan process appeared in 14.7% of the applications. This process encompasses planning activities to create the actions that will best achieve the requirements of the Make, Delivery, and Return processes. The studies dealing with the Make process totaled 11.8% of the sample, including activities dealing with the transformation of products or the execution of services. Finally, the Delivery process, which involves order, transportation, and distribution management, appeared in just one application (2.9%). Similarly, the Return process, which is related to receiving products that are returned for any reason [48], was also associated with just a single application.

Figure 6b displays the results of the classification of these studies by the type of SCM strategy adopted by the company where the model was applied (RQ2.3). While 58.8% of the studies did not clearly identify the type of SCM strategy employed, 20.6% of the studies involved companies that adopted a Sustainable strategy. In general, sustainable SCM studies are based on triple-bottom-line dimensions and include decision-making criteria related to economic, environmental, and social aspects. Applications in companies employing green supply chains represented 8.8% of the sample and were oriented toward minimizing environmental pollution and improving the environmental performance of products and processes. They were followed by 5.9% of the companies with digital (or smart) supply chains, which are focused on promoting digitalization, automation, and the integration of operations throughout the supply chain. The Lean strategy, which focuses on a reduction in costs and the elimination of waste, was represented by one application (2.9%). Similarly, the LARG strategy, which combines aspects of the Lean, Agile, Resilient, and Green strategies, was also represented by just one application. We did not find applications devoted exclusively to agile or resilient supply chains.

The predominance of applications in sustainable and green SCM is due to the importance that environmental management and social responsibility have achieved in recent decades. Companies began adapting to deal with stricter regulations and customers who were more concerned about the socio-environmental impacts of business activities [8,76]. As the use of environmental and social performance indicators is relatively recent in most companies and these indicators usually involve qualitative aspects or variables for which there is no performance history, the use of HFLTS techniques and HFLTS extensions is quite appropriate [19].

Table 7. Classification of the Studies according to their Application Objectives.

Application Objective	Author(s)	SCM Process	Total
Supplier selection	Liao et al. [49]; Zhang et al. [71]; Krishankumar et al. [77]; Krishankumar et al. [78]; Liu et al. [49]; Dolatabad et al. [50]; Qu et al. [74]; Krishankumar et al. [64].	Source	8
Failure evaluation	Chang et al. [59]; Zolfaghari and Mousavi [79]; Dai et al. [66]; Duan et al. [67].	Make	4
Performance evaluation	Tüysüz and Şimşek [54]; Büyüközkan and Güler [56]; Divsalar et al. [73].	Enable	3
Risk evaluation	Wu et al. [58]; Shen and Liu [65].	Enable	2
Logistics service provider selection	Wang et al. [72]; Zhang et al. [70].	Source	2
Personnel selection	Wu et al. [75]; Yalçın et al. [61].	Enable	2
Location selection	Wu et al. [80]; Ren et al. [81].	Enable	2
Barrier assessment	Erol et al. [53].	Plan	1
Traffic congestion assessment	Wang et al. [68].	Delivery	1
Critical success factor evaluation	Zheng et al. [57].	Plan	1
Lean tools evaluation	Pérez-Domínguez et al. [55].	Plan	1
Supplier evaluation	Lima-Junior and Hsiao [52].	Source	1
Supplier segmentation	Borges et al. [4].	Source	1
Emergency logistics plan selection	Qin et al. [9].	Plan	1
Packaging design selection	Lima et al. [82].	Plan	1
Recycling partner selection	Li et al. [69].	Return	1
Supplier development program selection	Finger and Lima-Junior [8].	Source	1
SC performance indicator selection	Osiro et al. [19].	Enable	1

Finally, Figure 7 displays the classification of these studies according to the sector of the participating company (RQ2.4). The automotive sector stood out with 10 applications (29.4%). It was followed by the health and electro-electronic sectors with five and four applications, respectively. The food, glass, high-technology, infrastructure, manufacturing, retail, and transportation sectors were represented by just one application apiece. Five of the studies did not specify the company's sector. The large number of company applications seemed to be related to the fact that these chains are of great economic importance and have a complex structure. Assembly companies purchase many components externally, for which there is often more than one supplier. Moreover, many automotive companies have traditionally been among the early adopters of management tools and practices [83,84].

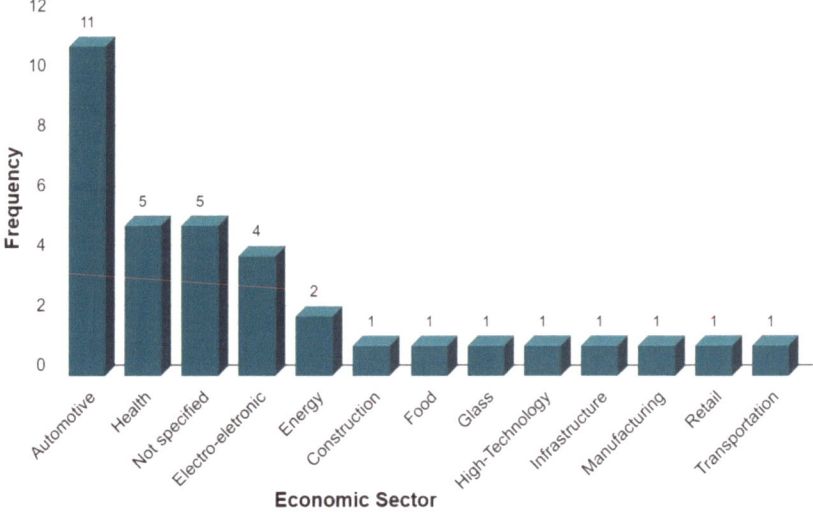

Figure 7. Classification of the Studies according to Economic Sector.

4.4. Aspects Related to Decision Making and HFLTSs

This section deals with subjects related to research questions RQ3.1 to RQ3.8. In the graph displayed in Figure 8, the selected studies are classified according to the main decision-making problem that they dealt with (RQ3.1). The results indicated that half of the applications were devoted to ranking problems, or in other words, problems with ranking alternatives by global preference. This was followed by 32.4% of the applications, which dealt with choice problems, in which there was a desire to choose a subgroup of alternatives within the available options. Finally, in 17.6% of the applications, we found studies that dealt with sorting issues, or in other words, problems in which the objective was to classify each alternative in a predetermined category.

In Figure 9, the studies were classified according to the adopted HFLTS approach (RQ3.2). The HFLTS approach was used in 23 studies (67.6%), while 15 studies (44.1%) adopted an HFLTS extension. Among the 13 HFLTS extensions discussed in Section 2, we found applications based on just 7 of them: DHHFLTS (7 studies); PLTS (3 studies); and DHFLTS, EHFLTS, HPFLS, IVHFLTS, and MHFLTS (with 1 study apiece). Four studies combined HFLTSs with an HFLTS extensions such as PLTSs and DHFLTSs.

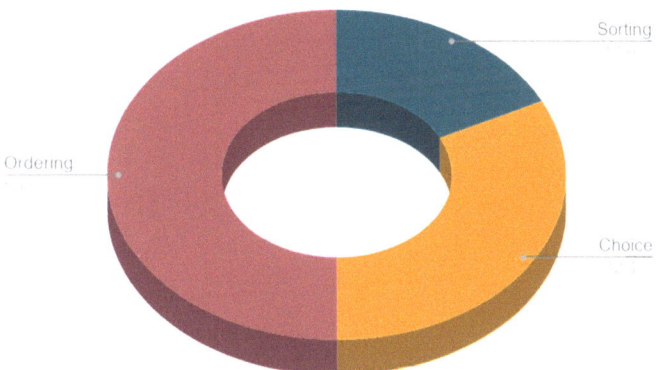

Figure 8. Classification of the Studies by Type of Decision-Making Problem.

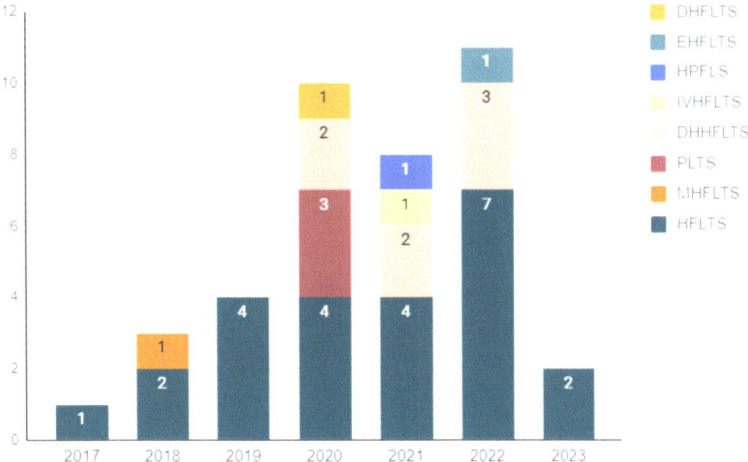

Figure 9. Frequency of Use of HFLTSs and HFLTS Extensions.

The great frequency of the HFLTS approach originally proposed by Rodríguez et al. [14] may be related to its ease of use by DMs and the ease of calculations. Meanwhile, the use of DHHFLTSs may be associated with the possibility of using complex linguistic expressions, which combine two scales of linguistic terms in each judgment. In addition to increasing the flexibility of DM preferences, the use of DHHFLTSs increases the possibilities of values that can be attributed to the evaluated object [40]. Furthermore, the interest of using PLTSs lies in the fact that this approach takes into account a possibility degree for each linguistic term selected by a DM. For each judgment provided by a DM, the sum of the possibility degrees must be equal to 1. For example, if a DM selects only the term "s_3: Medium" without hesitation, their judgment will be quantified as (s_3, 1.0). When selecting "between s_4: good and s_6: extremely good", the DM's judgment will be quantified as (s_4, 0.33; s_5, 0.33; s_6, 0.33). Since the values of the degrees of possibility act as weighting factors for the linguistic terms, the terms chosen in situations of greater hesitation will have less influence on the final result [37].

Table 8 displays the techniques that were applied in each of the analyzed studies. It also describes how the criteria weights were determined and which techniques were employed in calculating these weights. In terms of research question RQ3.3, we verified that all of the studies proposed the integration of HFLTS or HFLTS extension techniques with other techniques, which included Multicriteria Decision-Making (MCDM) methods, quality management, and statistical techniques. Integration occurred through the hybridization or sequential application of these techniques.

Table 8. Decision-Making Techniques in Each Study.

	Proposed By	Decision Technique(s)	Method for Obtaining Criteria Weights	Method for Calculating Criteria Weights
	Tüysüz & Şimşek [54]	HFLTS-AHP	Weights not applied	N/A
	Osiro et al. [19]	HFLTS-QFD	Calculation	HFLTS-QFD
	Yalçın et al. [61]	HFLTS-CODAS	Calculation	HFLTS-CODAS
	Qu et al. [74]	DHFLTS, HFLTS, and Regret and Rejoice Theory	Calculation	HFLTS and Regret and Rejoice Theory
	Ren et al. [81]	Programming models based on Incomplete Hesitant Fuzzy Linguistic Preference Relation	Calculation	Programming models
	Wang et al. [68]	DHHFLTS-ORESTE	Calculation	DHHFL-ORESTE
Hybrid techniques (13)	Zhang et al. [70]	HFLTS-PLTS and Ratio Index-Based Probabilistic Linguistic Ranking Method	Calculation	CoCoSo
	Liu et al. [51]	HFLTS and Dempster–Shafer Theory	Assigned by DMs	N/A
	Lima-Junior & Hsiao [52]	HFLTS-TOPSIS	Assigned by DMs	N/A
	Borges et al. [4]	HFLTS-TOPSIS	Calculation	HFLTS-TOPSIS
	Finger and Lima-Junior [8]	HFLTS-QFD	Calculation	HFLTS-QFD
	Qin et al. [9]	HFLTS and Swaps Method	Calculation	HFLTS, and Swaps Method
	Zheng et al. [57]	HFLTS-DEMATEL	Calculation	HFLTS-DEMATEL
	Chang et al. [59]	HFLTS-FMEA, DEMATEL, and Ordered Weighted Geometric Average (OWGA)	Calculation	OWGA
	Wang et al. [20,72]	MHFLTES, Heronian Mean (HM), and Prioritized Average Operator	Calculation	MHFLTE and HM
	Liao et al. [59]	HFLTS-BMW and HFLTS-ARAS	Calculation	HFLTS-BWM
	Pérez-Domínguez et al. [55]	AHP and HFLTS-TOPSIS	Calculation	AHP
	Wu et al. [58]	HFLTS, Fuzzy Synthetic Method, Eigenvalue Method, and Triangular Fuzzy Number (TFN)	Calculation	Eigenvalue Method
	Zhang et al. [71]	HFLTS, PLTS, Modified BWM, and CoCoSo	Calculation	HFLTS-BWM
Combined techniques (21)	Krishankumar et al. [77]	DHHFLTS, Generalized Maclaurin Symmetric Mean (GMSM), and Borda Method	Calculation	DHHFLTS and SV
	Li et al. [69]	HFLTS, PLTS, TFN, DEMATEL, and Generalized Weighted Ordered Weighted Average (GWOWA)	Calculation	PLTS-DEMATEL
	Büyüközkan and Güler [56]	HFLTS-AHP and HFLTS-MULTIMOORA	Calculation	HFLTS-AHP

Table 8. Cont.

	Proposed By	Decision Technique(s)	Method for Obtaining Criteria Weights	Method for Calculating Criteria Weights
	Shen and Liu [65]	DHHFLTS-FMEA, DHHFLTS-COPRAS, and Kemeny Median Method (KEM)-SWARA	Calculation	KEM-SWARA
	Krishankumar et al. [78]	DHHFLTS, GMSM, TODIM, and Cronbach's Alpha Coefficient	Calculation	Mathematical model
	Wu et al. [75]	HPFLS-TOPSIS and Weighted Cross-Entropy	Calculation	HPFLS-TOPSIS, and Weighted Cross-Entropy
	Zolfaghari and Mousavi [79]	IVHFLS-FMEA, MULTIMOORA, and TPOP	Calculation	MULTIMOORA
	Duan et al. [67]	DHHFLTS-FMEA, K-Means Clustering, and Maximizing Deviation Method	Calculation	Maximizing Deviation Method
	Dai et al. [66]	DHHFLTS-FMEA, K-Means Clustering, and Entropy Weight Method	Calculation	K-means Clustering
	Divsalar et al. [73]	Fuzzy Delphi, EHFLTS-VIKOR and IVHF-DANP	Calculation	IVHF-DANP
	Erol et al. [53]	HFLTS-QFD, HFLTS-Delphi, HFLTS-ANP, and HFLTS-TOPSIS	Calculation	HFLTS-ANP
	Krishankumar et al. [64]	DHHFLTS combined with the Attitudinal-CRITIC (Criteria Importance Through Intercriteria Correlation) Approach and the Weighted Distance Approximation (WDA) Algorithm	Calculation	Attitudinal-CRITIC Approach
Combined techniques (21)	Wu et al. [80]	HFLTS and the DEA-Based TODIM Method	Calculation	DEA-Based TODIM Method
	Lima et al. [82]	HFLTS-AHP, HFLTS-QFD, and HFLTS-PROMETHEE	Calculation	HFLTS-AHP
	Dolatabad et al. [50]	HFLTS-VIKOR and Fuzzy Cognitive Map	Calculation	Fuzzy Cognitive Map

Figure 10 presents the frequency with which each technique was utilized in the integrated methods. We identified 36 distinct techniques. The most frequently utilized methods were FMEA and TOPSIS, which were present in five of the studies. These were followed by AHP, DEMATEL, and QFD with four applications apiece.

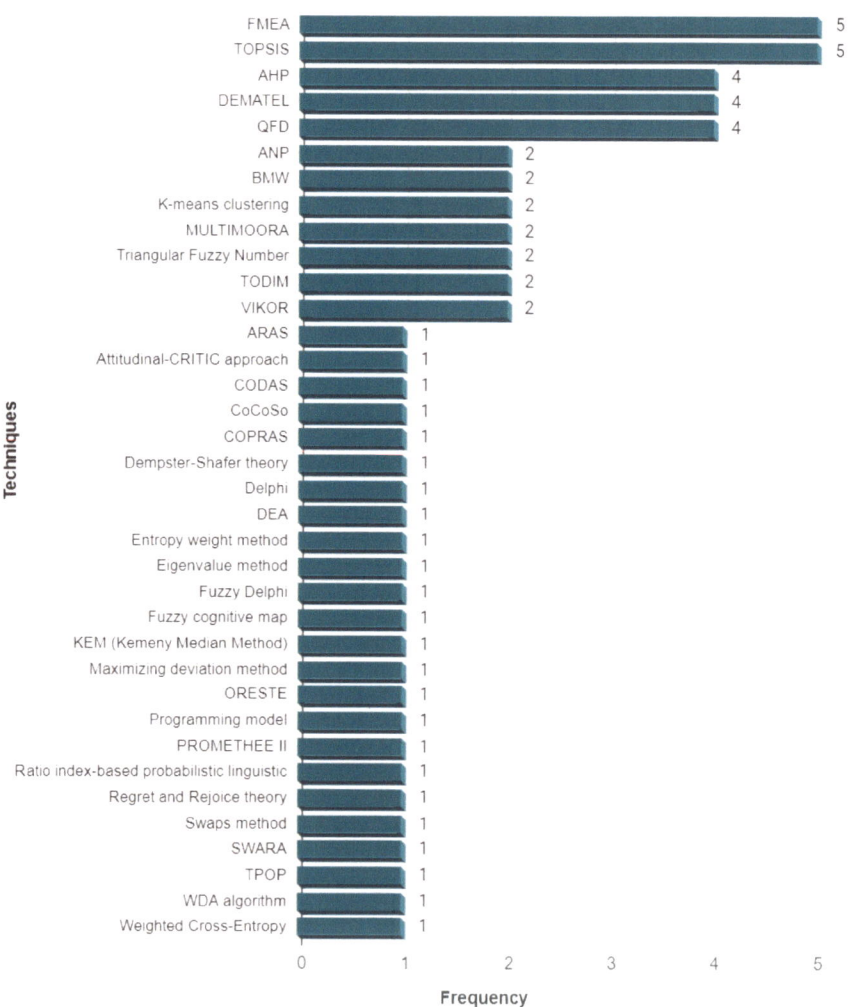

Figure 10. Frequency of Techniques that have been Integrated with HFLTSs and Extensions.

Table 9 describes the techniques that were found in at least two applications. In this table, the techniques are positioned according to their frequency of use. We believe that the great frequency of the use of FMEA is related to the fact that this method is recognized worldwide as an efficient risk evaluation tool. Combining it with HFLTS techniques makes it possible to overcome deficiencies in the original version of FMEA, such as the inability to attribute weights to the criteria and the need to review occasional erroneous evaluations of failures utilizing exact numerical values [65]. On the other hand, the methods resulting from the combination of TOPSIS and HFLTS approaches provide greater flexibility to the alternative evaluation process [4].

Table 9. Description of the Techniques that are Most Frequently Combined with HFLTSs and HFLTS Extensions.

Technique	Description
FMEA	A risk analysis method that is traditionally used to prioritize failure modes. Each failure mode receives a numerical value based on the "Severity", "Occurrence" and "Detection" criteria. In traditional FMEA, the values range from 0 to 10, and the criteria weights are not considered. Combining FMEA with HFLTSs or HFLTS extensions enables DMs to use linguistic expressions to evaluate failures. It also makes it possible to consider criteria weights and support group decision making under conditions of uncertainty and hesitation [65].
TOPSIS	An MCDM method that ranks alternatives according to their proximity to the ideal solutions. While the positive ideal solution is formed by the highest numerical values achieved by the alternatives in each criterion, the negative ideal solution consists of the lowest numerical values for each criterion. The traditional version of TOPSIS only uses numerical values and is not appropriate for decision making under conditions of uncertainty. On the other hand, combining TOPSIS with HFLTSs or HFLTS extensions enables DMs to use linguistic expressions to evaluate alternatives and in some cases criteria weights as well [4].
AHP	An MCDM method in which decision-making problems are represented as a hierarchy composed of the problem objective, the criteria, and the alternatives. The criteria weights and the values for the alternatives are defined based on pairwise comparisons conducted by the DMs. This method requires the verification of the consistency of the DM judgments and also requires a greater number of judgments than TOPSIS and VIKOR [54,56].
DEMATEL	A very useful technique for identifying the cause-and-effect relationships among a group of system elements. In the decision-making area, this method is often used to identify the inter-relationships among the criteria. The evaluations of the individual relationships between the elements serve as the inputs for this method. The main output is a structural model that provides a global view of the inter-relationships among the criteria, which makes it possible to discover which of them are the most influential [57].
QFD	A quality management method traditionally applied to the product development area. It is made up of a "what" matrix that contains the prioritization of the customer requirements and a "how" matrix that relates the requirements to the desirable technical characteristics of the product. While traditional QFD uses numerical values, the versions based on HFLTSs and HFLTS extensions use linguistic terms and expressions provided by the DMs. There are also versions of HFLTS-QFD that can take into account the degree of risk in the decision-making problem [8,19].
ANP	An MCDM method based on AHP. The decision-making problem is represented by a network composed of various nodes. Like AHP, comparisons between pairs of alternatives and criteria serve as the input data for this method. ANP also requires the verification of the consistency of the DMs' paired judgments. A distinguishing characteristic of ANP is that this method takes into account the inter-relationships among decision-making criteria [53]
BWM	An MCDM method that ranks alternatives from the best to the worst solution. Initially, the DMs should indicate which are the most important and least important criteria. The other criteria are compared in terms of the best and worst criteria. The same principle is applied in the evaluation of the alternatives. In addition to requiring fewer comparisons than AHP, the BWM method presents an advantage in that it does not require tests to verify its consistency [59,71].
K-Means Clustering	An unsupervised learning technique which makes it possible to partition n observations into k groups. Initially, this method randomly selects k alternatives and associates each alternative with a cluster. For each remaining alternative, the similarity is calculated between the analyzed alternative and the center of each cluster (centroid). Then, each alternative is associated with the closest cluster. The centroids of the clusters are updated and the alternatives are reclassified until they achieve the convergence of the algorithm [66,67].
MULTI-MOORA	A compensatory method that works with independent criteria and does not use the weights of these criteria. This method seeks to simultaneously optimize two or more objectives while taking into account a group of restrictions. The output of MULTIMOORA is a ranking obtained through the aggregation of the results of the following methods: the Ratio System, the Reference Point Approach, and the Full Multiplicative Form [56,79].
TFNs	Triangular Fuzzy Numbers are used to represent imprecise numerical values or qualitative linguistic evaluations. Each TFN is made up of a triangular membership function in which the central vertex indicates the value with the greatest membership degree, while the vertices of the extremities indicate values of lesser membership degrees. One of the main advantages of TFNs is the simplicity of the calculations and the low computational complexity involved in manipulating their values [58,69].
TODIM	An MCDM method that is based on prospect theory. TODIM works with quantitative criteria as well as qualitative criteria. It measures the dominance degree of each alternative over the other alternatives based on the value overall. TODIM presents resources to eliminate inconsistencies in the DM judgments. A distinguishing characteristic of this method is that it tests more specific forms of profit and loss functions [78].
VIKOR	An MCDM method for determining a compromise solution that is closer to the ideal solution. One benefit of the VIKOR method is its capacity to generate rankings that consider judgments related to all of the criteria (group utility) or the criteria that have the worst evaluations (individual regret) [50,73].

In addition to the approaches used to model DM preferences, another factor that directly influences decision making is the weights (degrees of relative importance) attributed to the criteria. The definition of these weights allows DMs to decide which aspects should have priority in evaluating alternatives for the solution of a problem. The literature features various ways of obtaining the values of criteria weights. As shown in Table 8, in most of the studies (31), the weights were calculated by a decision-making technique or an aggregation operator based on the DM judgments (RQ3.4). The most often applied techniques to accomplish this were HFLTS-QFD, HFLTS-AHP, and HFLTS-BWM. In two studies, the weights were attributed directly by the DMs. One study did not attribute weights to the criteria.

Table 10 presents the classification of the studies based on factors related to their support of GDM (RQ3.5). We found that all of the studies applied methods to support GDM, even though the application presented by Finger and Lima-Junior [8] considered only one DM. Figure 11 presents the frequency of given numbers of participating DMs in the analyzed applications. The average number of participating DMs in the analyzed applications was 6.4 and the mode was 3. Applications with three or four participating DMs have been also quite frequent. In nine studies (26.5%), the weights of the DMs were taken into account in obtaining the results (RQ3.6). Among these, seven studies (20.6%) used methods to calculate the weight values of the DMs, and in two studies (5.9%) numerical values were attributed directly to weight the DM opinions.

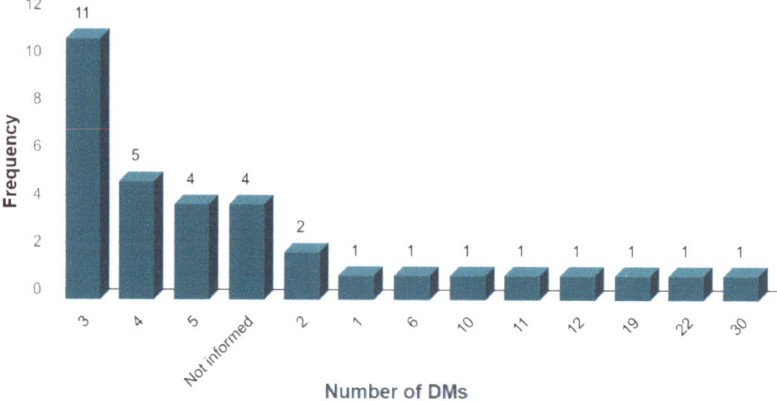

Figure 11. Frequency of Studies with Given Numbers of DMs in the Analyzed Applications.

In terms of the approaches adopted to manipulate the individual preferences of the DMs (RQ3.7), 31 (88.2%) applied a decision-making technique or a mathematical operator to aggregate DM preferences; for example, VHFLPWA, Interval Weighed Geometric Aggregation (IWGA), OWA, HFLWA, DHHLWA, GWOWA, and GMSM. Three studies (11.8%) presented a decision-making matrix obtained by consensus without specifying how this consensus among the various DM evaluations was achieved. Only four studies (11.8%) applied iterative methods in the search for consensus. These methods made it possible to suggest modifications to the DM evaluations by calculating measures of consensus during the evaluation rounds. Examples of iterative methods seeking consensus were presented in Liu et al. [51], Wu et al. [75], Divsalar et al. [73], and Erol et al. [53].

Table 10. Classification of the Studies in terms of Group Decision-Making (GDM).

Author(s)	Supports GDM?	Number of DMs	Allows Weighting of DM Opinions?	How Were DM Weights Assigned?	Method Used to Calculate DM Weights	Aggregates DM Opinions?	Procedure Used to Aggregate DM Opinions	Applies Iterative Consensus Method?
Tüysüz and Şimşek [54]	Yes	3	No	N/A	N/A	Yes	HFLTS-AHP	No
Osiro et al. [19]	Yes	3	Yes	N/A	N/A	Yes	HFLTS-QFD	No
Wang et al. [11]	Yes	3	No	N/A	N/A	Yes	HM	No
Chang et al. [59]	Yes	4	No	N/A	N/A	Yes	OWGA	No
Wu et al. [58]	Yes	30	No	N/A	N/A	Yes	Fuzzy arithmetic mean	No
Liao et al. [49]	Yes	4	Yes	Attributed numeric values	N/A	Yes	IWGA	No
Pérez-Domínguez et al. [55]	Yes	6	No	N/A	N/A	Yes	HFLTS-TOPSIS	No
Yalçın et al. [61]	Yes	5	No	N/A	N/A	Yes	Ordered Weighted Average (OWA)	No
Krishankumar et al. [77]	Yes	3	No	N/A	N/A	Yes	GMSM	No
Li et al. [69]	Yes	5	No	N/A	N/A	Yes	GWOWA	No
Qu et al. [74]	Yes	3	No	N/A	N/A	Yes	Degree of group satisfaction and the regret theory	No
Wang et al. [68]	Yes	Not informed	No	N/A	N/A	Yes	DHHFLTS-ORESTE	No
Ren et al. [81]	Yes	4	No	N/A	N/A	Yes	Programming model	No
Zhang et al. [70]	Yes	Not informed	No	N/A	N/A	Yes	Ratio index-based probabilistic linguistic	No
Zhang et al. [71]	Yes	5	No	N/A	N/A	Yes	HFLTS-BWM	No
Krishankumar et al. [78]	Yes	3	Yes	Calculated	Mathematical model	Yes	GMSM	No
Liu et al. [51]	Yes	4	Yes	Attributes numeric values	N/A	Yes	Based on degrees of hesitancy and similarity	Yes
Büyüközkan and Güler [56]	Yes	3	No	N/A	N/A	Yes	AHP and OWA	No
Lima-Junior and Hsiao [52]	Yes	2	No	N/A	N/A	Yes	HFLTS-TOPSIS	No
Zolfaghari and Mousavi [79]	Yes	3	No	N/A	N/A	Yes	Interval-valued Hesitant Fuzzy Linguistic Prioritized Weighted Average	No

Table 10. *Cont.*

Author(s)	Supports GDM?	Number of DMs	Allows Weighting of DM Opinions?	How Were DM Weights Assigned?	Method Used to Calculate DM Weights	Aggregates DM Opinions?	Procedure Used to Aggregate DM Opinions	Applies Iterative Consensus Method?
Shen and Liu [65]	Yes	Not informed	No	N/A	N/A	Yes	DHHFLTS-COPRAS	No
Wu et al. [75]	Yes	10	No	N/A	N/A	No	N/A	Yes
Finger and Lima-Junior [8]	Yes	1	No	N/A	N/A	Yes	HFLTS-QFD	No
Erol et al. [53]	Yes	19	No	N/A	N/A	Yes	HFLTS-Delphi and HFLTS-ANP	Yes
Qin et al. [9]	Yes	Not informed	No	N/A	N/A	No	N/A	No
Divsalar et al. [73]	Yes	12	No	N/A	N/A	No	N/A	Yes
Borges et al. [4]	Yes	2	No	N/A	N/A	Yes	HFLTS-TOPSIS	No
Duan et al. [67]	Yes	5	Yes	Calculated	Non-linear programming and genetic algorithm	Yes	Double Hierarchy Hesitant Linguistic Weighted Average (DHHLWA)	No
Dai et al. [66]	Yes	3	Yes	Calculated	Entropy Weight Method	Yes	Entropy Weight Method	No
Wu et al. [80]	Yes	3	Yes	Calculated	Optimization model	Yes	Optimization model	No
Lima et al. [82]	Yes	4	No	N/A	N/A	Yes	HFLTS-AHP and OWA	No
Krishankumar et al. [64]	Yes	3	Yes	Calculated	WDA algorithm	Yes	Attitudinal-CRITIC approach	No
Dolatabad et al. [50]	Yes	11	Yes	Calculated	Method not defined	Yes	Hesitant Fuzzy Linguistic Weighted Average (HFLWA)	No
Zheng et al. [57]	Yes	22	Yes	Calculated	Maximizing consensus approach	Yes	HFLTS-DEMATEL	No

Finally, another important aspect of these applications has to do with how their results were validated. As presented in Figure 12, the results in 12 studies (35.3%) were validated by comparing them with the results of other decision-making methods applied to the same problem (RQ3.8). In general, this comparison is based on the ranking or categorization (sorting) supplied for each analyzed method. In 11 studies (23.5%), the authors compared the obtained results with those furnished by other methods and also conducted sensitivity analysis tests. Sensitivity analyses were utilized in an isolated manner in four studies (11.8%). The main purpose of sensitivity analyses is to verify alterations in the outputs furnished by the model when the input parameter values are varied. In three studies (8.8%), the results were validated through a combination of sensitivity analysis, comparisons with other methods, and statistical tests. In addition, three studies featured just one application without specifying how the results were validated. Just one study validated the results by comparing them with real data.

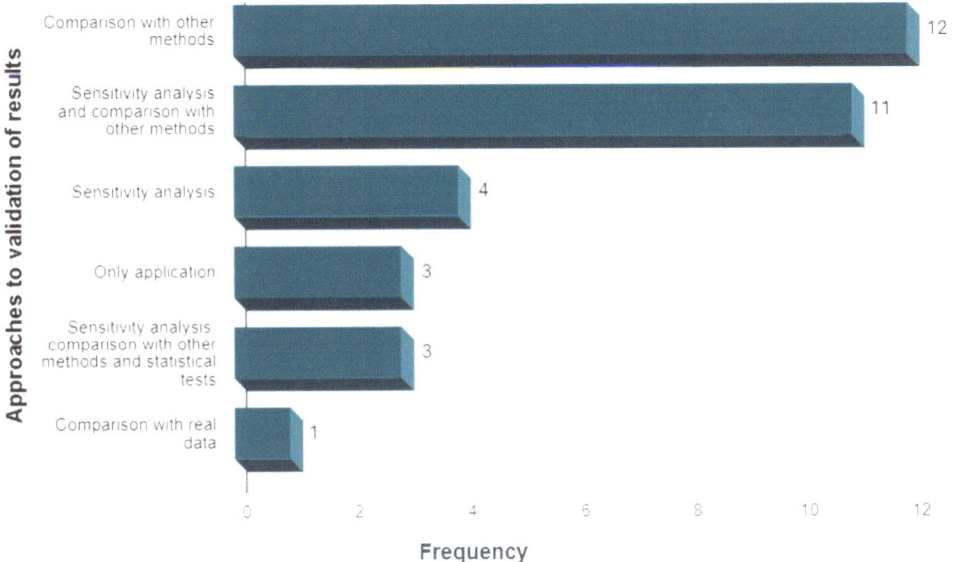

Figure 12. Approaches Utilized to Validate the Analyzed Study Results.

4.5. Comparison of the Results of Previous Studies

The results of this study were compared to those of previous literature review studies on subjects related to our objective. Thus, as in Yu et al. [24], which analyzed 1080 studies based on HFS and HFS extensions, this study found that the HFLTS approach proposed by Rodríguez et al. [14] was the most utilized in the analyzed studies. In terms of the type of SCM strategy, our results were similar to the study by Lima-Junior and Carpinetti [6], which investigated decision-making models to evaluate supply chain performance. Both studies indicated the predominance of sustainable and green supply chains. In terms of the procedures adopted to validate their results, Lima-Junior and Carpinetti [6] related the prevailing use of sensitivity analysis followed by statistical techniques and comparisons with other methods, while our study concluded that comparisons with other methods have been the most utilized validation procedure followed by sensitivity analysis combined with comparisons with other methods.

The results of this study indicated that the most frequently adopted techniques in integrated models have been TOPSIS, FMEA, AHP, DEMATEL, and QFD; while in Lima-Junior and Carpinetti's study [6], the AHP and Data Envelopment Analysis (DEA) techniques and linear programming predominated. In terms of the SCM processes that had the most applications, in analyzing artificial intelligence techniques in SCM, Yang et al. [5] found that the Return and Make processes had fewer applications, while our study found that the least studied processes have been Return and Delivery. On the other hand, while Yang et al. [5] found that the Enable and Plan processes had the most applications, this study found that the Source and Enable processes have had the most applications.

Finally, in terms of the economic sectors of the participating companies that have received these applications, Yang et al.'s study [5] noted the predominance of retail, food, and manufacturing; while our study showed a prevalence in the automotive, health, and electro-electronic industries. These results were somewhat similar to those found by Lima-Junior and Carpinetti [6], who related that the sectors that received the most applications of decision-making models were the automotive, food, and electro-electronic industries.

5. Recommendations for Future Research

The results of our systematic review of the literature identified various research gaps in this subject. Based on such gaps, as well as those in SCC [48], Lima-Junior and Carpinetti [6], Wang et al. [11], Kabak [47], and Yang et al. [5], we proposed a framework that seeks to help researchers and managers develop future studies on this subject. The framework presented in Figure 13 encompasses pertinent recommendations regarding the following topics: innovative applications of techniques; new combinations of decision-making techniques; comparative studies of decision-making techniques; GDM issues; and validation procedures for new decision-making support models.

As indicated in Figure 13, we recommend taking into account the participating economic sector and the type of SCM strategy of the company participating in the application for various types of future studies, since this has a strong influence on the choice of criteria and the assessment of their weights. Since there is no single decision-making method, each technique has benefits and limitations that depend on the context of the application [26]. Thus, the choice of MCDM methods to be applied must take into account aspects such as: the purpose of the application; the type of decision-making problem; the number of alternatives, the criteria, and the DMs involved; the nature of the criteria; the format for representing DM preferences; and the need to weigh DM opinions among other factors. Thus, choosing a suitable decision-making method requires a detailed analysis of the features of the techniques and the problem in question. Proposing new hybrid techniques has the potential to bring together the benefits of HFLTS approaches with techniques traditionally used in companies, which will thus generate more appropriate solutions for each context.

5.1. Innovative Applications

The results of this study indicated that the use of HFLTS and HFLTS extension techniques has still not been tested in various decision-making problems that are important to ensure effective SCM. Problems related to the delivery and return processes have been less studied until now. There are also various types of industries that have not participated in HFLTS and HFLTS extension applications.

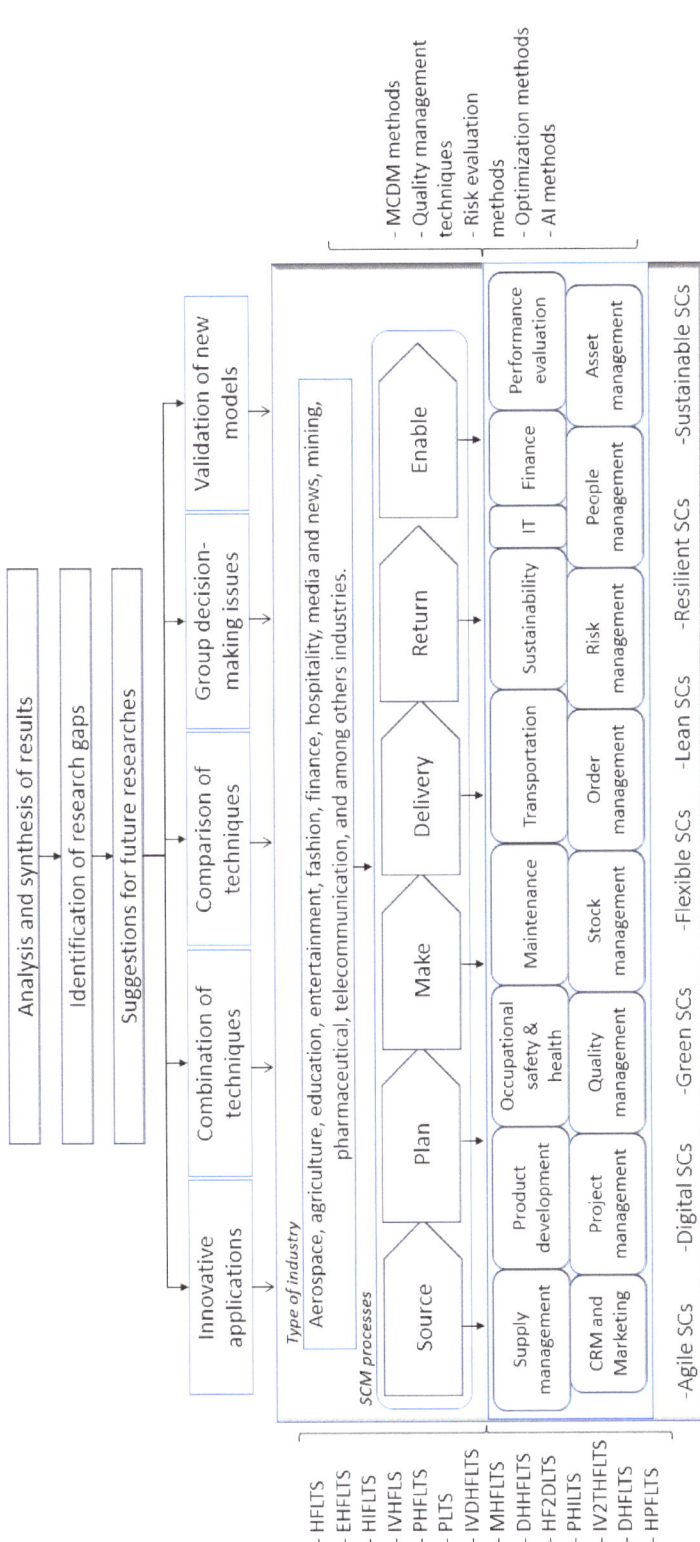

Figure 13. Framework to Guide the Development of Future Studies on this Subject.

As Figure 13 illustrates, SCM processes span various business areas that present multiple-criteria decision-making problems for which few or no applications have been applied. The use of HFLTS and HFLTS extension techniques has great potential to contribute to applications dealing with these problems due to the capacity of these techniques to provide support for GDM under conditions of uncertainty. Based on the research opportunities that we have identified, below is a list of SCM problems that involve selecting, ordering, and categorizing that can be explored in future works. It includes problems associated with various business areas that involve strategic, tactical, and/or operational decision making:

1. Asset management: prioritization of asset management investments, strategic asset allocation, selecting the location of new installations, and layout selection;
2. Customer relationship management (CRM) and marketing: marketplace selection, marketing strategy selection, market segmentation, customer satisfaction analysis, and customer relationship management software selection;
3. Finance: credit risk evaluation, corporate financial performance analysis, investment appraisal, budget allocations, and the evaluation of financial plans;
4. Information technology (IT): information system selection, computer workstation selection, software quality evaluation, IT service provider selection, and disruptive Industry 4.0 technology evaluation;
5. Maintenance: maintenance strategy selection, maintenance service provider selection, maintenance machine selection, and the prioritization of maintenance activities;
6. Occupational safety and health: accident risk evaluation, the selection of key indicators for improving the occupational safety system, the prioritization of emergency plans, individual protection equipment selection, and system reliability evaluation;
7. Order management: prioritization of production orders, order delivery evaluation, and the prioritization of plans to improve order management;
8. Product development: product development strategy selection, new product material selection, prototype evaluation, and product portfolio evaluation;
9. Project management: project proposal selection, project risk evaluation, project performance indicator selection, project management practice maturity evaluation, and program and/or project success evaluation;
10. Quality management: Six Sigma project selection, benchmarking, product or service requirement prioritization, selection of the certifying body for the implementation of ISO 9001 certification, and prioritization of continual improvement actions;
11. Risk management: risk evaluation tool selection, organizational risk evaluation, and supply chain risk evaluation;
12. Stock management: stock management strategy selection, ABC classification of stocks, warehouse location selection, and warehouse structure selection;
13. Supply management: make or buy, supplier performance monitoring, supplier segmentation, and supplier development program evaluation;
14. Personnel management: organizational climate evaluation, personnel selection, position evaluation, and skills and qualifications evaluation;
15. Sustainability: waste treatment alternative evaluation, prioritization of sanitary landfill location selection actions, prioritization of actions designed to promote sustainability, evaluation of the barriers to the adoption of sustainable practices, and product lifecycle evaluation;
16. Transportation: route selection, modes of transport evaluation, logistics service provider selection, vehicle selection, and geographic information system selection.

The suggested applications open a gamut of possibilities for new studies. On one hand, one can explore problems that still have not been addressed with applications, such as those related to asset management, finance, IT, maintenance, marketing, occupational safety and health, order management, and stock management, among other areas. On the other hand, one can test the use of techniques that still have not been applied to problems that have received more attention, such as supplier selection, failure evaluation, and performance evaluation. These applications can involve companies in sectors that have not been very studied until now or firms in sectors that have not had any applications, such as aerospace, agriculture, education, entertainment, fashion, finance, hospitality, media and news, mining, pharmaceuticals, and telecommunications. It is also important to take SCM strategies into account in each case. Applications related to agile, flexible, and resilient supply chains have received less attention in recent studies. Different combinations of SCM strategies could also be studied in future studies.

5.2. Technique Integration

The results indicated that the choice, ordering, or sorting of alternatives has rarely been conducted using HFLTSs or one of their extensions alone. Instead, HFLTSs and HFLTS extensions have usually been combined with other methods. However, there are various combinations of HFLTS and HFLTS extension techniques and other types of methods that still have not been tested in SCM problems.

Although HFLTSs and their extensions can be used to calculate criteria weights, in many cases techniques based on paired comparisons have been used. The most frequent of these is the AHP method, which requires a greater amount of judgment from DMs and also requires consistency tests. Therefore, there are several opportunities for using HFLTS approaches in defining criteria weights and evaluating possible inter-relationships between criteria. The use of HFLTS extensions in these cases can bring greater flexibility to representing criteria weights and avoid a loss of information in the initial stages of the decision-making problem.

Given the low frequency of applications that employed the DHFLTS, EHFLTS, HPFLS, IVHFLTS, and MHFLTS techniques, as well as the absence of applications based on HI-FLTS, PHFLTS, IVDHFLTS, HF2DLTS, PHILTS, IV2THFLTS techniques, future studies could test new combinations of these approaches with MCDM methods, quality management techniques, risk evaluation techniques, optimization methods, and/or artificial intelligence techniques.

Although the EHFLTS and MHFLTS approaches are especially useful for GDM problems in which it is necessary to aggregate the preferences of several DMs, repeated linguistic terms of the global set of evaluations are not accounted for, which can lead to a loss of information. In view of this, the adoption of PHFLTSs, PLTSs, and PHILTSs is more appropriate because they take into account each term that appears in the individual or subgroup evaluation of the DMs and at the same time assign greater weights to the terms that appear more frequently.

Since HIFLTSs, IVDHFLTSs, DHFLTSs, and HPFLTSs take into account the number of membership and non-membership degrees, these approaches can be applied in GDM that involves high complexity and risk. On the other hand, IVHFLSs are appropriate when the group's opinion is given according to the limits of the valued intervals and are not appropriate for preserving the individual preferences of the DMs. Thus, IVHFLSs can be used in problems with more homogeneous groups and are not suitable for situations in which there is no consensus.

New hybrid methods can be created based on MCDM methods, which have great potential for integration with HFLTS approaches but have rarely or never been used, such as ORESTHE, ARAS, CODAS, CoCoSo, Delphi, ELECTRE (ÉLimination Et Choix Traduisant la REalité in French), Measurement of Alternatives and Ranking according to Compromise Solution (MARCOS), Measuring Attractiveness by a Categorical Based Evaluation Technique (MACBETH), ORESTE, PROMETHEE, the Qualitative Flexible Multiple Criteria Method (QUALIFLEX), the Simple Multi-Attribute Rating Technique (SMART), and SWARA.

There are also quality management techniques such as QFD, Service Quality Measurement (SERVQUAL), and the GUT matrix (a process prioritization matrix based on Gravity, Urgency, and Tendency) that are multicriteria in nature and can be integrated with HFLTS extensions to create new MCDM methods. Similarly, FMEA, risk matrix, and fault tree analysis can be combined with these approaches to generate new methodologies for risk evaluation.

In cases in which a problem step seeks to order, select, or sort alternatives while another step seeks to optimize resources, one can use combinations with optimization methods such as linear programming, non-linear programming, stochastic programming, and dynamic programming. On the other hand, for problems that seek to classify patterns or predictions and/or group values and/or require the analysis of a large quantity of data, HFLTS approaches can be combined with AI methods such as artificial neural networks, neuro-fuzzy systems, genetic algorithms, and case-based logic.

5.3. Comparison of Techniques

The results of our study also demonstrated the need to conduct comparative studies involving MCDM techniques based on HFLTSs and HFLTS extensions. Even though some of the analyzed studies compared the numeric outputs of distinct techniques when applied to the same problem, the literature offered few comparative studies that discussed the benefits and limitations of HFLTS and HFLTS extension techniques in specific problem domains.

The realization of these comparative studies would contribute to a greater understanding of the technical characteristics being compared and could assist researchers and practitioners in choosing the most appropriate techniques for given SCM problems. In addition to comparing the outputs generated by each technique, it is recommended that these studies take into account comparison factors such as computational complexity, limitations in terms of the number of input variables, the effect of variations in criteria and alternatives, support for GDM, and agility in the decision-making process [26].

The realization of studies that compare various HFLTS extensions could also be valuable in mapping the advantages in use as well as the similarities and differences among these approaches. These comparative studies could take into account factors such as the complexity of modeling and processing, the effect of differences in the representation of DM preferences, and the appropriateness of each approach in dealing with various types of uncertainty. The behavior of various aggregation operators for HFLTS and HFTLS extension information could also be analyzed.

5.4. GDM Issues

There are some topics related to GDM in SCM that have been little studied and deserve more attention. Even though all of the analyzed studies provided support for GDM, we verified that there are few methods that make it possible to attribute weights to the DMs. This may be especially useful when one wants to weight the opinion of DMs based on their level of experience, positions, and/or knowledge of a problem. In addition, since methods that allow the attribution of weights to DMs using linguistic expressions were not found, we suggest the development of methods that make this possible to deal appropriately with uncertainty in the definition of DM weights.

An important emergent research topic is large GDM problems. These problems are a special case in terms of GDM processes in which the opinions of a large number of people are collected. There are various large GDM problems inherent to SCM that could be investigated in future works; for example, strategic decisions that involve DMs from various departments or organizations or product or service evaluations conducted by a gamut of customers. In cases in which the number and diversity of DMs are large, we recommend the adoption of EHFLTS techniques, which make it possible to organize DMs in subgroups and avoid losses of information in situations in which there is no consensus among the DMs.

Finally, another relevant topic regarding GDM that requires more investigation has to do with models based on the consensus-reaching process. There are a variety of opportunities to develop new models of this type that can be explored through a combination of iterative methods with HFLTS and/or HFLTS extension techniques. One of them consists of the development of new approaches that combine the Delphi method with HFLTS extensions. In addition, it would be interesting to test new consensus models that propose modifications to DM preferences as well as models based on adaptive consensus strategies that automatically update DM weights with each iteration.

5.5. Validation of New Decision-Making Models

Future studies could use validation procedures that have been little explored to evaluate the reliability of the results of new decision-making models. Given that most studies currently conduct sensitivity analysis tests and compare their results with other methods, it is plausible to adopt statistical techniques such as hypothesis tests, variance analysis, and error measurements that analyze the obtained results. The use of similarity measures is also a useful way to compare the results generated by different techniques. In addition, the realization of factor analysis experiments would make it possible to identify which input variables have the greatest influence on the results.

To verify the consistency of the obtained results, it is important to conduct tests that consider a larger number of application cases in order to evaluate the performance of these techniques under distinct scenarios while varying the number of alternatives, criteria, and linguistic terms. We also recommend verifying the usability of HFLTS and HFLTS extension techniques by users who are not specialists in dealing with these techniques. To accomplish this, we suggest that future studies develop software with graphical interfaces based on these techniques to verify their usability in various organization areas.

6. Conclusions

This study presented a systematic review of the literature on applications of HFLTS and HFLTS extension techniques in SCM decision-making problems. In order to answer a series of research questions regarding this subject, the selected studies were characterized in accordance with a group of factors related to their origin, SCM, HFLTSs, and decision making. The results demonstrated that this research subject is quite recent and that there has been substantial growth in the number of publications about this topic. The applications we identified provide support for a wide variety of decision-making problems; their main focuses were on supplier selection, failure evaluation, and performance evaluation. The results reinforced the high applicability of HFLTSs and their extensions to business practices, since companies from any economic sector can adopt these techniques in their decision-making processes no matter whether they are strategic, tactical, or operational.

We verified the predominance of the use of HFLTS, TOPSIS, and FMEA techniques. Among HFLTS extensions, we can highlight DHHFLTS and PLTS applications. Applications in automotive firms and sustainable supply chains have received the most attention. It was confirmed that all of the analyzed models are appropriate for providing support for GDM, even though few of them permit the attribution of distinct weights to DMs. There were also few models designed to obtain a consensus among DMs. The results of this study demonstrated that even though there is a wide variety of HFLTS extensions, we did not find SCM applications for around half of them. There are also various types of SCM strategies, industries, and decision-making problems that deserve greater attention from researchers and practitioners. A challenge for real applications is to computationally implement HFLTS techniques and their extensions, since decision-making software products based on these techniques are still rare. Although most of them can be implemented using spreadsheet software such as MS Excel, the development of new forms of software with a graphical interface can simplify their use and contribute to the greater adoption of these techniques by users who are not specialists in HFLTSs.

The main contribution of this study consists of presenting an overview of the use of HFLTSs and HFLTS extensions in SCM in practice, highlighting trends and research opportunities. Our study presented a wide array of directions for future studies that encompass topics related to innovative applications, combinations of techniques, comparisons of techniques, GDM issues, and validation procedures for new decision-making models. To our knowledge, this was the first study to present a systematic review that focused on real applications of HFLTS and HFLTS extension techniques. It was also the first study to analyze applications of decision-making techniques that deal with hesitation in SCM. The analyzed techniques can be applied to various fields in SCM. That being said, we believe that this study can contribute to the dissemination of the use of HFLTSs and HFLTS extensions to minimize the effects of uncertainty on the results.

Finally, a limitation of this study was that there may be works that present HFLTS or HFLTS extension applications that were not identified in our searches. Even though we consulted various databases, this list was not exhaustive. In addition, we opted to include only articles in English and did not include gray literature or non-realistic numerical applications. Future studies can complement the results of this systematic review of the literature by including new works in the study sample. Other reviews can also be conducted that consider applications of techniques derived from HFSs and HFS extensions in various areas of knowledge such as the engineering, health, construction, and energy fields.

Author Contributions: Conceptualization, F.R.L.-J. and M.E.B.d.O.; Methodology, F.R.L.-J. and M.E.B.d.O.; Formal Analysis, C.H.L.R. and M.E.B.d.O.; Investigation, C.H.L.R. and M.E.B.d.O.; Data Curation, C.H.L.R. and F.R.L.-J.; Writing—Original Draft Preparation, F.R.L.-J.; Writing—Review and Editing, F.R.L.-J.; Visualization, C.H.L.R.; Supervision, F.R.L.-J.; Funding Acquisition, F.R.L.-J. All authors have read and agreed to the published version of the manuscript.

Funding: This research was funded by the National Council for Scientific and Technological Development (CNPq) (Code 409529/2021-4).

Data Availability Statement: The data generated in this study have been presented in Section 4. Data sharing is not applicable to this article.

Acknowledgments: We thank the journal *Mathematics* for the invitation to publish a review paper. We also thank the anonymous reviewers for their suggestions for improving this work.

Conflicts of Interest: The authors declare no conflict of interest.

References

1. Lambert, D.M.; Enz, M.G. Issues in Supply Chain Management: Progress and Potential. *Ind. Mark. Manag.* **2017**, *62*, 1–16. [CrossRef]
2. Rinaldi, M.; Murino, T.; Gebennini, E.; Morea, D.; Bottani, E. A Literature Review on Quantitative Models for Supply Chain Risk Management: Can They Be Applied to Pandemic Disruptions? *Comput. Ind. Eng.* **2022**, *170*, 108329. [CrossRef]
3. Lima-Junior, F.R.; Carpinetti, L.C.R. An Adaptive Network-Based Fuzzy Inference System to Supply Chain Performance Evaluation Based on SCOR® Metrics. *Comput. Ind. Eng.* **2020**, *139*, 106191. [CrossRef]
4. Borges, W.V.; Lima Junior, F.R.; Peinado, J.; Carpinetti, L.C.R. A Hesitant Fuzzy Linguistic TOPSIS Model to Support Supplier Segmentation. *Rev. Adm. Contemp.* **2022**, *26*, e210533. [CrossRef]
5. Yang, M.; Lim, M.K.; Qu, Y.; Ni, D.; Xiao, Z. Supply Chain Risk Management with Machine Learning Technology: A Literature Review and Future Research Directions. *Comput. Ind. Eng.* **2023**, *175*, 108859. [CrossRef]
6. Lima-Junior, F.R.; Carpinetti, L.C.R. Quantitative Models for Supply Chain Performance Evaluation: A Literature Review. *Comput. Ind. Eng.* **2017**, *113*, 333–346. [CrossRef]
7. Resende, C.H.L.; Geraldes, C.A.S.; Lima Junior, F.R.; Lima, F.R. Decision Models for Supplier Selection in Industry 4.0 Era: A Systematic Literature Review. *Procedia Manuf.* **2021**, *55*, 492–499. [CrossRef]
8. Finger, G.S.W.; Lima-Junior, F.R. A Hesitant Fuzzy Linguistic QFD Approach for Formulating Sustainable Supplier Development Programs. *Int. J. Prod. Econ.* **2022**, *247*, 108428. [CrossRef]
9. Qin, R.; Liao, H.; Jiang, L. An Enhanced Even Swaps Method Based on Prospect Theory with Hesitant Fuzzy Linguistic Information and Its Application to the Selection of Emergency Logistics Plans under the COVID-19 Pandemic Outbreak. *J. Oper. Res. Soc.* **2022**, *73*, 1227–1239. [CrossRef]
10. Wu, Z.; Xu, J.; Jiang, X.; Zhong, L. Two MAGDM Models Based on Hesitant Fuzzy Linguistic Term Sets with Possibility Distributions: VIKOR and TOPSIS. *Inf. Sci.* **2019**, *473*, 101–120. [CrossRef]
11. Wang, H.; Xu, Z.; Zeng, X.-J. Hesitant Fuzzy Linguistic Term Sets for Linguistic Decision Making: Current Developments, Issues and Challenges. *Inf. Fusion* **2018**, *43*, 1–12. [CrossRef]
12. Wang, H.; Xu, Z.; Zeng, X.-J. Modeling Complex Linguistic Expressions in Qualitative Decision Making: An Overview. *Knowl.-Based Syst.* **2018**, *144*, 174–187. [CrossRef]
13. Riahi, Y.; Saikouk, T.; Gunasekaran, A.; Badraoui, I. Artificial Intelligence Applications in Supply Chain: A Descriptive Bibliometric Analysis and Future Research Directions. *Expert Syst. Appl.* **2021**, *173*, 114702. [CrossRef]

14. Rodriguez, R.M.; Martinez, L.; Herrera, F. Hesitant Fuzzy Linguistic Term Sets for Decision Making. *IEEE Trans. Fuzzy Syst.* **2012**, *20*, 109–119. [CrossRef]
15. Wang, H. Extended Hesitant Fuzzy Linguistic Term Sets and Their Aggregation in Group Decision Making. *Int. J. Comput. Intell. Syst.* **2015**, *8*, 14. [CrossRef]
16. Chen, Z.-S.; Chin, K.-S.; Li, Y.-L.; Yang, Y. Proportional Hesitant Fuzzy Linguistic Term Set for Multiple Criteria Group Decision Making. *Inf. Sci.* **2016**, *357*, 61–87. [CrossRef]
17. Si, G.; Liao, H.; Yu, D.; Llopis-Albert, C. Interval-Valued 2-Tuple Hesitant Fuzzy Linguistic Term Set and Its Application in Multiple Attribute Decision Making. *J. Intell. Fuzzy Syst.* **2018**, *34*, 4225–4236. [CrossRef]
18. Beg, I.; Rashid, T. Hesitant Intuitionistic Fuzzy Linguistic Term Sets. *Notes Intuit. Fuzzy Sets* **2014**, *20*, 53–64.
19. Osiro, L.; Lima-Junior, F.R.; Carpinetti, L.C.R. A Group Decision Model Based on Quality Function Deployment and Hesitant Fuzzy for Selecting Supply Chain Sustainability Metrics. *J. Clean. Prod.* **2018**, *183*, 964–978. [CrossRef]
20. Lu, K.; Liao, H.; Zavadskas, E.K. An Overview of Fuzzy Techniques in Supply Chain Management: Bibliometrics, Methodologies, Applications and Future Directions. *Technol. Econ. Dev. Econ.* **2021**, *27*, 402–458. [CrossRef]
21. De, A.; Singh, S.P. Analysis of Fuzzy Applications in the Agri-Supply Chain: A Literature Review. *J. Clean. Prod.* **2021**, *283*, 124577. [CrossRef]
22. Liao, H.; Gou, X.; Xu, Z. A Survey of Decision Making Theory and Methodologies of Hesitant Fuzzy Linguistic Term Set. *Syst. Eng. Theory Pract.* **2017**, *37*, 35–48. [CrossRef]
23. Morente-Molinera, J.A.; Pérez, I.J.; Ureña, M.R.; Herrera-Viedma, E. On Multi-Granular Fuzzy Linguistic Modeling in Group Decision Making Problems: A Systematic Review and Future Trends. *Knowl.-Based Syst.* **2015**, *74*, 49–60. [CrossRef]
24. Yu, D.; Sheng, L.; Xu, Z. Knowledge Diffusion Trajectories in the Hesitant Fuzzy Domain in the Past Decade: A Citation-Based Analysis. *Int. J. Fuzzy Syst.* **2022**, *24*, 2382–2396. [CrossRef]
25. Zadeh, L.A. Fuzzy Sets. *Inf. Control* **1965**, *8*, 338–353. [CrossRef]
26. Lima Junior, F.R.; Osiro, L.; Carpinetti, L.C.R. A Comparison between Fuzzy AHP and Fuzzy TOPSIS Methods to Supplier Selection. *Appl. Soft Comput.* **2014**, *21*, 194–209. [CrossRef]
27. Malik, M.G.A.; Bashir, Z.; Rashid, T.; Ali, J. Probabilistic Hesitant Intuitionistic Linguistic Term Sets in Multi-Attribute Group Decision Making. *Symmetry* **2018**, *10*, 392. [CrossRef]
28. Torra, V. Hesitant Fuzzy Sets. *Int. J. Intell. Syst.* **2010**, *25*, 529–539. [CrossRef]
29. Rodríguez, R.M.; Martínez, L.; Herrera, F.; Martínez, L.; Herrera, F. A Group Decision Making Model Dealing with Comparative Linguistic Expressions Based on Hesitant Fuzzy Linguistic Term Sets. *Inf. Sci.* **2013**, *241*, 28–42. [CrossRef]
30. Liu, P.; Shi, L. The Generalized Hybrid Weighted Average Operator Based on Interval Neutrosophic Hesitant Set and Its Application to Multiple Attribute Decision Making. *Neural Comput. Appl.* **2015**, *26*, 457–471. [CrossRef]
31. Kong, M.; Ren, F.; Park, D.-S.; Hao, F.; Pei, Z. An Induced Hesitant Linguistic Aggregation Operator and Its Application for Creating Fuzzy Ontology. *KSII Trans. Internet Inf. Syst.* **2018**, *12*, 4952–4975. [CrossRef]
32. Zhang, Z.; Wu, C. Hesitant Fuzzy Linguistic Aggregation Operators and Their Applications to Multiple Attribute Group Decision Making. *J. Intell. Fuzzy Syst.* **2014**, *26*, 2185–2202. [CrossRef]
33. Liao, H.; Xu, Z.; Zeng, X.J. Distance and Similarity Measures for Hesitant Fuzzy Linguistic Term Sets and Their Application in Multi-Criteria Decision Making. *Inf. Sci.* **2014**, *271*, 125–142. [CrossRef]
34. Liu, D.; Liu, Y.; Chen, X. The New Similarity Measure and Distance Measure of a Hesitant Fuzzy Linguistic Term Set Based on a Linguistic Scale Function. *Symmetry* **2018**, *10*, 367. [CrossRef]
35. Ren, P.; Wang, X.; Xu, Z.; Zeng, X.-J. Hesitant Fuzzy Linguistic Iterative Method for Consistency and Consensus-Driven Group Decision Making. *Comput. Ind. Eng.* **2022**, *173*, 108673. [CrossRef]
36. Wang, J.; Wu, J.; Wang, J.; Zhang, H.; Chen, X. Interval-Valued Hesitant Fuzzy Linguistic Sets and Their Applications in Multi-Criteria Decision-Making Problems. *Inf. Sci.* **2014**, *288*, 55–72. [CrossRef]
37. Pang, Q.; Wang, H.; Xu, Z. Probabilistic Linguistic Term Sets in Multi-Attribute Group Decision Making. *Inf. Sci.* **2016**, *369*, 128–143. [CrossRef]
38. Qi, X.; Liang, C.; Zhang, J. Multiple Attribute Group Decision Making Based on Generalized Power Aggregation Operators under Interval-Valued Dual Hesitant Fuzzy Linguistic Environment. *Int. J. Mach. Learn. Cybern.* **2016**, *7*, 1147–1193. [CrossRef]
39. Wang, J.; Wang, J.; Zhang, H.; Chen, X. Multi-Criteria Group Decision-Making Approach Based on 2-Tuple Linguistic Aggregation Operators with Multi-Hesitant Fuzzy Linguistic Information. *Int. J. Fuzzy Syst.* **2016**, *18*, 81–97. [CrossRef]
40. Gou, X.; Liao, H.; Xu, Z.; Herrera, F. Double Hierarchy Hesitant Fuzzy Linguistic Term Set and MULTIMOORA Method: A Case of Study to Evaluate the Implementation Status of Haze Controlling Measures. *Inf. Fusion* **2017**, *38*, 22–34. [CrossRef]
41. Liu, X.; Ju, Y.; Qu, Q. Hesitant Fuzzy 2-Dimension Linguistic Term Set and Its Application to Multiple Attribute Group Decision Making. *Int. J. Fuzzy Syst.* **2018**, *20*, 2301–2321. [CrossRef]
42. Zhang, R.; Li, Z.; Liao, H. Multiple-Attribute Decision-Making Method Based on the Correlation Coefficient between Dual Hesitant Fuzzy Linguistic Term Sets. *Knowl.-Based Syst.* **2018**, *159*, 186–192. [CrossRef]
43. Yang, L.; Wu, X.-H.; Qian, J. A Novel Multicriteria Group Decision-Making Approach with Hesitant Picture Fuzzy Linguistic Information. *Math. Probl. Eng.* **2020**, *2020*, 1–19. [CrossRef]
44. Ghadikolaei, A.S.; Madhoushi, M.; Divsalar, M. Extension of the VIKOR Method for Group Decision Making with Extended Hesitant Fuzzy Linguistic Information. *Neural Comput. Appl.* **2018**, *30*, 3589–3602. [CrossRef]

45. Page, M.J.; McKenzie, J.E.; Bossuyt, P.M.; Boutron, I.; Hoffmann, T.C.; Mulrow, C.D.; Shamseer, L.; Tetzlaff, J.M.; Akl, E.A.; Brennan, S.E.; et al. The PRISMA 2020 Statement: An Updated Guideline for Reporting Systematic Reviews. *BMJ* **2021**, *372*, n71. [CrossRef]
46. Zimmer, K.; Fröhling, M.; Schultmann, F. Sustainable Supplier Management—A Review of Models Supporting Sustainable Supplier Selection, Monitoring and Development. *Int. J. Prod. Res.* **2016**, *54*, 1412–1442. [CrossRef]
47. Kabak, Ö.; Ervural, B. Multiple Attribute Group Decision Making: A Generic Conceptual Framework and a Classification Scheme. *Knowl.-Based Syst.* **2017**, *123*, 13–30. [CrossRef]
48. SCC—Supply Chain Council. *Supply Chain Operations Reference (SCOR)*; SCC: Cypress, TX, USA, 2012.
49. Liao, H.; Wen, Z.; Liu, L. Integrating Bwm and Aras Under Hesitant Linguistic Environment For Digital Supply Chain Finance Supplier Section. *Technol. Econ. Dev. Econ.* **2019**, *25*, 1188–1212. [CrossRef]
50. Dolatabad, A.H.; Heidary Dahooie, J.; Antucheviciene, J.; Azari, M.; Razavi Hajiagha, S.H.; Dolatabad, A.H.; Heidary Dahooie, J.; Antucheviciene, J.; Azari, M.; Razavi Hajiagha, S.H. Supplier Selection in the Industry 4.0 Era by Using a Fuzzy Cognitive Map and Hesitant Fuzzy Linguistic VIKOR Methodology. *Environ. Sci. Pollut. Res.* **2023**, *30*, 52923–52942. [CrossRef]
51. Liu, P.; Zhang, X.; Pedrycz, W. A Consensus Model for Hesitant Fuzzy Linguistic Group Decision-Making in the Framework of {Dempster}–{Shafer} Evidence Theory. *Knowl.-Based Syst.* **2021**, *212*, 106559. [CrossRef]
52. Lima Junior, F.R.; Hsiao, M. A Hesitant Fuzzy Topsis Model to Supplier Performance Evaluation. *DYNA* **2021**, *88*, 126–135. [CrossRef]
53. Erol, I.; Murat Ar, I.; Peker, I.; Searcy, C. Alleviating the Impact of the Barriers to Circular Economy Adoption Through Blockchain: An Investigation Using an Integrated MCDM-Based QFD With Hesitant Fuzzy Linguistic Term Sets. *Comput. Ind. Eng.* **2022**, *165*, 107962. [CrossRef]
54. Tüysüz, F.; Şimşek, B. A Hesitant Fuzzy Linguistic Term Sets-Based AHP Approach for Analyzing the Performance Evaluation Factors: An Application to Cargo Sector. *Complex Intell. Syst.* **2017**, *3*, 167–175. [CrossRef]
55. Pérez-Domínguez, L.; Luviano-Cruz, D.; Valles-Rosales, D.; Hernández Hernández, J.; Rodríguez Borbón, M.; Hernández, J.I.H.J.I.H.; Borbón, M.I.R.M.I.R.; Hernández Hernández, J.; Rodríguez Borbón, M. Hesitant Fuzzy Linguistic Term and TOPSIS to Assess Lean Performance. *Appl. Sci.* **2019**, *9*, 873. [CrossRef]
56. Büyüközkan, G.; Güler, M. A Combined Hesitant Fuzzy MCDM Approach for Supply Chain Analytics Tool Evaluation. *Appl. Soft Comput.* **2021**, *112*, 107812. [CrossRef]
57. Zheng, C.; Peng, B.; Zhao, X.; Wei, G.; Wan, A.; Yue, M. A Large Group Hesitant Fuzzy Linguistic DEMATEL Approach for Identifying Critical Success Factors in Public Health Emergencies. *Aslib J. Inf. Manag.* **2022**. ahead of print. [CrossRef]
58. Wu, Y.; Jia, W.; Li, L.; Song, Z.; Xu, C.; Liu, F. Risk Assessment of Electric Vehicle Supply Chain Based on Fuzzy Synthetic Evaluation. *Energy* **2019**, *182*, 397–411. [CrossRef]
59. Chang, K.-H.; Wen, T.-C.; Chung, H.-Y. Soft Failure Mode and Effects Analysis Using the OWG Operator and Hesitant Fuzzy Linguistic Term Sets. *J. Intell. Fuzzy Syst.* **2018**, *34*, 2625–2639. [CrossRef]
60. Wu, H.; Xu, Z. Cognitively Inspired Multi-Attribute Decision-Making Methods Under Uncertainty: A State-of-the-Art Survey. *Cognit. Comput.* **2022**, *14*, 511–530. [CrossRef]
61. Yalçin, N.; Pehlivan, N.Y. Application of the Fuzzy CODAS Method Based on Fuzzy Envelopes for Hesitant Fuzzy Linguistic Term Sets: A Case Study on a Personnel Selection Problem. *Symmetry* **2019**, *11*, 493. [CrossRef]
62. Krishankumar, R.; Ravichandran, K.S.; Shyam, V.; Sneha, S.V.; Kar, S.; Garg, H. Multi-Attribute Group Decision-Making Using Double Hierarchy Hesitant Fuzzy Linguistic Preference Information. *Neural Comput. Appl.* **2020**, *32*, 14031–14045. [CrossRef]
63. Krishankumar, R.; Ravichandran, K.S.; Kar, S.; Gupta, P.; Mehlawat, M.K. Double-Hierarchy Hesitant Fuzzy Linguistic Term Set-Based Decision Framework for Multi-Attribute Group Decision-Making. *Soft Comput.* **2021**, *25*, 2665–2685. [CrossRef]
64. Krishankumar, R.; Pamucar, D.; Pandey, A.; Kar, S.; Ravichandran, K.S. Double Hierarchy Hesitant Fuzzy Linguistic Information Based Framework for Personalized Ranking of Sustainable Suppliers. *Environ. Sci. Pollut. Res.* **2022**, *29*, 65371–65390. [CrossRef] [PubMed]
65. Shen, M.; Liu, P. Risk Assessment of Logistics Enterprises Using FMEA Under Free Double Hierarchy Hesitant Fuzzy Linguistic Environments. *Int. J. Inf. Technol. Decis. Mak.* **2021**, *20*, 1221–1259. [CrossRef]
66. Dai, J.; Pang, J.; Luo, Q.; Huang, Q. Failure Evaluation of Electronic Products Based on Double Hierarchy Hesitant Fuzzy Linguistic Term Set and K-Means Clustering Algorithm. *Symmetry* **2022**, *14*, 2555. [CrossRef]
67. Duan, C.-Y.; Chen, X.-Q.; Shi, H.; Liu, H.-C. A New Model for Failure Mode and Effects Analysis Based on k-Means Clustering Within Hesitant Linguistic Environment. *IEEE Trans. Eng. Manag.* **2022**, *69*, 1837–1847. [CrossRef]
68. Wang, X.; Gou, X.; Xu, Z. Assessment of Traffic Congestion with ORESTE Method under Double Hierarchy Hesitant Fuzzy Linguistic Environment. *Appl. Soft Comput.* **2020**, *86*, 105864. [CrossRef]
69. Li, P.; Liu, J.; Wei, C. Factor Relation Analysis for Sustainable Recycling Partner Evaluation Using Probabilistic Linguistic DEMATEL. *Fuzzy Optim. Decis. Mak.* **2020**, *19*, 471–497. [CrossRef]
70. Zhang, X.; Su, T.; Xin, B. The Dominance Degree-Based Heterogeneous Linguistic Decision-Making Technique for Sustainable 3PRLP Selection. *Complexity* **2020**, *2020*, 6102036. [CrossRef]
71. Zhang, Z.; Liao, H.; Al-Barakati, A.; Zavadskas, E.K.; Antuchevičienė, J. Supplier Selection for Housing Development by an Integrated Method with Interval Rough Boundaries. *Int. J. Strateg. Prop. Manag.* **2020**, *24*, 269–284. [CrossRef]

72. Wang, J.; Wang, J.; Tian, Z.; Zhao, D. A Multihesitant Fuzzy Linguistic Multicriteria Decision-Making Approach for Logistics Outsourcing with Incomplete Weight Information. *Int. Trans. Oper. Res.* **2018**, *25*, 831–856. [CrossRef]
73. Divsalar, M.; Ahmadi, M.; Nemati, Y. A SCOR-Based Model to Evaluate LARG Supply Chain Performance Using a Hybrid MADM Method. *IEEE Trans. Eng. Manag.* **2020**, *69*, 1101–1120. [CrossRef]
74. Qu, G.; Xue, R.; Li, T.; Qu, W.; Xu, Z. A Stochastic Multi-Attribute Method for Measuring Sustainability Performance of a Supplier Based on a Triple Bottom Line Approach in a Dual Hesitant Fuzzy Linguistic Environment. *Int. J. Environ. Res. Public Health* **2020**, *17*, 2138. [CrossRef] [PubMed]
75. Wu, X.-H.; Yang, L.; Qian, J. Selecting Personnel with the Weighted Cross-Entropy TOPSIS of Hesitant Picture Fuzzy Linguistic Sets. *J. Math.* **2021**, *2021*, 1–26. [CrossRef]
76. Karmaker, C.L.; Al Aziz, R.; Palit, T.; Bari, A.B.M.M. Analyzing Supply Chain Risk Factors in the Small and Medium Enterprises under Fuzzy Environment: Implications towards Sustainability for Emerging Economies. *Sustain. Technol. Entrep.* **2023**, *2*, 100032. [CrossRef]
77. Krishankumar, R.; Ravichandran, K.S.; Liao, H.; Kar, S. An Integrated Decision Framework for Group Decision-Making with Double Hierarchy Hesitant Fuzzy Linguistic Information and Unknown Weights. *Int. J. Comput. Intell. Syst.* **2020**, *13*, 624–637. [CrossRef]
78. Krishankumar, R.; Arun, K.; Kumar, A.; Rani, P.; Ravichandran, K.S.; Gandomi, A.H. Double-Hierarchy Hesitant Fuzzy Linguistic Information-Based Framework for Green Supplier Selection with Partial Weight Information. *Neural Comput. Appl.* **2021**, *33*, 14837–14859. [CrossRef]
79. Zolfaghari, S.; Mousavi, S.M. A New Risk Evaluation Methodology Based on FMEA, MULTIMOORA, TPOP, and Interval-Valued Hesitant Fuzzy Linguistic Sets with an Application to Healthcare Industry. *Kybernetes* **2021**, *50*, 2521–2547. [CrossRef]
80. Wu, P.; Zhou, L.; Martínez, L. An Integrated Hesitant Fuzzy Linguistic Model for Multiple Attribute Group Decision-Making for Health Management Center Selection. *Comput. Ind. Eng.* **2022**, *171*, 108404. [CrossRef]
81. Ren, P.; Hao, Z.; Wang, X.; Zeng, X.-J.; Xu, Z. Decision-Making Models Based on Incomplete Hesitant Fuzzy Linguistic Preference Relation with Application to Site Selection of Hydropower Stations. *IEEE Trans. Eng. Manag.* **2022**, *69*, 904–915. [CrossRef]
82. Lima, B.P.; da Silva, A.F.; Marins, F.A.S. New Hybrid AHP-QFD-PROMETHEE Decision-Making Support Method in the Hesitant Fuzzy Environment: An Application in Packaging Design Selection. *J. Intell. Fuzzy Syst.* **2022**, *42*, 2881–2897. [CrossRef]
83. Kumar Singh, R.; Modgil, S. Assessment of Lean Supply Chain Practices in Indian Automotive Industry. *Glob. Bus. Rev.* **2023**, *24*, 68–105. [CrossRef]
84. Dang, T.-T.; Nguyen, N.-A.-T.; Nguyen, V.-T.; Dang, L.-T.-H. A Two-Stage Multi-Criteria Supplier Selection Model for Sustainable Automotive Supply Chain under Uncertainty. *Axioms* **2022**, *11*, 228. [CrossRef]

Disclaimer/Publisher's Note: The statements, opinions and data contained in all publications are solely those of the individual author(s) and contributor(s) and not of MDPI and/or the editor(s). MDPI and/or the editor(s) disclaim responsibility for any injury to people or property resulting from any ideas, methods, instructions or products referred to in the content.

Article

A New Method for Commercial-Scale Water Purification Selection Using Linguistic Neural Networks

Saleem Abdullah [1], Alaa O. Almagrabi [2,*] and Nawab Ali [1]

1. Department of Mathematics, Abdul Wali Khan University Mardan, Mardan 23200, Pakistan; saleemabdullah@awkum.edu.pk (S.A.)
2. Department of Information Systems, Faculty of Computing and Information Technology, King Abdulaziz University, Jeddah 21589, Saudi Arabia
* Correspondence: aalmagrabi3@kau.edu.sa

Citation: Abdullah, S.; Almagrabi, A.O.; Ali, N. A New Method for Commercial-Scale Water Purification Selection Using Linguistic Neural Networks. *Mathematics* **2023**, *11*, 2972. https://doi.org/10.3390/math11132972

Academic Editor: Francisco Rodrigues Lima Junior

Received: 5 April 2023
Revised: 16 May 2023
Accepted: 31 May 2023
Published: 3 July 2023

Copyright: © 2023 by the authors. Licensee MDPI, Basel, Switzerland. This article is an open access article distributed under the terms and conditions of the Creative Commons Attribution (CC BY) license (https://creativecommons.org/licenses/by/4.0/).

Abstract: A neural network is a very useful tool in artificial intelligence (AI) that can also be referred to as an ANN. An artificial neural network (ANN) is a deep learning model that has a broad range of applications in real life. The combination and interrelationship of neurons and nodes with each other facilitate the transmission of information. An ANN has a feed-forward neural network. The neurons are arranged in layers, and each layer performs a particular calculation on the incoming data. Up until the output layer, which generates the network's ultimate output, is reached, each layer's output is transmitted as an input to the subsequent layer. A feed-forward neural network (FFNN) is a method for finding the output of expert information. In this research, we expand upon the concept of fuzzy neural network systems and introduce feed-forward double-hierarchy linguistic neural network systems (FFDHLNNS) using Yager–Dombi aggregation operators. We also discuss the desirable properties of Yager–Dombi aggregation operators. Moreover, we describe double-hierarchy linguistic term sets (DHLTSs) and discuss the score function of DHLTSs and the distance between any two double-hierarchy linguistic term elements (DHLTEs). Here, we discuss different approaches to choosing a novel water purification technique on a commercial scale, as well as some variables influencing these approaches. We apply a feed-forward double-hierarchy linguistic neural network (FFDHLNN) to select the best method for water purification. Moreover, we use the extended version of the Technique for Order Preference by Similarity to Ideal Solution (extended TOPSIS) method and the grey relational analysis (GRA) method for the verification of our suggested approach. Remarkably, both approaches yield almost the same results as those obtained using our proposed method. The proposed models were compared with other existing models of decision support systems, and the comparison demonstrated that the proposed models are feasible and valid decision support systems. The proposed technique is more reliable and accurate for the selection of large-scale water purification methods.

Keywords: double-hierarchy linguistic term set; Dombi t-norms; artificial neural network; decision-making

MSC: 94D05; 90C70

1. Introduction

1.1. A Brief Review of the Development of Neural Networks and Their Types

Classical statistical methods have been widely used in various industries for decades, particularly in fields such as quality control, experimental design, and process optimization. However, in recent years, neural networks (NNs) [1] have emerged as powerful tools for solving complex problems in various fields, including finance, engineering [2], medicine [3], and computer science [4]. NNs have gained popularity due to their ability to handle large and complex data sets, learn patterns and relationships in the data, and make accurate predictions or classifications. In comparison to classical statistical methods [5], NNs have the advantage of being able to model nonlinear relationships and capture complex interactions between variables. In the field of pattern recognition, NNs have been shown to outperform classical statistical methods, particularly when dealing with complex and high-dimensional data. In prediction and classification tasks, NNs have also been successful, achieving high accuracy rates in fields such as image [6,7] and speech recognition [8], natural language processing [9,10], and sentiment analysis [11,12]. Overall, while classical statistical methods remain useful in many applications, NNs have become a popular and powerful tool for solving complex problems in various fields. Regression [13] and NNs are often seen as competing model-building methods, as they can both be used for modeling and predicting outcomes based on input variables. However, while regression is a linear method that requires the assumption of linearity between the input and output variables, NNs are capable of modeling nonlinear relationships and do not require such assumptions. The structure and operation of the brain served as the first inspiration for neural networks, which are created to resemble the behavior of organic neurons. As a result, they share some performance characteristics with human neural biology. Layers of linked nodes or neurons that are arranged into input, hidden, and output layers are used in NNs to replicate the behavior of the human brain. The input layer receives data and sends them to the hidden layers for processing, while the output layer produces the final result. In order to increase the model's accuracy, the hidden layers are in charge of observing patterns and correlations in the data and modifying the weights of the connections between neurons. One of the key features of neural networks is their ability to learn from data and store that knowledge in the form of learned parameters. This allows them to make predictions [14,15] and classifications [16,17] based on previously seen examples and to generalize that knowledge to new, unseen examples. Neural networks are also capable of identifying patterns [18] in data, even in the presence of noise [19] or other sources of variability. This makes them useful for a variety of applications, including natural language processing, image and speech recognition, and predictive modeling [20,21]. In addition, neural networks are capable of taking past experiences into consideration and using that information to make inferences and judgments about new situations. This is known as "contextual learning," and it allows neural networks to adapt to changing conditions and make more accurate predictions over time. Overall, NNs are a powerful tool for processing and analyzing complex data, and their performance characteristics make them well-suited to a wide range of tasks in fields such as machine learning [22], artificial intelligence [23,24], and cognitive science [25]. There are different kinds of NNs, each with its own unique architecture and characteristics. Among the most common types, we can find feed-forward neural networks [26], recurrent neural networks (RNNs) [27], convolutional neural networks (CNNs) [28], auto-encoder neural networks [29], generative adversarial networks [30], etc. In this paper, we discuss feed-forward neural networks in detail.

1.2. A Brief Review of Feed-Forward Neural Networks and Their Uses

The connections between the nodes in feed-forward neural networks (FFNNs) do not cycle, and data only move from the input layer to the output layer in one direction. This makes FFNNs a relatively simple type of neural network compared to others such as RNNs and CNNs. Numerous applications have made use of FFNNs, such as natural language processing [10,11], image [7,8] and speech recognition [9], and financial forecasting [31]. They have been among the most successful learning algorithms and have been the basis for many other types of neural networks. However, FFNNs are a method for determining experts' output, similar to techniques such as the Technique for Order Preference by Similarity to Ideal Solution (TOPSIS) method [32] and the grey relational analysis (GRA) method [33], among others. FFNNs are machine learning models that learn to make predictions or classifications based on input data. They do not rely on explicit expert knowledge or rules but rather on patterns and relationships in the data themselves. TOPSIS and GRA, on the other hand, are decision-making methods used in multi-criteria decision analysis (MCDA) [34]. These methods involve comparing alternatives based on multiple criteria and weighing their importance to reach a decision. They do not involve machine learning or neural networks.

1.3. A Brief Review of Activation Function and Its Importance

Both the output layer and hidden layers of a FFNN can use activation functions [35]. Activation functions are essential in neural networks because they introduce nonlinearity, which allows the network to simulate complicated interactions between input and output variables. The rectified linear unit (ReLU) [36] and the family of sigmoid functions (including the logistic sigmoid function [37], hyperbolic tangent [38], and arctangent function [38]) are popular activation functions used in FFNNs. ReLU is widely used in the hidden layers because of its simplicity and effectiveness in training deep networks. It outputs zero for negative input and increases linearly for positive input. On the other hand, the logistic sigmoid function is commonly used in the output layer to transform the output of the network into a probability between 0 and 1. The logistic sigmoid function has a smooth and differentiable curve, which makes it easy to compute the gradient during back-propagation. The logistic sigmoid function's output is constantly between 0 and 1, making it appropriate for binary classification tasks. However, it is susceptible to the vanishing gradient problem, particularly in deep neural networks [39], which can result in delayed training and poor performance. Overall, choosing the appropriate activation function for a neural network depends on the specific problem at hand and the structure of the network.

The main objectives of this research are:

1. There are numerous aggregation operators, such as Einstein, Yager, and Dombi. In this research, we combine the Yager [40–42] and Dombi [43] aggregation operators to make a new aggregation operator called the Yager–Dombi aggregation operator and explain its desirable properties;
2. We expand the concept of a feed-forward neural network to incorporate a feed-forward double-hierarchy linguistic neural network using Yager–Dombi operators;
3. We develop a fuzzy neural network for the selection of water purification methods using a double-hierarchy linguistic neural network and use it for the selection of water purification methods;
4. We extend the Yager–Dombi operations to aggregate a double hierarchy for fuzzy information.

1.4. Motivation behind the Study

According to the study analysis above, there has been no extensive usage of double-hierarchy linguistic term sets (DHLTSs) in the field of FFNNs or of the Yager–Dombi aggregation operator.

The primary objectives of this study are as follows:

1. To extend the concept of fuzzy neural networks to incorporate double-hierarchy linguistic neural networks;
2. To combine existing aggregation operators to create a new aggregation operator;
3. To develop a fuzzy neural network for the selection of water purification methods;
4. To extend the Yager–Dombi operations to aggregate double-hierarchy fuzzy information.

1.5. Contribution of the Study

In this paper, we combine two t-norms (Yager and Dombi t-norms) and apply them to linguistic neural networks, develop a new model of linguistic neural networks, and solve the selection problem of the water purification procedure. The contribution of this paper can be summarized as follows:

1. We develop new t-norms and their operations by using the Yager and Dombi t-norms and discuss their relationships;
2. The developed t-norms are further expanded to aggregation operators to develop a new set of double-hierarchy linguistic terms;

3. The proposed aggregation is necessary for artificial neural networks. Therefore, we integrate the proposed aggregation operators into the hidden layers of a linguistic neural network;
4. We develop a new approach to linguistic neural networks and linguistic decision models using linguistic neural networks;
5. The proposed linguistic decision model, based on a linguistic neural network, is applied to water-purification procedure-selection problems;
6. The proposed models are verified and compared with other models in Sections 7 and 8 for validation.

The remaining sections of the paper are structured as follows: In Section 2, we discuss some fundamental definitions. In Section 3, we define the Yager–Dombi t-norm and t-conorm and discuss the DHLTYDWA, DHLTYDOWA, and DHLTYDHWA operators and their desirable properties. We also discuss the score function and the distance between any two DHLTSs. In Section 4, we discuss activation functions and feed-forward neural networks. In Section 5, we discuss the output of the feed-forward neural network using the Yager–Dombi operator. Section 6 illustrates the selection of the best method for water purification. In Section 7, we use the extended TOPSIS [44] and GRA [45] methods for verification. In Section 8, we compare the extended TOPSIS and GRA methods with our proposed method and rank the output using different Yager–Dombi aggregation operators. In Section 9, the authors conclude their study and elaborate on its value and development direction.

2. Fundamental Concept

This section discusses fuzzy sets, intuitionistic fuzzy sets, a hierarchy linguistic term set, and a double-hierarchy linguistic term set.

If Y is a non-empty set, then an object having the form (L.ZADEH [46])

$$Z = \{\langle y, U_Z(y)\rangle | y \in Y\} \qquad (1)$$

is called a fuzzy set, and $U_Z(y) \in [0,1]$ denotes the degree of membership of Y in Z. An object having the form

$$Z = \{\langle y, U_Z(y), V_Z(y)\rangle | y \in Y\} \qquad (2)$$

is called an intuitionistic fuzzy set, with $U_Z(y) \in [0,1]$ denoting the degree of membership and $V_Z(y) \in [0,1]$ representing the degree of non-membership of Y in Z. Additionally, the following condition is also satisfied:

$$U_Z(y) + V_Z(y) \in [0,1],$$

Fuzzy sets are a mathematical framework for representing and manipulating uncertainty and imprecision in information. In contrast to traditional sets, which define membership in a binary, all-or-nothing way, fuzzy sets allow for partial membership, meaning that an element can belong to a set to a certain degree. The usefulness of fuzzy sets lies in their ability to model and reason about real-world problems that involve uncertainty, ambiguity, and vagueness. For example, in many domains such as decision-making, control systems, and artificial intelligence, it is often difficult or impossible to make precise distinctions between different categories or values. Fuzzy sets provide a flexible and intuitive way to represent and reason about such situations. They allow for a more nuanced and realistic representation of uncertainty and imprecision, which can lead to better decision-making and more robust system design. Several terms that define the variable in a natural language manner, such as "low", "medium", and "high", compose a fuzzy linguistic term set. The degree to which a variable belongs to a given term is described by a fuzzy set that is assigned to each term. Let S be a non-empty set. Then, the linguistic term set is defined as:

$$S = \{s_\alpha | \alpha \in [-\beta, \beta]\} \qquad (3)$$

Suppose $S = \{s_\alpha | \alpha \in [-\beta, \beta]\}$ is a FHLT (first-hierarchy linguistic term set) and $O = \{o_{\mathcal{P}} | \mathcal{P} \in [-k, k]\}$ is a SHLTS (second-hierarchy linguistic term set), then the double-hierarchy linguistic term set (DHLTS) is defined as:

$$S_0 = \{S_\alpha \langle O_{\mathcal{P}} \rangle | \alpha \in [-\beta, \beta], \mathcal{P} \in [-k, k]\} \quad (4)$$

Gou et al. suggested two transformed functions between the DHLTs subscript and its numerical scale in order to deal with DHLTSs [47] more effectively. $S_0 = S_\alpha \langle O_{\mathcal{P}} \rangle$ is an unbroken DHLTS. The transformed functions f and f^{-1} between the numerical value and the subscript $[\alpha, \mathcal{P}]$ of the DHLT $S_\alpha \langle O_{\mathcal{P}} \rangle$ are given below:

$$f : [-\beta, \beta] \times [-k, k] \to [0, 1],$$

$$= \frac{\mathcal{P} + (\alpha + \beta)k}{2\beta k} = \Upsilon \quad (5)$$

$$f^{-1} : [0, 1] \to [-\beta, \beta] \times [-k, k],$$

$$f^{-1}(\Upsilon) = S_{[2\beta\Upsilon - \Upsilon]} \langle O_{k\{(2\beta\Upsilon - \Upsilon) - [2\beta\Upsilon - \Upsilon]\}} \rangle \quad (6)$$

where $[2\beta\Upsilon - \Upsilon]$ represents the integer part of the value $2\beta\Upsilon - \Upsilon$.

3. Yager–Dombi Operators for DHLTSs

This section gives the basic concepts of Yager–Dombi t-norm, Yager–Dombi t-conorm, Yager–Dombi operators, and Yager–Dombi aggregation operators for DHLTSs.

If G and H are any two numbers, then the Yager–Dombi t-norm and Yager–Dombi t-conorm are given as follows:

$$T(G, H) = \frac{1}{2 + \left\{ \left\{ \left\{ \frac{1 - \{1 - \min(1,(1-G)^t)\}}{\min(1,(1-G)^t)} \right\}^k + \left\{ \frac{1 - \{1 - \min(1,(1-h)^t)\}}{\min(1,(1-h)^t)} \right\}^k \right\}^{\frac{1}{t}} \right\}^{\frac{1}{k}}} \quad (7)$$

$$T''(G, H) = 1 - \frac{1}{1 + \left\{ \left\{ \frac{\min(1,(G^t + H^t))}{1 - \min(1,(G^t + H^t))} \right\}^{\frac{k}{t}} + \left\{ \frac{\min(1,(G^t + H^t))}{1 - \min(1,(G^t + H^t))} \right\}^{\frac{k}{t}} \right\}^{\frac{1}{k}}} \quad (8)$$

where $k, t \geq 1$ and $G, H \in [0, 1]$.

To define operation rules based on Yager–Dombi t-norms for this, let $S_1 = F\left(S_{\alpha 1} \langle O_{\mathcal{P}1} \rangle\right)$ and $S_2 = F(S_{\alpha 2} \langle O_{\mathcal{P}2} \rangle)$ be any two DHLTSs. Let $k, t \geq 1$, and $\beta > 0$ be any real numbers. Then, we establish the Yager–Dombi operators for DHLTSs, which are given below:

I. $$S_1 \oplus S_2 = 1 - \frac{1}{1 + \left\{ \left\{ \frac{\min\left(1, F\left(S_{\alpha 1} \langle O_{\mathcal{P}1} \rangle\right)^t\right)}{1 - \min\left(1, F\left(S_{\alpha 1} \langle O_{\mathcal{P}1} \rangle\right)^t\right)} \right\}^{\frac{k}{t}} + \left\{ \frac{\min\left(1, F(S_{\alpha 2} \langle O_{\mathcal{P}2} \rangle)^k\right)}{1 - \min\left(1, F(S_{\alpha 2} \langle O_{\mathcal{P}2} \rangle)^k\right)} \right\}^{\frac{k}{t}} \right\}^{\frac{1}{k}}}$$

II. $$S_1 \otimes S_2 = \frac{1}{2 + \left\{ \left\{ \frac{1 - \{1 - \min(1,(1 - F(S_{\alpha 1} \langle O_{\mathcal{P}1} \rangle))^t)\}}{\min(1,(1 - F(S_{\alpha 1} \langle O_{\mathcal{P}1} \rangle))^t)} \right\}^k + \left\{ \frac{1 - \{1 - \min(1,(1 - F(S_{\alpha 2} \langle O_{\mathcal{P}2} \rangle))^t)\}}{\min(1,(1 - F(S_{\alpha 2} \langle O_{\mathcal{P}2} \rangle))^t)} \right\}^k \right\}^{\frac{1}{t}} \right\}^{\frac{1}{k}}}$$

III. $\mathcal{B} \odot S_1 = 1 - \dfrac{1}{1+\left\{(\mathcal{B})\left\{\dfrac{\min\left(1, F\left(S_{\alpha 1}\langle O_{\mathcal{P}1}\rangle\right)^t\right)}{1-\min\left(1, F\left(S_{\alpha 1}\langle O_{\mathcal{P}1}\rangle\right)^t\right)}\right\}^k\right\}^{\frac{1}{k}}}$

IV. $(S_1)^{\mathcal{B}} = \dfrac{1}{2+\left\{(\mathcal{B})\left\{\left\{\dfrac{1-\{1-\min(1,(1-G)^t)\}}{\min(1,(1-G)^t)}\right\}^k\right\}^{\frac{1}{t}}\right\}^{\frac{1}{k}}}$

If $S_1 = F\left(S_{\alpha 1}\langle O_{\mathcal{P}1}\rangle\right)$ and $S_2 = F(S_{\alpha 2}\langle O_{\mathcal{P}2}\rangle)$ are any two DLTEs, then the distance between $S_1 = F\left(S_{\alpha 1}\langle O_{\mathcal{P}1}\rangle\right)$ and $S_2 = F(S_{\alpha 2}\langle O_{\mathcal{P}2}\rangle)$ is defined as:

$$d(S_1, S_2) = \left| F\left(S_{\alpha 1}\langle O_{\mathcal{P}1}\rangle\right) - F\left(S_{\alpha 2}\langle O_{\mathcal{P}2}\rangle\right) \right| \tag{9}$$

where $d(S_1, S_2) \in [0, 1]$.

Suppose $S_i = \{F(S_{\alpha i}\langle O_{\mathcal{P}i}\rangle) | i = 1, 2, 3 \ldots n\}$ is a group of DHLTSs, then the double-hierarchy linguistic term Yager–Dombi weighted averaging (DHLTYDWA) operator is a mapping $\mathbb{Q}^n \to \mathbb{Q}$ such that:

$$DHLTYDWA(S_1, S_2, S_3, S_4, \ldots S_i) = \bigoplus_{i=1}^{n} W_i S_i$$

$W_i = (W_1, W_2, W_3, W_4 \ldots W_i)^l$ is the weight vector of $S_i = \{F(S_{\alpha i}\langle O_{\mathcal{P}i}\rangle) | i = 1, 2, 3 \ldots n\}$ that fulfills the condition that $\sum_{i=1}^{n} W_i = 1$.

In the following theorem, we prove that the aggregated values obtained from different DHLTSs by using the proposed aggregation operators are again DHLTSs. This means that the aggregation operators are valid. Additionally, we provide validation for the proposed aggregation operators using Yager–Dombi t-norms.

Theorem 1. *Suppose $S_i = \{F(S_{\alpha i}\langle O_{\mathcal{P}i}\rangle) | i = 1, 2, 3 \ldots n\}$ is a group of DHLTSs, then the aggregated value of DHLTSs using the DHLTYDWA operator is also a DHLTS, that is:*

$$DHLTYDWA(S_1, S_2, S_3, S_4, \ldots S_i) = \bigoplus_{i=1}^{n} W_i S_i = 1 - \dfrac{1}{1+\left\{\left(\sum_{i=1}^{n} W_i\right)\left\{\dfrac{\min\left(1, F\left(S_{\alpha i}\langle O_{\mathcal{P}i}\rangle\right)^t\right)}{1-\min\left(1, F\left(S_{\alpha i}\langle O_{\mathcal{P}i}\rangle\right)^t\right)}\right\}^k\right\}^{\frac{1}{k}}} \tag{10}$$

$W_i = (W_1, W_2, W_3, W_4 \ldots W_i)^l$ is the weight vector of $S_i = \{F(S_{\alpha i}\langle O_{\mathcal{P}i}\rangle) | i = 1, 2, 3 \ldots n\}$ that fulfills the condition that $\sum_{i=1}^{n} W_i = 1$.

Proof. To prove this theorem, we employ the mathematical induction approach.

For $n = 1$, we have:

$$DHLTYDWA(S_1) = W_1 S_1 = 1 - \dfrac{1}{1+\left\{(W_1)\left\{\dfrac{\min\left(1, F\left(S_{\alpha 1}\langle O_{\mathcal{P}1}\rangle\right)^t\right)}{1-\min\left(1, F\left(S_{\alpha 1}\langle O_{\mathcal{P}1}\rangle\right)^t\right)}\right\}^k\right\}^{\frac{1}{k}}}$$

As a result, the statement holds true for $n = 1$.
Assume that the above result holds true for $n = h$.

$$DHLTYDWA(S_1, S_2, S_3, \ldots S_h) = \bigoplus_{i=1}^{h} W_i S_i = 1 - \cfrac{1}{1 + \left\{ \left(\sum_{i=1}^{h} W_i\right) \left\{ \cfrac{\min\left(1, F\left(S_{\alpha i}\langle O_{\mathcal{P} i}\rangle\right)^t\right)}{1 - \min\left(1, F\left(S_{\alpha i}\langle O_{\mathcal{P} i}\rangle\right)^t\right)} \right\}^{\frac{k}{t}} \right\}^{\frac{1}{k}}}$$

Next, we are going to show that the result is true for $n = h + 1$:

$$DHLTYDWA(S_1, S_2, S_3, S_4, \ldots S_{h+1}) = \bigoplus_{i=1}^{h+1} W_i S_i$$

$$= \left(1 - \cfrac{1}{1 + \left\{ \left(\sum_{i=1}^{h} W_i\right) \left\{ \cfrac{\min\left(1, F\left(S_{\alpha i}\langle O_{\mathcal{P} i}\rangle\right)^t\right)}{1 - \min\left(1, F\left(S_{\alpha i}\langle O_{\mathcal{P} i}\rangle\right)^t\right)} \right\}^{\frac{k}{t}} \right\}^{\frac{1}{k}}} \right) \oplus \left(1 - \cfrac{1}{1 + \left\{ (W_{h+1}) \left\{ \cfrac{\min\left(1, F\left(S_{\alpha i}\langle O_{\mathcal{P} i}\rangle\right)^t\right)}{1 - \min\left(1, F\left(S_{\alpha i}\langle O_{\mathcal{P} i}\rangle\right)^t\right)} \right\}^{\frac{k}{t}} \right\}^{\frac{1}{k}}} \right)$$

$$= 1 - \cfrac{1}{1 + \left\{ \left(\sum_{i=1}^{h+1} W_i\right) \left\{ \cfrac{\min\left(1, F\left(S_{\alpha i}\langle O_{\mathcal{P} i}\rangle\right)^t\right)}{1 - \min\left(1, F\left(S_{\alpha i}\langle O_{\mathcal{P} i}\rangle\right)^t\right)} \right\}^{\frac{k}{t}} \right\}^{\frac{1}{k}}}$$

The result is valid for $n = h + 1$. Moreover, the DHLTYDWA operator can easily hold its idempotence, boundedness, and monotonicity properties.

Suppose $S_i = \{F(S_{\alpha i}\langle O_{\mathcal{P} i}\rangle) | i = 1, 2, 3 \ldots n\}$ is a group of DHLTSs, then the double-hierarchy linguistic term Yager–Dombi order weighted averaging (DHLTYDOWA) operator is a mapping $\mathbb{Q}^n \to \mathbb{Q}$ such that:

$$DHLTYDOWA(S_1, S_2, S_3 \ldots S_i) = \bigoplus_{i=1}^{n} W_i S_{R(i)}$$

$W_i = (W_1, W_2, W_3, W_4 \ldots W_i)^t$ is the weight vector of $S_i = \{F(S_{\alpha i}\langle O_{\mathcal{P} i}\rangle) | i = 1, 2, 3 \ldots n\}$ that fulfills the condition that $\sum_{i=1}^{n} W_i = 1$; $(R(1), R(2), R(3) \ldots R(n))$ are the permutations of $(i = 1, 2, 3 \ldots n)$ for which $S_{R(i-1)} \geq S_{R(i)}$ for all $i = 1, 2, 3 \ldots n$.

In the following theorem, we prove that the aggregated values obtained from different DHLTSs by using the proposed ordered weighted aggregation operators are again DHLTSs. This means that the aggregation operators are valid. Additionally, we provide validation for the proposed ordered weighted aggregation operators using Yager–Dombi t-norms. □

Theorem 2. *Suppose $S_i = \{F(S_{\alpha i}\langle O_{\mathcal{P} i}\rangle) | i = 1, 2, 3 \ldots n\}$ is a group of DHLTSs, then the aggregated value of DHLTSs using the DHLTYDOWA operator is also a DHLTS, that is:*

$$DHLTYDOWA(S_1, S_2, \ldots S_i) = \bigoplus_{i=1}^{n} W_i S_{R(i)} = 1 - \cfrac{1}{1 + \left\{ \left(\sum_{i=1}^{n} W_i\right) \left\{ \cfrac{\min\left(1, F\left(S_{\alpha i}\langle O_{\mathcal{P} i}\rangle\right)^t\right)}{1 - \min\left(1, F\left(S_{\alpha i}\langle O_{\mathcal{P} i}\rangle\right)^t\right)} \right\}^{\frac{k}{t}} \right\}^{\frac{1}{k}}} \quad (11)$$

Proof. The proof of this result is similar to the proof of Theorem 1. Moreover, the DHLTYDWA operator can easily hold its idempotence, boundedness, and monotonicity properties. Suppose $S_i^h = \{F(S_{\alpha i}\langle O_{\mathcal{P} i}\rangle) | i = 1, 2, 3 \ldots n\}$ is a collection of DHLTSs, then the double-hierarchy linguistic term Yager–Dombi hybrid weighted averaging (DHLTYDHWA) operator is a mapping $\mathbb{Q}^n \to \mathbb{Q}$ such that:

$$DHLTYDHWA\left(S_1{}^h, S_2{}^h, S_3{}^h, S_3{}^h, \ldots S_i{}^h\right) = \bigoplus_{i=1}^n W_i S_i{}^h$$

$W_i = (W_1, W_2, W_3, W_4 \ldots W_i)^l$ is the weight vector of $S_i{}^o = \{F(S_{\alpha i}\langle O_{\mathcal{P} i}\rangle)|i = 1, 2, 3 \ldots n\}$ that fulfills the condition that $\sum_{i=1}^n W_i = 1$. □

In the following theorem, we prove that the aggregated values obtained from different DHLTSs by using the proposed hybrid weighted aggregation operators are again DHLTSs. This means that the aggregation operators are valid. Additionally, we provide validation for the proposed ordered weighted aggregation operators using Yager–Dombi t-norms.

Theorem 3. *Suppose $S_i{}^h = \{F(S_{\alpha i}\langle O_{\mathcal{P} i}\rangle)|i = 1, 2, 3 \ldots n\}$ is a group of DHLTSs, then the aggregated value of DHLTSs using the DHLTYDHWA operator is also a DHLTS, that is:*

$$DHLTYDHWA\left(S_1{}^h, S_2{}^h, \ldots S_i{}^h\right) = \bigoplus_{i=1}^n W_i S_i{}^h = 1 - \frac{1}{1 + \left\{(\sum_{i=1}^n W_i)\left\{\frac{\min\left(1, F\left(S_{\alpha i}\langle O_{\mathcal{P} i}\rangle\right)^t\right)}{1 - \min\left(1, F\left(S_{\alpha i}\langle O_{\mathcal{P} i}\rangle\right)^t\right)}\right\}^{\frac{k}{t}}\right\}^{\frac{1}{k}}} \quad (12)$$

Proof. The proof of this result is similar to the proof of Theorem 1. Moreover, the DHLTY-DWA operator can easily hold its idempotent, bounded, and monotonicity properties. □

4. Activation Functions and Neural Network Systems

An activation function [48] is a function that gives the output of a node. Additionally, it is known as the "transfer function." It is used to determine if the output of a neural network is a yes or no response. Depending on the function, it can transfer the output values to a range between 0 and 1 or between −1 and 1, among others. An activation function can be either linear or nonlinear. The terms monotonic function and derivative are essential for understanding nonlinear functions. The range or curves of the nonlinear activation functions are the major factors used to categorize them. The sigmoid or logistic activation function that we use is defined as follows:

$$f(x) = Y = \frac{e^x}{e^x + 1} \quad (13)$$

where $x = 1 \pm 2 \pm 3 \pm 4 \ldots \psi$.

The sigmoid activation function, which successfully accomplishes its duty and has a range of 0 to 1, is commonly employed in decision-making because it is effective at performing what it is supposed to perform. Since it has the lowest range and generates the most precise predictions, we therefore employ this activation function anytime we need to determine an outcome. The function could take many different shapes. As a consequence, we are able to determine the slope of the sigmoid curve between any two positions. Although the derivative of the function is not monotonic, the function itself is. The logistic sigmoid function has the effect of making it possible for a neural network to encounter an impasse during training. Neural networks were first developed in the 1950s as a means of addressing this issue. Programming everyday computers to function similarly to a network of brain cells is the process used to create ANNs. Artificial neural networks employ a complex mathematical algorithm to make sense of the data they are fed. A typical artificial neural network is made up of hundreds to millions of units, commonly referred to as artificial neurons, arranged in an order of layers. The input layer receives a variety of outside data types. The network's goal is to process or comprehend these data. After leaving the input layer, the data go through one or more hidden units. The hidden unit is in charge of transforming the incoming data into a format that the output unit can use. Most neural networks have complete interconnections between layers. These connections, similarly to the human brain, are weighted; the higher the number, the greater the impact one unit has on another. Every unit in the network gains knowledge as the data pass through it. The output units, which are placed on the other side of the network, provide the network with the data that have been received and processed. ANNs come in different types, but in this paper, we only discuss FFNNs in detail.

A feed-forward neural network (FFNN) is one in which there is no cycling of the connections between the nodes. The opposite of a FFNN is a feed-backward neural network (FBNN), which cycles through certain routes. The feed-forward model is the most fundamental type of neural network since incoming data are only ever processed in one way. In a feed-backward neural network, some of the input data come back to the input layer from hidden layers. However, regardless of the number of hidden nodes the data may pass through, they never flow backward and always move ahead. The input layer, hidden layer, and output layer are only a few of the layers that make up this collection of fundamental processing units. Each unit in the layer below it is linked to every other unit in the layer above it. Because they are not all made equally, each of these connections may have a different weight or strength. A feed-forward neural network is a function such that:

$$\beta = f\left(\sum_{j=1}^{n} W_j a_{ij}\right) \tag{14}$$

where the activation function is denoted by f; a_{ij} is the input signal; W_j is the criteria weight; and β is the single output. The output of feed-forward neural network is shown in Figure 1.

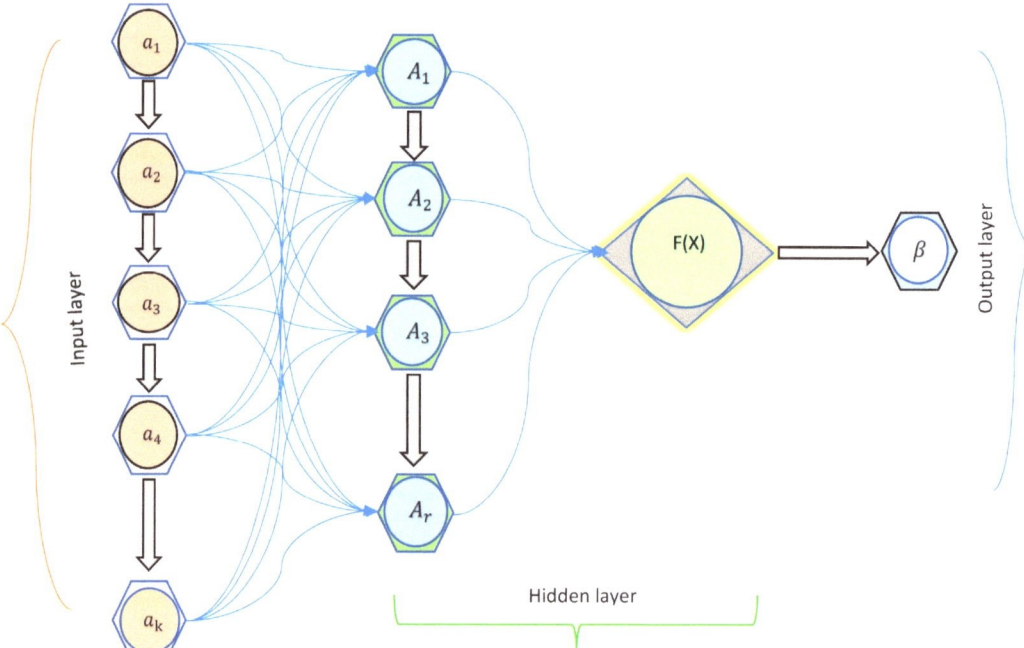

Figure 1. Output of a feed-forward neural network.

Figure 2 shows that the Artificial neural networks and fuzzy logic are both used in neuro-fuzzy systems, a sort of hybrid intelligent system that is effective at processing complicated data. The following figure depicts a fuzzy neuron, a processing element of a hybrid neural net:

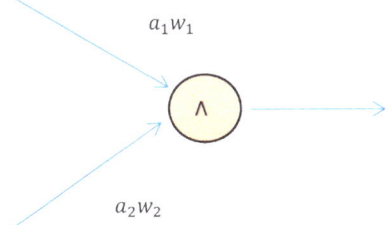

Figure 2. Simple neuro-fuzzy system.

where a_1, a_2 are the input signals and w_1, w_2 are the corresponding weights of the input signals, respectively. $\wedge$ is the AND fuzzy neuron.

5. The Output of Neural Networks Using Yager–Dombi Operators

In this section, we put forward a new extension to the fuzzy neural system to develop a fuzzy neural system that is based on DHLTSs. For this, let $P_j = \{p_1, p_2, p_3, p_4 \ldots p_k\}$ be the arrangement of attributes that will be evaluated, and let $Q_i = \{q_1, q_2, q_3, q_4 \ldots q_z\}$ be the discrete set of z alternatives that must be selected. Suppose $W_j = (W_1, W_2, W_3, W_4 \ldots W_k)^l$ is the attribute weight vector, where $W_j\{j = 1, 2, 3, 4 \ldots k\}$ are all real numbers such that $W_j > 0$ and $\sum_{j=1}^{z} W_j = 1$. Let there be n number of experts, $DM_r\{r = 1, 2, 3, 4 \ldots n\}$, to express their views on the z alternatives in reference to the k criterion in the context of DHLTSs. The data given by the nth expert in the form matrix appear as follows:

$$\vartheta_r = \begin{bmatrix} a_{11} & a_{12} & \cdots & a_{1k} \\ a_{21} & a_{22} & \cdots & a_{2k} \\ \vdots & \vdots & \ddots & \vdots \\ a_{z1} & a_{z2} & \cdots & a_{zk} \end{bmatrix}$$

To determine the output of the feed-forward fuzzy neural system, we go through different phases. In phase 1, we determine the criteria weight vector of each matrix provided by the experts corresponding to each input information or signal. In phase 2, we find the hidden layer using Equations (18) and (19). In phase 3, the activation function is used to calculate the output of the input data.

To determine the feed-forward fuzzy neural system's output in a DHLTS environment, we suggest the following algorithm:

Phase 1: The following are the steps in Phase 1:

Step 1: Recognize the expert data or information presented in the form of a matrix $\vartheta_r = \{A_{ij}\}_{z \times k}$, where $j = (1, 2, 3, 4 \ldots k)$ and $i = (1, 2, 3, 4 \ldots z)$ represent the jth criterion value in relation to the ith alternative, respectively;

Step 2: Determine the criteria weight vector of each matrix provided by the experts corresponding to each input information or signal using the entropy measure method, which consists of the following steps:

1. To evaluate the entropy of the information provided by the experts in the form of matrices, we use the following equation:

$$n_j = \frac{1}{i}\sum_{i=1}^{z}\left(\frac{1}{j}\sum_{j=1}^{k}\left(\left\{\sqrt{2}\cos\pi\left(\frac{F(S_{\alpha i}\langle O_{\mathcal{P}_i}\rangle) - (F(S_{\alpha i}\langle O_{\mathcal{P}_i}\rangle))^c}{4}\right) - 1\right\} \times \frac{1}{\sqrt{2}-1}\right)\right) \quad (15)$$

where $j = (1, 2, 3, \ldots, k)$ and $i = (1, 2, 3, \ldots, z)$ denote the jth alternative and ith criterium, respectively, and $F(S_{\alpha i}\langle O_{\mathcal{P}i}\rangle)$ denotes the DHLTS, $(F(S_{\alpha i}\langle O_{\mathcal{P}i}\rangle))^c$ is the complement of the given DHLTEs, and is determined as:

$$(F(S_{\alpha i}\langle O_{\mathcal{P}i}\rangle))^c = F(S_{-\alpha i}\langle O_{-\mathcal{P}i}\rangle) \tag{16}$$

2. We obtain the final criteria weight vector as follows:

$$W_j = \frac{n_j}{\sum_{j=1}^{k} n_j} \tag{17}$$

Phase 2: The steps in Phase 2 are listed below:

Step 1: Find the scalar product of the weight vector and input signal using the Yager–Dombi aggregation operator as follows:

$$\mathcal{M} = \sum_{r=1}^{n} (W_j a_j), \tag{18}$$

where $W_j = (W_1, W_2, W_3, \ldots W_k)$ and $r = (1, 2, 3, 4 \ldots n)$ denote the weight vector and the experts, respectively. $a_j = (a_1, a_2, a_3, \ldots, a_j)$ denotes the input signals;

Step 2: Take the minimum weight and multiply it with $\mathcal{M}$ using the Yager–Dombi averaging operator.

$$W_j = \min\{\vartheta_1(w_j), \vartheta_2(w_j), \vartheta_3(w_j), \ldots, \vartheta_n(w_j)\} \tag{19}$$

Phase 3: The steps taken in Phase 3 are as follows:

Step 1: Apply the logistic activation function to the outcome of Step 2 in Phase 2 to obtain the final output β. The logistic activation function is described as follows:

$$f(x) = Y = \frac{e^x}{e^x + 1}$$

Step 2: Rank the possible outcomes for each β in descending order.

Output of Feed-Forward Double-Hierarchy Linguistic Term Neural Networks

An artificial neural network of this sort, known as a feed-forward double-hierarchy linguistic term neural network (FFDHLTNN), only allows information to travel in one direction: from the input layer using one or more hidden layers to the output layer. It is called feed-forward because data move forward through the network without looping back on themselves. In a typical FFDHLTNN architecture, each layer consists of a set of neurons or nodes that perform a simple mathematical operation on the inputs and output the result to the next layer. The network's ultimate output is produced by the output layer once the input layer has received the input data. As demonstrated in Figure 3, the hidden layers execute intricate modifications on the input data to provide characteristics that are beneficial for the output layer.

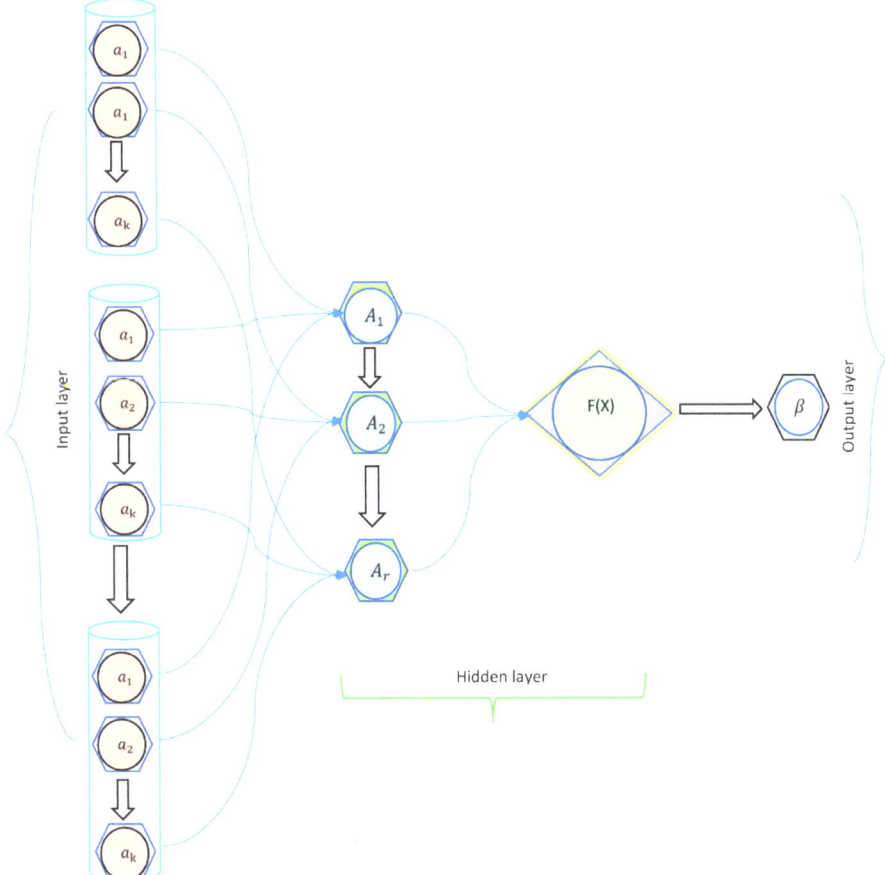

Figure 3. Output of a feed-forward double-hierarchy linguistic neural network.

Step 1: The decision-making matrix given by the decision-makers in the LTS environment must be identified, and all evaluations must be based on the LTS that the DM has provided:

$$S_o = \left(S_\alpha \langle O_P \rangle\right) \begin{bmatrix} S_{-4} = extremely\ low, S_{-3} = very\ low, S_{-2} = low, S_{-1} = slightly\ low, \\ S_0 = medium \\ S_1 = slightly\ high, \quad S_2 = high, \quad S_1 = very\ high, \quad S_3 = extremely\ high \\ O_{-4} = far\ from, \quad O_{-4} = scarcely, O_{-4} = only\ a\ little, O_{-4} = a\ little \\ O_{-4} = just\ right \\ O_1 = much, O_2 = very\ much, O_3 = exremely\ much, O_4 = entierly \end{bmatrix}$$

Step 2: Find the criteria weight for each matrix provided by the experts in the environment of the DHLTEs using the entropy measure method;

Step 3: Find the scalar product of input information or signals and their corresponding weights, and add the scalar product of the input signals using Equations (18) and (19);

Step 4: Apply the logistic activation function to the outcome of Step 2 in Phase 2 to obtain the final output β;

Step 5: Rank the possible outcomes for each β in descending order.

6. Numerical Example

Water is a clear, flavorless, and odorless liquid that is necessary for all life on Earth. It is a chemical substance with the chemical formula H_2O, consisting of two hydrogen atoms and one oxygen atom. The three states of matter that water may exist in are solid (ice), liquid (water), and gas (water vapors). It is the only substance on Earth that can naturally exist in all three states at normal temperatures and pressures. Water plays a crucial role in many aspects of our lives, including hydration, agriculture, industry, transportation, and recreation. However, due to the growth of industries, access to clean and safe drinking water has become a serious concern for many people throughout the world. Water pollution causes various diseases, some of which lead to death. As a result, supplying clean water to domestic areas is the most crucial responsibility of governments. In this section, we will look at different commercial-scale water purification systems and the factors that influence them.

The following techniques are used to purify water on a commercial scale:

(1) q_1: **Chlorination**: Chlorination is a process that uses chlorine to disinfect water. Chlorine is added to water, which kills the harmful bacteria and viruses present in it. This method is effective in killing most of the disease-causing pathogens;

(2) q_2: **Reverse Osmosis**: In the reverse osmosis (RO) process, a semi-permeable membrane is utilized to filter out dissolved particles, contaminants, and minerals from water. It is a highly effective method of water purification and is commonly used in households and industries;

(3) q_3: **Ultraviolet Purification**: UV purification uses ultraviolet light to kill bacteria, viruses, and other microorganisms present in water. It is an effective method of water purification and does not use any chemicals;

(4) q_4: **Filtration**: Filtration is a process that removes impurities from water by passing it through a porous material. Sediments, dirt, and other bigger particles can be effectively removed from water with this technique;

(5) q_5: **Coagulation and sedimentation:** Chemicals such as alum are added to water to cause impurities to clump together and settle at the bottom of a tank, which can then be removed through sedimentation;

(6) q_6: **Boiling:** By bringing water to a boil for at least one minute, the majority of disease-causing organisms can be killed;

(7) q_7: **Distillation:** Water is heated during distillation, and the steam is subsequently condensed back into water. Minerals, chemicals, and bacteria are just a few of the impurities that can be removed by using this method.

The factors that affect the methods of water purification are:

1. p_1: **Economic factors:** Economic factors can have a significant impact on water purification, as the process of treating and purifying water can be expensive and require significant investments in infrastructure, technology, and human resources. One major economic factor that can affect water purification is the availability and cost of resources such as energy, chemicals, and materials needed for the purification process;

2. p_4: **Socio-political factors:** The socio-political environment can have a significant impact on water purification. Governments have an obligation to make sure that their populations have access to safe drinking water since access to clean water is a fundamental human right. However, the provision of clean water can be influenced by a variety of socio-political factors, such as social factors, public health, and political instability, among others;

3. p_3: **Environmental factors:** There are many environmental factors that can affect water purification, including temperature, chemicals, turbidity, and climate change. Overall, environmental factors can have a significant impact on water purification and must be taken into account when designing and implementing water purification systems;

4. p_5: **Type of Contaminants:** The type and concentration of contaminants present in the water will also affect the purification process. Different treatment processes are better

suited for removing different types of contaminants, such as chemicals, microbes, or sediments;

5. p_5: **Water Quality Standards:** The level of purity required for the final product will affect the purification process. Different industries and applications have different standards for water quality, which will influence the choice of treatment process and the extent of purification required.

For this, let $(q_1, q_2, q_3, q_4, q_5, q_6, q_7)$ represent the set of alternatives for commercial-scale water purification. Let $(p_1, p_2, p_3, p_4, p_5)$ be the five variables influencing these procedures. The network's ultimate output is produced by the output layer once the input layer has received the input data. Between the input and output layers, the hidden layers carry out complicated modifications to the data to provide characteristics that are beneficial to the output layer. The linguistic information is considered input signals, interacting with weight vectors in the input layer and using the Yager–Dombi t-norms to produce the product of input signals as a linguistic variable. Then, the aggregated information of input signals is calculated using Yager–Dombi t-conorms. The hidden layer information, after performing the Yager–Dombi t-norms and t-conorms, is given in Table 1. Apply the activation function to the hidden layer signals and determine the output layer signals of the linguistic feed-forward neural network. The output layer information is given in Table 2.

Table 1. Hidden layer of feed-forward neural network for DHLTSs.

	p_1	p_2	p_3	p_4	p_5
$\mathscr{g}_1$	0.738521429	0.326423333	0.212368533	0.136978263	0.1258662
$\mathscr{g}_2$	0.969678667	0.368916545	0.333768273	0.154811222	0.095050909
$\mathscr{g}_3$	0.570108667	0.487515667	0.185739733	0.104634095	0.087066391
$\mathcal{M}$	2.278308762	1.182855545	0.731876539	0.396423581	0.3079835
	0.694964668	0.541884482	0.422591636	0.283884909	0.235464362

Table 2. Output layer of feed-forward neural network for DHLTSs.

	q_1	q_2	q_3	q_4	q_5	q_6	q_7
DHLTYDWA	0.61904	0.66270	0.64047	0.63911	0.66773	0.63961	0.62284
DHLTYDOWA	0.62006	0.66972	0.64923	0.63942	0.67406	0.64974	0.62513
DHLTYDHWA	0.51688	0.52162	0.51856	0.51912	0.52149	0.51857	0.51747

According to experts, the best solution is reverse osmosis, whereas filtration is effective for basic water purification tasks, including chlorine and sediment removal. Reverse osmosis removes contaminants across a wider range. All pathogens in the water are eliminated by alternative methods. However, the dead bacteria still float in the water. On the other hand, an RO water purifier eliminates bacteria by filtering out their floating, dead corpses after they have been killed. As a result, RO-purified water is cleaner.

7. Verification of Our Proposed Method

In this section, we consider the extended TOPSIS approach and the GRA method to verify the effectiveness and validity of our proposed approach. The extended TOPSIS and GRA methods are used for verification in decision-making because they provide a more comprehensive and objective analysis of multiple criteria and are effective in handling uncertainty, imprecision, and complex decision-making scenarios. For this verification, we consider the information given by the experts in the form of three matrices in the environment of DHLTSs. The matrices provided by the experts consist of five attributes related to seven alternatives. We use the Yager–Dombi aggregation operator to aggregate the information given by the experts in the form of matrices in the environment of DHLTSs. Then, we apply the entropy approach to determine the criteria weight vector of the aggregated matrices. Finally, we apply the extended TOPSIS method to obtain the required output, which is given in Table 3.

Table 3. Output of extended TOPSIS method for DHLTSs.

	q_1	q_2	q_3	q_4	q_5	q_6	q_7
d_{ij}^+	0.21584	0.03857	0.14964	0.14933	0.04855	0.14578	0.22714
d_{ij}^-	0.07570	0.27066	0.1419	0.14222	0.24299	0.14576	0.0644
output	0.25967	0.87526	0.48672	0.48781	0.83347	0.49997	0.22089
raking	0.87526	0.83347	0.49997	0.48781	0.48672	0.25967	0.22089

Similarly, we use the Yager–Dombi aggregation operator to aggregate the experts' information and then apply the entropy measure method to determine the criteria weight vector. Finally, we apply the GRA method to obtain the required output, as shown in Table 4.

Table 4. Output of GRA method for DHLTSs.

	q_1	q_2	q_3	q_4	q_5	q_6	q_7
d_{ij}^+	0.58842	0.87869	0.642025	0.70632	0.86793	0.66403	0.60736
d_{ij}^-	0.79424	0.52724	0.73448	0.70624	0.57306	0.677098	0.85416
output	0.42557	0.62499	0.46642	0.50003	0.60231	0.49513	0.41556
raking	0.62499	0.60231	0.50003	0.49513	0.46642	0.42557	0.41556

From Tables 2–4, we can see that the results obtained by using both methods (the extended TOPSIS method and the GRA method) are almost the same as the results obtained by using our proposed method. This ensures that our proposed method is valid. Reverse osmosis is therefore the ideal option for experts to choose, while filtering is appropriate for simple water jobs such as chlorine and sediment removal. Reverse osmosis removes contaminants across a wider range. All pathogens in the water are eliminated by alternative methods. However, the dead bacteria still float in the water. An RO water purifier, on the other hand, eliminates germs by killing them and filtering away their floating, lifeless bodies. Therefore, RO-purified water is cleaner.

8. Discussion and Comparison

In this section, we compare our suggested approach with the GRA method. For this, let $P_j = \{p_1, p_2, p_3, p_4 \ldots p_k\}$ be the arrangement of attributes that will be evaluated, and let $Q_i = \{q_1, q_2, q_3, q_4 \ldots q_z\}$ be the discrete set of z alternatives that must be selected. Suppose $W_j = (W_1, W_2, W_3, W_4 \ldots W_k)^l$ is the attribute weight vector, where $W_j\{j = 1, 2, 3, 4 \ldots k\}$ are all real numbers satisfying $W_j > 0$ and $\sum_{j=1}^{z} W_j = 1$. Let there be an n number of experts, $DM_r\{r = 1, 2, 3, 4 \ldots n\}$, expressing their views on the z alternatives with respect to the k criterion in the context of DHLTSs. Here, the experts gave data in the form of three matrices in the environment of DHLTSs. The matrices provided by the experts consist of five attributes related to seven alternatives. First, we find the expert weight vector using the entropy measure method. We use the Yager–Dombi aggregation operator to aggregate the information provided by the experts in the form of matrices in the environment of DHLTSs and then apply the entropy measure method to calculate the criteria weight vector of the aggregated matrix, which is $W_j = (0.214625973, 0.201184071, 0.167857403, 0.204584654, 0.211747899)^l$. Next, we determine the aggregated matrix's PIS and NIS. Finally, we apply the GRA method to determine the output and rank the possible outcomes of each output, as shown in Table 4. For this comparison, we will also use the extended TOPSIS approach. Both the extended TOPSIS and TOPSIS methods are MCDM techniques that seek to isolate the best alternative from a group of alternatives based on several criteria. The number of criteria that are taken into account is the fundamental distinction between TOPSIS and extended TOPSIS. Extended TOPSIS can accommodate many sets of criteria, whereas TOPSIS only takes into account one set of criteria. Similarly, in the extended TOPSIS method, we find the expert weight vector using the entropy measure method. We use the Yager–Dombi aggregation operator to aggregate the information provided by the experts in the form of matrices in the environment of DHLTSs and then apply the entropy measure method to determine the criteria weight vector of the aggregated matrix. Next, we find the positive and negative ideal solutions of the aggregated matrix. Then, the expert information is converted to one set of criteria, and we finally apply the TOPSIS method to determine the output and rank the possible outcomes of each output, as shown in Table 3. The graphical representation of the comparison between the proposed method and other similar methods can be observed in Figure 4. The comparison of detail information is given in Table 5.

Figure 4. Steps of GRA method, extended TOPSIS method, and our proposed method.

Table 5. Output ranking of feed-forward neural network for DHLTSs.

DHLTYDWA	$q_5 > q_2 > q_3 > q_6 > q_4 > q_7 > q_1$
DHLTYDOWA	$q_5 > q_2 > q_6 > q_3 > q_4 > q_7 > q_1$
DHLTYDHWA	$q_2 > q_5 > q_4 > q_6 > q_3 > q_7 > q_1$

In our suggested approach, we take the experts' information in the environment of DHLTSs and find the expert weight using the entropy measure method. After that, we use the Yager–Dombi aggregation operator to find the hidden layer from the experts' matrices, apply the activation function to generate the output of our proposed method, and finally rank the output. The comparison between the GRA method, the extended TOPSIS method, and our proposed method is shown in Figure 5. Both methods yield the same results as our proposed method. Filtration is useful for simple water tasks such as chlorine and sediment removal, but reverse osmosis is the best choice according to experts. A wider range of contaminants may be removed via reverse osmosis. Other techniques eliminate all germs in the water. However, the dead microorganisms still float in the water. An RO water purifier, on the other hand, filters out the dead bacteria that are floating in the water and eliminates them. It follows that RO-purified water is more sanitary. As a result, we conclude that our suggested technique is valid and superior to both methods (the extended TOPSIS method and the GRA method), since it produces the output in fewer steps, saving experts' time.

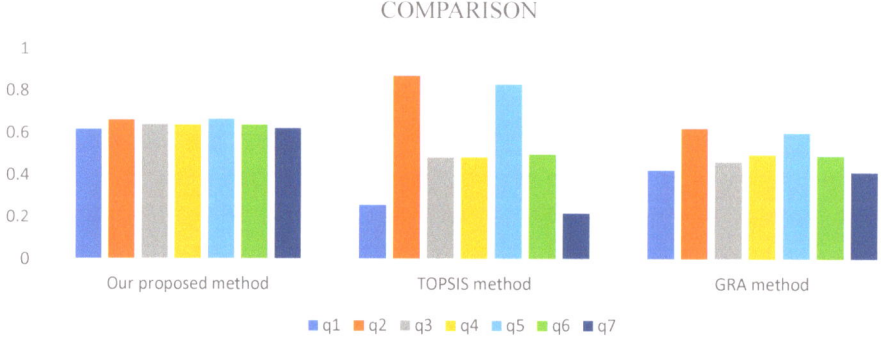

Figure 5. Comparison of Extended TOPSIS Method, GRA Method, and Our Proposed Method.

9. Conclusions

The neural network is a very important topic in machine learning, and artificial neural networks have garnered significant attention in the decision-making process. Linguistic information is a useful tool for describing uncertainty in information science and decision-making. Therefore, this paper applies the linguistic term set to artificial neural networks and develops a decision-making model based on linguistic neural networks. First, using the concepts of Yager t-norms and Dombi t-norms, we define a new hybrid t-norm known as the Yager–Dombi t-norm. We further develop some operational rules for double-hierarchy linguistic term sets and generalize them to incorporate more than two double-hierarchy linguistic term sets. The Yager–Dombi aggregation operator has been developed for double-hierarchy linguistic terms. We discuss the desirable properties of the DHLTYDWA operator, the DHLTYDOWA operator, and the DHLTYDHWA operator. Furthermore, we extend the concept of fuzzy neural network systems to feed-forward double-hierarchy linguistic neural network systems. Experts provide information in the environment of DHLTSs, and to determine the expert weight, we use the entropy measure method. We then obtain the hidden layer data using the DHLTYDWA operator and apply FFNNs to the hidden layer to derive the output of the information provided by the experts. In addition, a real-life MADM problem has been formulated. Filtration is beneficial for simple water tasks such as chlorine and sediment removal, but reverse osmosis is the best choice, according to experts, because a wider range of contaminants can be removed with this method. Other techniques eliminate all germs in the water. However, the dead microorganisms still float in it. An RO water purifier, on the other hand, filters out the dead bacteria that are floating in the water and eliminates them. It follows that RO-purified water is more sanitary. Moreover, we use the extended TOPSIS approach and the GRA approach for the verification of our proposed method, and both methods yield almost the same results as our proposed method.

A comparison analysis has been carried out, as seen in Figure 4, to demonstrate the validity and viability of our suggested method in comparison to other existing methods. We affirm that our proposed method is significantly better than other existing methods because it produces expert information output in a shorter time.

In future work, we will apply our proposed technique to intuitionistic fuzzy sets for decision-making problems and also use it in TWDs, Pythagorean fuzzy sets.

Author Contributions: Methodology, S.A. and N.A.; Formal analysis, A.O.A.; Data curation, A.O.A.; Writing—original draft, S.A. and N.A. All authors have read and agreed to the published version of the manuscript.

Funding: This research work was funded by Institutional Fund Projects under grant no. IFPIP: 414-611-1443. The authors gratefully acknowledge the technical and financial support provided by the Ministry of Education and King Abdulaziz University, DSR, Jeddah, Saudi Arabia.

Data Availability Statement: Not applicable.

Conflicts of Interest: The authors declare no conflict of interest.

References

1. Razi, M.A.; Athappilly, K. A comparative predictive analysis of neural networks (NNs), nonlinear regression and classification and regression tree (CART) models. *Expert Syst. Appl.* **2005**, *29*, 65–74. [CrossRef]
2. Coats, P.K.; Fant, L.F. Recognizing financial distress patterns using a neural network tool. *Financ. Manag.* **1993**, *1993*, 142–155. [CrossRef]
3. Ceylan, H.; Bayrak, M.B.; Gopalakrishnan, K. Neural Networks Applications in Pavement Engineering: A Recent Survey. *Int. J. Pavement Res. Technol.* **2014**, *7*, 434–444.
4. Sarvamangala, D.R.; Kulkarni, R.V. Convolutional neural networks in medical image understanding: A survey. *Evol. Intell.* **2022**, *15*, 1–22. [CrossRef]
5. Fang, C.; Dong, H.; Zhang, T. Mathematical models of overparameterized neural networks. *Proc. IEEE* **2021**, *109*, 683–703. [CrossRef]
6. Mitchell, J.M.O. Classical statistical methods. *Mach. Learn. Neural Stat. Classif.* **1994**, *1994*, 17–28.
7. Koh, J.; Lee, J.; Yoon, S. Single-image deblurring with neural networks: A comparative survey. *Comput. Vis. Image Underst.* **2021**, *203*, 103134. [CrossRef]
8. Alshehri, S.A. Neural network technique for image compression. *IET Image Process.* **2016**, *10*, 222–226. [CrossRef]
9. Yen, Y.; Fanty, M.; Cole, R. Speech recognition using neural networks with forward-backward probability generated targets. In Proceedings of the 1997 IEEE International Conference on Acoustics, Speech, and Signal Processing, Munich, Germany, 21–24 April 1997; Volume 4, pp. 3241–3244.
10. Collobert, R.; Weston, J. A unified architecture for natural language processing: Deep neural networks with multitask learning. In Proceedings of the 25th International Conference on Machine Learning, Helsinki, Finland, 5–9 July 2008; pp. 160–167.
11. Ma, Q. Natural language processing with neural networks. In Proceedings of the Language Engineering Conference, Hyderabad, India, 13–15 December 2002; pp. 45–56.
12. Rani, S.; Kumar, P. Deep learning based sentiment analysis using convolution neural network. *Arab. J. Sci. Eng.* **2019**, *44*, 3305–3314. [CrossRef]
13. Chen, P.; Sun, Z.; Bing, L.; Yang, W. Recurrent attention network on memory for aspect sentiment analysis. In Proceedings of the 2017 Conference on Empirical Methods in Natural Language Processing, Copenhagen, Denmark, 7–11 September 2017; pp. 452–461.
14. Draper, N.R.; Smith, H. *Applied Regression Analysis*; John Wiley & Sons: Hoboken, NJ, USA, 1998; Volume 326.
15. Yoo, P.D.; Kim, M.H.; Jan, T. Machine learning techniques and use of event information for stock market prediction: A survey and evaluation. In Proceedings of the International Conference on Computational Intelligence for Modelling, Control and Automation and International Conference on Intelligent Agents, Web Technologies and Internet Commerce (CIMCA-IAWTIC'06), Sydney, Australia, 29 November–1 December 2006; Volume 2, pp. 835–841.
16. Adya, M.; Collopy, F. How effective are neural networks at forecasting and prediction? A review and evaluation. *J. Forecast.* **1998**, *17*, 481–495. [CrossRef]
17. Zaghloul, W.; Lee, S.M.; Trimi, S. Text classification: Neural networks vs support vector machines. *Ind. Manag. Data Syst.* **2009**, *109*, 708–717. [CrossRef]
18. Naseer, M.; Minhas, M.F.; Khalid, F.; Hanif, M.A.; Hasan, O.; Shafique, M. Fannet: Formal analysis of noise tolerance, training bias and input sensitivity in neural networks. In Proceedings of the 2020 Design, Automation & Test in Europe Conference & Exhibition (DATE), Grenoble, France, 9–13 March 2020; pp. 666–669.
19. Shin, K.S.; Lee, T.S.; Kim, H.J. An application of support vector machines in bankruptcy prediction model. *Expert Syst. Appl.* **2005**, *28*, 127–135. [CrossRef]

20. Sterling, A.J.; Zavitsanou, S.; Ford, J.; Duarte, F. Selectivity in organocatalysis—From qualitative to quantitative predictive models. *Wiley Interdiscip. Rev. Comput. Mol. Sci.* **2021**, *11*, e1518. [CrossRef]
21. Liu, Y.; Liu, S.; Wang, Y.; Lombardi, F.; Han, J. A survey of stochastic computing neural networks for machine learning applications. *IEEE Trans. Neural Netw. Learn. Syst.* **2020**, *32*, 2809–2824. [CrossRef]
22. Schwendicke, F.A.; Samek, W.; Krois, J. Artificial intelligence in dentistry: Chances and challenges. *J. Dent. Res.* **2020**, *99*, 769–774. [CrossRef]
23. Ossowska, A.; Kusiak, A.; Świetlik, D. Artificial intelligence in dentistry—Narrative review. *Int. J. Environ. Res. Public Health* **2022**, *19*, 3449. [CrossRef]
24. Drakopoulos, G.; Giannoukou, I.; Mylonas, P.; Sioutas, S. The converging triangle of cultural content, cognitive science, and behavioral economics. In *Artificial Intelligence Applications and Innovations, Proceedings of the AIAI 2020 IFIP WG 12.5 International Workshops: MHDW 2020 and 5G-PINE 2020, Neos Marmaras, Greece, 5–7 June 2020, Proceedings 16*; Springer International Publishing: Berlin/Heidelberg, Germany, 2020; pp. 200–212.
25. Bebis, G.; Georgiopoulos, M. Feed-forward neural networks. *IEEE Potentials* **1994**, *13*, 27–31. [CrossRef]
26. Medsker, L.R.; Jain, L.C. Recurrent neural networks. *Des. Appl.* **2001**, *5*, 64–67.
27. Gu, J.; Wang, Z.; Kuen, J.; Ma, L.; Shahroudy, A.; Shuai, B.; Liu, T.; Wang, X.; Wang, G.; Cai, J.; et al. Recent advances in convolutional neural networks. *Pattern Recognit.* **2018**, *77*, 354–377. [CrossRef]
28. Lange, S.; Riedmiller, M. Deep auto-encoder neural networks in reinforcement learning. In Proceedings of the 2010 International Joint Conference on Neural Networks (IJCNN), Barcelona, Spain, 18–23 July 2010; pp. 1–8.
29. Creswell, A.; White, T.; Dumoulin, V.; Arulkumaran, K.; Sengupta, B.; Bharath, A.A. Generative adversarial networks: An overview. *IEEE Signal Process. Mag.* **2018**, *35*, 53–65. [CrossRef]
30. Chao, J.; Shen, F.; Zhao, J. Forecasting exchange rate with deep belief networks. In Proceedings of the 2011 International Joint Conference on Neural Networks, San Jose, CA, USA, 31 July–5 August 2011; pp. 1259–1266.
31. Zavadskas, E.K.; Mardani, A.; Turskis, Z.; Jusoh, A.; Nor, K.M. Development of TOPSIS method to solve complicated decision-making problems—An overview on developments from 2000 to 2015. *Int. J. Inf. Technol. Decis. Mak.* **2016**, *15*, 645–682. [CrossRef]
32. Wei, G.W. GRA method for multiple attribute decision making with incomplete weight information in intuitionistic fuzzy setting. *Knowl.-Based Syst.* **2010**, *23*, 243–247. [CrossRef]
33. Guitouni, A.; Martel, J.M. Tentative guidelines to help choosing an appropriate MCDA method. *Eur. J. Oper. Res.* **1998**, *109*, 501–521. [CrossRef]
34. Elliott, D.L. *A Better Activation Function for Artificial Neural Networks*; Institute for Systems Research, Harvard University: College Park, MA, USA, 1993.
35. Schmidt-Hieber, J. Nonparametric Regression Using Deep Neural Networks with ReLU Activation Function. 2020. Available online: https://arxiv.org/abs/1708.06633 (accessed on 30 May 2023).
36. Yin, X.; Goudriaan, J.A.N.; Lantinga, E.A.; Vos, J.A.N.; Spiertz, H.J. A flexible sigmoid function of determinate growth. *Ann. Bot.* **2003**, *91*, 361–371. [CrossRef] [PubMed]
37. Zamanlooy, B.; Mirhassani, M. Efficient VLSI implementation of neural networks with hyperbolic tangent activation function. *IEEE Trans. Very Large Scale Integr. (VLSI) Syst.* **2013**, *22*, 39–48. [CrossRef]
38. Kamruzzaman, J. Arctangent activation function to accelerate backpropagation learning. *IEICE Trans. Fundam. Electron. Commun. Comput. Sci.* **2002**, *85*, 2373–2376.
39. Montavon, G.; Samek, W.; Müller, K.R. Methods for interpreting and understanding deep neural networks. *Digit. Signal Process.* **2018**, *73*, 1–15. [CrossRef]
40. Sideris, A.; Orita, K. Structured learning in feedforward neural networks with application to robot trajectory control. In Proceedings of the 1991 IEEE International Joint Conference on Neural Networks, Singapore, 18–21 November 1991; pp. 1067–1072.
41. Sharma, S.; Mehra, R. Implications of pooling strategies in convolutional neural networks: A deep insight. *Found. Comput. Decis. Sci.* **2019**, *44*, 303–330. [CrossRef]
42. Garg, H.; Shahzadi, G.; Akram, M. Decision-making analysis based on Fermatean fuzzy Yager aggregation operators with application in COVID-19 testing facility. *Math. Probl. Eng.* **2020**, *2020*, 7279027. [CrossRef]
43. Akram, M.; Khan, A.; Borumand Saeid, A. Complex Pythagorean Dombi fuzzy operators using aggregation operators and their decision-making. *Expert Syst.* **2021**, *38*, e12626. [CrossRef]
44. Ye, F. An extended TOPSIS method with interval-valued intuitionistic fuzzy numbers for virtual enterprise partner selection. *Expert Syst. Appl.* **2010**, *37*, 7050–7055. [CrossRef]
45. Zadeh, L.A. Fuzzy sets as a basis for a theory of possibility. *Fuzzy Sets Syst.* **1978**, *1*, 3–28. [CrossRef]
46. Atanassov, K. Intuitionistic fuzzy sets. *Int. J. Bioautomation* **2016**, *20*, 1.

47. Li, X.; Xu, Z.; Wang, H. Three-way decisions based on some Hamacher aggregation operators under double hierarchy linguistic environment. *Int. J. Intell. Syst.* **2021**, *36*, 7731–7753. [CrossRef]
48. Cheng, F.; Liang, H.; Niu, B.; Zhao, N.; Zhao, X. Adaptive neural self-triggered bipartite secure control for nonlinear MASs subject to DoS attacks. *Inf. Sci.* **2023**, *631*, 256–270. [CrossRef]

Disclaimer/Publisher's Note: The statements, opinions and data contained in all publications are solely those of the individual author(s) and contributor(s) and not of MDPI and/or the editor(s). MDPI and/or the editor(s) disclaim responsibility for any injury to people or property resulting from any ideas, methods, instructions or products referred to in the content.

Article

A Fuzzy Multi-Criteria Evaluation System for Share Price Prediction: A Tesla Case Study

Simona Hašková *, Petr Šuleř and Róbert Kuchár

Institute of Technology and Business in České Budějovice, Okružní 517/10, 370 01 Ceské Budejovice, Czech Republic; 24786@mail.vstecb.cz (P.Š.); kuchar@mail.vstecb.cz (R.K.)
* Correspondence: haskova@mail.vstecb.cz

Abstract: The article presents the predictive capabilities of a fuzzy multi-criteria evaluation system that operates on the basis of a non-fuzzy neural approach, but also one that is capable of implementing a learning paradigm and working with vague concepts. Within this context, the necessary elements of fuzzy logic are identified and the algebraic formulation of the fuzzy system is presented. It is with the help of the aforementioned that the task of predicting the short-term trend and price of the Tesla share is solved. The functioning of a fuzzy system and fuzzy neural network in the field of time series value prediction is discussed. The authors are inclined to the opinion that, despite the fact that a fuzzy neural network reacts in terms of applicability and effectiveness when solving prediction problems in relation to input data with a faster output than a fuzzy system, and is more "user friendly", a sufficiently knowledgeable and experienced solver/expert could, by using a fuzzy system, achieve a higher speed of convergence in the learning process than a fuzzy neural network using the minimum range of input data carrying the necessary information. A fuzzy system could therefore be a possible alternative to a fuzzy neural network from the point of view of prediction.

Keywords: fuzzy system; neural network; prediction; Tesla stock

MSC: 26E50

Citation: Hašková, S.; Šuleř, P.; Kuchár, R. A Fuzzy Multi-Criteria Evaluation System for Share Price Prediction: A Tesla Case Study. *Mathematics* **2023**, *11*, 3033. https://doi.org/10.3390/math11133033

Academic Editor: Francisco Rodrigues Lima Junior

Received: 26 May 2023
Revised: 30 June 2023
Accepted: 5 July 2023
Published: 7 July 2023

Copyright: © 2023 by the authors. Licensee MDPI, Basel, Switzerland. This article is an open access article distributed under the terms and conditions of the Creative Commons Attribution (CC BY) license (https:// creativecommons.org/licenses/by/ 4.0/).

1. Introduction

Predicting stock market prices is one of the basic objectives of scientific and business research. Historically, the key area of study has involved the modelling of time series [1]. Traditional methods are based on parametric models and include the autoregressive method, exponential smoothing, and structural models of time series (e.g., [2,3]). For example, Debnath and Srivastava [4] conducted a paired t-test for the significance of the difference between the average returns of the sectors in 2019 and 2021 to test the appropriateness of the stock selection methodology proposed.

Modern approaches to temporal data analysis include machine learning methods ([5], which provide procedures for learning temporal dynamics purely based on data series [6] or prediction algorithms based on fuzzy logic, as demonstrated in Castillo and Melin [7]).

Fuzzy logic makes it possible, to a certain extent, to solve the problem of insufficient knowledge. The reason being that it provides a simple way to reach a certain conclusion based on vague, ambiguous, imprecise, or missing input data [8]. Models combining fuzzy logic and dynamic modelling based on regression analysis [9] and time series analysis [10] have been used to this effect.

There is a merger of methods implementing the paradigm of learning with methods based on fuzzy-logic-formed predictive fuzzy neural networks [11]. In terms of cognitive science, these are artificially created systems capable of working with vague concepts and effectively utilising the implicit knowledge gained from the data presented in the learning process, either under the supervision of a teacher or without [12]. The basic structure of these systems is a tailor-made problem of a neural network type in terms of an allegorical model of the structure and the functioning of the human brain on the micro level [13]. The learning process of the fuzzy neural network is controlled by a fuzzy learning algorithm, which gradually selects each element of the training set and calculates the necessary changes in synaptic weights, so that the network generates an adequate right side for its left side, i.e., an output that differs as little as possible from the desired output [14]. The vagueness of the term "as little as possible" is one of the reasons why fuzzy logic, with fuzzy inference rules of the fuzzy implications type, which therefore become part of the instructions of the fuzzy learning algorithm, enters the game.

With a properly chosen fuzzy learning algorithm, the learning process, after some time, converges to such a synaptic weight setting, whereby the sum of the quadrates of output deviations from the desired values reaches its minimum and the running of the fuzzy learning algorithm stops. This means that all the available information, contained in the training set, is "incarnated" into the structure of the network, which responds to each new input not contained in the training set, with an output in terms of the learnt range [15].

Our aim is to introduce a prediction tool that is able to implement the learning paradigm and work with vague concepts with a minimum of input data carrying the necessary information; however, it will operate on a different basis than the aforementioned fuzzy neural approach. This prediction tool is in the general concept presented as a fuzzy system for multi-criteria evaluation (hereinafter "fuzzy system").

The conceptual framework of the fuzzy system is based on Kahneman's allegorical model of the functioning of the human mind in terms of the identification of the processes of the functioning of systems S_1 and S_2 of Kahneman's model, with the phases of the functioning of the fuzzy system [16]. S_1 and S_2 perform two cooperating subsystems represented by the conscious and subconscious components of thinking. The characteristic of S_1, among others, is intuitiveness. It is a system that operates unconsciously, producing spontaneous reactions to various stimuli. It is responsible for quick situation assessment and automatic responses, such as facial recognition or responses to danger, and so on. The characteristics of S_2, among others, are analytical thinking, intentionality, and awareness. S_2 is activated when tackling more complex tasks that require logical reasoning, analysis, and planning. It is responsible for critical thinking, decision making based on rational deliberation, and control over the impulses triggered by S_1. In contrast to vaguely defined intuitive concepts, into which the human mind categorizes more or less related details (S_1), there are vaguely defined supports of fuzzy sets, into which the fuzzy system "blurs" the more or less related values of the input stimuli. Just as the rational component of the human mind oversees the appropriateness of the reactions to stimuli (S_2), the processes of fuzzy logical inference, functioning within fuzzy logic, maintain the proper proportionality between the left and right sides of the inference rules.

A fuzzy system is a tool with a uniform structure, independent of the type of problem, in which the inference rules are not fuzzy implications, whereby simple rules of conventional situational control are applied.

In the text that follows, we summarize the current state of knowledge in relation to the issue of modern prediction methods. In the methodological part, we identify the necessary elements of fuzzy logic and present the algebraic formulation of the fuzzy system, with the help of which, the task of predicting the short-term trend of Tesla, Inc. (hereinafter "Tesla") shares is solved. As part of the discussion, the functioning of the fuzzy system and fuzzy neural networks in terms of time series value prediction is compared.

The key contribution of the article is the presentation of the fuzzy system and its verification as a possible alternative to a fuzzy neural network from the point of view of prediction ability based on the minimum required input data. The determination of the real capabilities of the fuzzy system, in terms of short-term trend prediction, is performed on the Tesla share price estimate.

2. Related Work

In the past, several methods were developed that utilised fuzzy logic in combination with artificial intelligence and machine learning. What follows maps some of these methods.

In the field of economics, fuzzy logic has been used as a prediction method for some time. In practice, fuzzy logic is used independently [17,18], but more often in combination with other procedures [19]. Fuzzy logic and fuzzy hybrid models are not just used for the prediction of stock price developments, but in many other fields too. For example, Maciel and Ballini [20] used a fuzzy model to analyse the problem of time within the framework of an interval time series of low and high asset prices. They compared their model to traditional econometric methods for time series and interval models based on statistical criteria. Their results revealed that the fuzzy model outperformed the standard sample of the competing method in the range of interval values. The high efficiency of this method was also pointed out by Kutlu et al. [21], who, through the Fuzzy Analytic Hierarchy Process (FAHP), tackled the issue of the optimal allocation of a renewable energy source from an economic point of view. Ben Jabeur et al. [22] investigated the effect of the 2008 financial crisis on the corporate performance of 805 French companies in the period of 2007–2009, using the method of a qualitative comparative analysis of fuzzy sets (fsQCA).

Similar models have been successfully used by traders for the fuzzy prediction of the exchange rates within the emissions allowance market [23]. Likewise, Hašková [24] dealt with predicting the profitability of investments in supported renewable energy sources—biogas stations; the given fuzzy model provided information about the level of investment security in terms of its resistance to possible loss. This was due, among other things, to the fact that the fuzzy approach expands the range of the sensitivity analysis performed by the evaluator when valuing projects and companies and enables decision makers to have more relevant information available for processes or predictions [25]. The connection between fuzzy predictive methods and sustainability was documented by Lo et al. [26]. They applied a regression analysis of least squares (OLS) and a qualitative comparative fuzzy set analysis (FsQCA) to obtain optimal sustainability models for a commercial firm. Research on fuzzy economic predictions in the form of an evaluation experiment has been used relatively rarely. An exception is represented by Shao et al. [27]. They successfully used the optimized intuitionistic fuzzy case-based reasoning (IFCBR) method to predict demand in crisis periods. Fuzzy logic can also be used as part of a pair trading strategy. It can partially solve the problem of the difficult prediction of this market-neutral arbitrage strategy and increase its returns, which has been proven by comparing it with traditional technical analysis methods for trading within the spread [28].

Wu et al. [29] used multivariate fuzzy logic relationships based on a technical analysis, affinity propagation (AP), clustering, and a support vector regression (SVR) model to predict the performance of the Taiwan Capitalization Weighted Stock Index (TAIEX), the Standard & Poor's 500 (S&P500), and the Dow Jones Industrial Average (DJIA) dataset. Chourmouziadis et al. [30] tested a model for the fuzzy prediction of the development of an investment portfolio on the Athens Stock Exchange, with the goal of outperforming the market (Buy and Hold strategy) in the medium and long terms. The approach was also proven to work during both bull and bear market periods. As part of their prediction of stock price developments, Mohamed et al. [31] applied an adaptive fuzzy neural model to 58 listed firms, examining the significance of four performance predictors—return on assets (ROA), return on equity (ROE), earnings per share (EPS), and profit margin (PM). Chang and Liu [32] developed a Takagi–Sugeno–Kang (TSK) Fuzzy Rule Based System for stock price prediction. The TSK fuzzy model applied a technical index as an input variable, with the subsequent part being a linear combination of the input variables. This was successfully tested on Taiwanese electronic stocks listed on the Taiwan Stock Exchange. Liu et al. [33] presented an application of type two fuzzy neural modelling to predict TAIEX and NASDAQ stock prices based on a given set of training data. Xie et al. [34] proposed an approach that integrated a fuzzy neural system with a Hammerstein–Wiener model that formed an indivisible five-layer network, whereby the implication of the fuzzy neural system was realised through a linear dynamic computation of the Hammerstein–Wiener model. The effectiveness of the model was evaluated on three data sets of financial stocks. An interesting contribution was published by Nasiri and Ebadzadeh [35], where a new multi-functional recurrent fuzzy neural network was introduced. It consisted of two fuzzy neural networks with Takagi–Sugeno–Kang fuzzy rules. There was a feedback loop between these two networks, which allowed the network to learn and memorize historical information from past observations, enabling it to learn multiple functions simultaneously.

The summary of the text, which includes the mentioned methods, has been addressed in several works. In this regard, a highly successful review work is "Systematic literature review of fuzzy logic based text summarization" by Kumar and Sharma [36], which presented a systematic review of the literature with the aim of gathering, analysing, and reporting the trends, gaps, and prospects of using fuzzy logic based on the findings in original studies. Another comprehensive study is "A comprehensive review of deep neuro-fuzzy system architectures and their optimization methods" by Talpur et al. [37]. The study aimed to assist researchers in understanding the various ways that deep neuro-fuzzy systems are developed through the hybridization of deep neural networks and reasoning aptitude from fuzzy inference systems, as well as gradient-based and metaheuristic-based optimization methods. In the context of the reviewed methods, it is worth mentioning the work "Applications of neuro fuzzy systems: A brief review and future outline" [38]. The work provided an overview of the development of neuro-fuzzy systems through the classification and a literature review of the articles from the decade of 2002–2012, aiming to explore how various neuro-fuzzy system methods were developed during this period.

Forecasting the price of Tesla shares has been covered by many authors. For example, Agrawal [39] proposed a non-linear regression method based on deep learning to predict the stock price. Barapatre et al. [40] proposed a machine learning artificial neural network (ANN) model for stock market price prediction that integrated a backpropagation algorithm that was used to train the ANN model, with the research implemented on the back of a Tesla stock price dataset. Aldhyani and Alzahrani [41] proposed a framework based on long short-term memory (LSTM) and a hybrid of a convolutional neural network with LSTM (CNN-LSTM) to predict the closing prices for Tesla, Inc. and Apple, Inc. Alkhatib et al. [42] used six deep learning models: MLP, GRU, LSTM, Bi-LSTM, CNN, and CNN-LSTM to predict the adjusted closing stock price. The results showed that the LSTM-based models improved with the new approach, with no model performing better or outperforming the other models.

Why is Tesla a frequent target for the analysis of techniques related to their ability to predict future share prices? Tesla is not only an exceptional company in the field of technology, but also in the field of economics. As a heavily loss-making company (until 2019), it has achieved very high market capitalization and brought profit to long-term equity investors. Tesla is an example of a company with a radical innovation strategy that, according to Czakon et al. [43], has remained relatively unaffected by the strong market uncertainty of recent years. The reason for the growth of Tesla's value was not only its value orientation focused on innovation and the future, but also the application of regular "air-to-air" software upgrades. Tesla undoubtedly timed its market entry to coincide with a period of rapid technological development, when the boundaries between products and the market became more dynamic [44]. Chen et al. [45] pointed out that Tesla, more than other brands, strives for a high degree of integrity in its product architecture, thereby employing significant vertical integration. The majority owner of the company, Elon Musk, also plays an important role. According to Kozinets [46], he is a charismatic, utopian entrepreneur. However, his power, given his significant involvement in social networks, is so great that he can effectively manipulate the markets.

3. Fuzzy Approach Methodology

The fuzzy approach has its foundations in various versions of fuzzy logic, which were created by adapting the binary numerical characteristics of propositional operators to the interval ⟨0, 1⟩. Here, fuzzy logic is used as a tool for the exact handling of fuzzy sets, the theory of which was published by Zadeh [47].

3.1. Principles of Fuzzy Set Theory

Let set U be the domain of consideration or discussion, let $\mu_A: U \rightarrow \langle 0, 1 \rangle$ be the so-called membership function, and let $\underline{A} = \{(y, \mu_A(y)): y \in U\}$ be the set of all pairs $(y, \mu_A(y))$, in which the numbers $0 \leq \mu_A(y) \leq 1$ indicate that, for a given $y \in U$, the degree of membership of the element y to the set $U_{\underline{A}} = \{y: 0 < \mu_A(y) \leq 1, y \in U\} \subset U$.

It follows that $\underline{A}$ is a fuzzy subset of universe U and $U_{\underline{A}}$ is its support. From the point of view of fuzzy logic, $\mu_A(y)$ is the truth value of the statement $y \in U_{\underline{A}}$. The element $y \in U$ with $\mu_A(y) = 0.5$ is called a crossover point in A. For values greater than 0.5, the element y rather belongs to $U_{\underline{A}}$, and for smaller values, it rather does not belong to it (for details see [48,49]).

3.2. Fuzzy System

The fuzzy system is shown schematically in Figure 1. It receives input data from the converter IN, which converts real values $x_i \in \mathcal{R}$ into internal input values $u_i \in \langle 0, 100 \rangle = U_i$. The internal output value $v_u \in \langle 0, 100 \rangle = V$ is then converted into the value of $y_x \in \mathcal{R}$ by the output converter OUT.

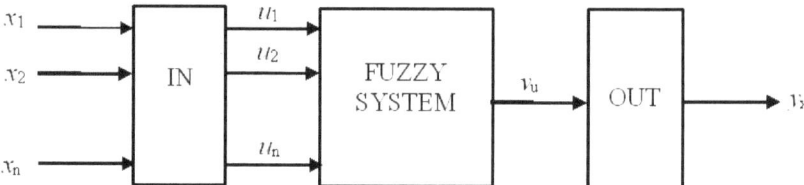

Figure 1. Fuzzy system.

From a conventional system operating on the basis of systems of algebraic or differential equations, the fuzzy system differs in terms of its internal mechanism for the transformation of u into v_u, i.e., what happens inside the "black box", i.e., what takes place in the process of solving problems within the fuzzy approach—see Section 3.3.

3.3. Framework Description of the Functioning of the Fuzzy Mechanism of the Fuzzy System

In the fuzzy system in Figure 1, there is a scale of internal numerical values u_i of universe $U_i = \langle 0, 100 \rangle$ of the input linguistic variable $\mathcal{U}_i$, $i = 1, 2, \ldots, n$, which is covered by three overlapping intervals at its edges by a knowledgeable expert. Each of these intervals is the support $U_{\underline{T}i}$ of one of three different fuzzy numbers $\underline{T}_i \in \{\underline{L}_i, \underline{M}_i, \underline{H}_i\} = S_i$, as specified by the membership function $\mu_{\underline{T}i}: U_i \rightarrow \langle 0,1 \rangle$, whose trapezoidal course is algebraically defined by the relations [50]:

$$\mu_{\underline{L}}(x) = 1 \text{ for } x < a, \ \mu_{\underline{L}}(x) = \frac{b-x}{(b-a)} \text{ for } a \leq x < b, \ \mu_{\underline{L}}(x) = 0 \text{ otherwise;}$$

$$\mu_{\underline{M}}(x) = \frac{x-a}{b-a} \text{ for } a \leq x < b, \ \mu_{\underline{M}}(x) = 1 \text{ for } b \leq x < c, \ \mu_{\underline{M}}(x) = \frac{d-x}{d-c} \text{ for } c \leq x < d, \ \mu_{\underline{M}}(x) = 0 \text{ otherwise;} \quad (1)$$

$$\mu_{\underline{H}}(x) = 0 \text{ for } x < c, \ \mu_{\underline{H}}(x) = \frac{x-c}{d-c} \text{ for } c \leq x < d, \ \mu_{\underline{H}}(x) = 1 \text{ otherwise.}$$

The same applies to universe $V = \langle 0, 100 \rangle$ of the output linguistic variable $\mathcal{V}$ with fuzzy numbers $\underline{T} \in \{\underline{L}, \underline{M}, \underline{H}\} = S$. The fuzzy numbers $\underline{L}_i, \underline{M}_i,$ and $\underline{H}_i$, respectively, $\underline{L}, \underline{M},$ and $\underline{H}$, formally represent the terms "low value", "common value", and "high value" of the linguistic variable $\mathcal{U}_i$, respectively, $\mathcal{V}$. It therefore acquires values on two levels: on the level of the numerical values u_i in universe U_i, respectively, the numerical values v in universe V, and on the level of the fuzzy numbers $\underline{T}_i$ (i.e., terms) in universe $S_i = \{\underline{L}_i, \underline{M}_i, \underline{H}_i\}$, respectively, and the fuzzy numbers $\underline{T}$ in universe $S = \{\underline{L}, \underline{M}, \underline{H}\}$ [51].

A significant role In the exact description of what happens inside the "black box" is played by the binary fuzzy coincidence relation $\underline{C}_i = \{((u_i, \underline{T}_i), \mu_{\underline{C}i}(u_i, \underline{T}_i)): (u_i, \underline{T}_i) \in U_i \times S_i\}$, $\mu_{\underline{S}i}: U_i \times S_i \to \langle 0, 1 \rangle$ from universe U_i to universe S_i of the input linguistic variable $\mathcal{U}I_i$, $i = 1,2,\ldots, n$. The relational matrix of this fuzzy relationship is presented in Table 1. In it, a corresponding line is assigned to each value $u_i \in U_i$, depending on which decomposition class of universe U_i it belongs to. In it, the general number $\mu_{\underline{C}i}(u_i, \underline{T}_i) = \mu_{\underline{T}i}(u_i) = |u_i \in U_{\underline{T}i}|$ is written in the field below $\underline{T}_i$, indicating the truth value of the statement "the element u_i from universe U_i is an element of the support $U_{\underline{T}i}$ of the fuzzy subset $\underline{T}_i$ on U_i". This is interpreted as the degree of coincidence of an element from the level of the numerical values with an element from the level of the terms of the linguistic variable $\mathcal{U}_i$. Positive values in a given row state on which fuzzy subsets $\underline{T}_i$ the singleton $\{(u_i,1)\}$ will be fuzzified. Table 1 (the relational matrix of the binary fuzzy relation $\underline{C}_i$) is therefore called a fuzzification table.

Table 1. Fuzzification table.

$U_i \backslash S_i$	$\underline{L}_i$	$\underline{M}_i$	$\underline{H}_i$
$u_i < a_i$	1	0	0
$a_i \leq u_i < b_i$	$(b_i - u_i)/(b_i - a_i)$	$(u_i - a_i)/(b_i - a_i)$	0
$b_i \leq u_i < c_i$	0	1	0
$c_i \leq u_i < d_i$	0	$(d_i - u_i)/(d_i - c_i)$	$(u_i - c_i)/(d_i - c_i)$
$u_i \geq d_i$	0	0	1

The fuzzification table assigns to each value $u_i \in U_i$ a three-element fuzzy set $S_i = \{(\underline{L}_i, \mu_{\underline{L}i}(u_i)), (\underline{M}_i, \mu_{\underline{M}i}(u_i)), (\underline{H}_i, \mu_{\underline{H}i}(u_i))\}$, with a one-element or two-element support. The Cartesian product $\mathcal{S} = S_1 \times S_2 \ldots \times S_n$ then defines the n-ary fuzzy relation $\underline{S}$ as $\underline{S} = S_1 \times S_2 \ldots \times S_n = \{((\underline{T}_1, \underline{T}_2, \ldots, \underline{T}_n), \mu_{\underline{S}}(\underline{T}_1, \underline{T}_2, \ldots, \underline{T}_n)): (\underline{T}_1, \underline{T}_2, \ldots, \underline{T}_n) \in S_1 \times S_2 \ldots \times S_n\}$, which is decomposable according to the components $S_1, S_2, \ldots, S_n$ with support $\mathcal{S}_{\underline{S}} = \{(\underline{T}_1, \underline{T}_2, \ldots, \underline{T}_n): \mu_{\underline{S}}(\underline{T}_1, \underline{T}_2, \ldots, \underline{T}_n) > 0, (\underline{T}_1, \underline{T}_2, \ldots, \underline{T}_n) \in S_1 \times S_2 \ldots \times S_n\}$, which, by means of projection F, whose elements (pair $(((\underline{T}_1, \underline{T}_2, \ldots, \underline{T}_n), \underline{T})$, where $\underline{T} = F(\underline{T}_1, \underline{T}_2, \ldots, \underline{T}_n))$, form a set of inference rules, is transferred from universe $\mathcal{S}$ to universe S in accordance with the extension principle.

The number of elements of the n-ary fuzzy relation $\underline{S}$ and the number of inference rules corresponding to it is 3^n; the number of elements of its support $\mathcal{S}_{\underline{S}}$, containing only those tuples of terms $(\underline{T}_1, \underline{T}_2, \ldots, \underline{T}_n) \in S_1 \times S_2 \ldots \times S_n$, with which the input vector $u = (u_1, u_2, \ldots, u_n)$ coincides and is therefore blurred into them, amounts to $2^{n-\alpha}$, where α is the number of prototypical elements (i.e., components u_i with $\mu_{\underline{T}i}(u_i) = 1$). For example, in the case where $n = 3$, $3^n = 27$, while $1 \leq 2^{n-\alpha} \leq 8$. Therefore, when applying fuzzy inference, it is not effective to take into account all inference rules, but sufficient to take into account only those whose left sides match with the elements of support $\mathcal{S}_{\underline{S}}$.

The process of the transformation of vector $x = (x_1, x_2, \ldots, x_n)$ into y_x then gradually passes through phases 1–5, as shown in Figure 2, where fuzzy inference begins with fuzzification and ends with aggregation [52]:

1. In the fuzzification phase, the entered point values are converted into membership values for individual fuzzy sets using the membership function. The "fuzzification" block in the fuzzy system also includes the IN converter (see Figure 1), which converts real values $x_i \in \mathcal{R}$ into internal input values $u_i \in \langle 0, 100 \rangle = U_i$. By applying the relevant fuzzification tables, the input vector $u = (u_1, u_2, \ldots, u_n)$ is then blurred into a tuple of terms $(\underline{T}_1, \underline{T}_2, \ldots, \underline{T}_n) \in S_1 \times S_2 \ldots \times S_n$, with which it coincides, even to a small extent. The set $\mathcal{S}_{\underline{S}}$ is then assembled from them, with each element $(\underline{T}_1, \underline{T}_2, \ldots, \underline{T}_n) \in \mathcal{S}_{\underline{S}}$ being evaluated according to its degree of coincidence with vector u, which is the number $\min\{\mu_{\underline{T}1}(u_1), \mu_{\underline{T}2}(u_2), \ldots, \mu_{\underline{T}n}(u_n)\}$.

2. In the previous stage, sharp sets of variables enter the system, which are subsequently fuzzified, i.e., converted into input fuzzy sets. The next steps include the derivation of the output fuzzy sets using an inference mechanism based on the rule base. In the "application of inference rules", the relation $F^{-1}(\underline{T}) = \{(\underline{T}_1, \underline{T}_2, \ldots, \underline{T}_n): F(\underline{T}_1, \underline{T}_2, \ldots, \underline{T}_n) = \underline{T}, (\underline{T}_1, \underline{T}_2, \ldots, \underline{T}_n) \in \mathcal{S}_{\underline{S}}\}, \underline{T} \in S$, is inversed to projection F and set $\mathcal{S}_{\underline{S}}$ is decomposed into the mutually disjoint classes $F^{-1}(\underline{L})$, $F^{-1}(\underline{M})$ and $F^{-1}(\underline{H})$. If any of the resulting decomposition classes are empty ($F^{-1}(\underline{T}) = \emptyset$), they are characterised by the number $\mathcal{M}_{\underline{T}} = 0$; otherwise ($F^{-1}(\underline{T}) \neq \emptyset$) the characteristic number of class $F^{-1}(\underline{T})$ is the number $\mathcal{M}_{\underline{T}} = \max\{\min\{\mu_{\underline{T}1}(u_1), \mu_{\underline{T}2}(u_2), \ldots, \mu_{\underline{T}n}(u_n)\}: (\underline{T}_1, \underline{T}_2, \ldots, \underline{T}_n) \in F^{-1}(\underline{T})\}$. The number $\mathcal{M}_{\underline{T}}$ is interpreted as the strength of the reaction $\underline{T}$ "lent" to it by vector $u = (u_1, u_2, \ldots, u_n)$.

3. In the "result processing" phase, the fuzzy mechanism of the fuzzy system generates on V fuzzy subsets $\underline{T}^* = \{(v, \mu_{\underline{T}}^*(v)): v \in V\}$, where $\mu_{\underline{T}}^*(v) = \min\{\mu_{\underline{T}}(v), \mathcal{M}_{\underline{T}}\}$, $\underline{T} \in S$. These are mostly fuzzy subsets of $\underline{T} \in \{\underline{L}, \underline{M}, \underline{H}\}$ with membership functions of $\mu_{\underline{T}}(v)$, bound from above by $\mathcal{M}_{\underline{T}}$ constants, so that $\mu_{\underline{T}}^*: V \to \langle 0, \mathcal{M}_{\underline{T}} \rangle$. In other words, in this phase, the evaluation of the rule takes place with the determination of the values of the causes, creating the minimum result value of the entire rule.

4. In the "aggregation" phase, the output values of all the activated rules for each linguistic variable are united into one fuzzy set, i.e., the fuzzy subsets $\underline{L}^*, \underline{M}^*$, and $\underline{H}^*$ are united, resulting in a fuzzy set $\underline{R} = \{(v, \mu_{agg}(v)): v \in V\} = \underline{L}^* \cup \underline{M}^* \cup \underline{H}^*$, in which $\mu_{agg}(v) = \max\{\mu_{\underline{L}}^*(v), \mu_{\underline{M}}^*(v), \mu_{\underline{H}}^*(v)\}, v \in V$.

5. Finally, in the "defuzzification" phase, the fuzzy mechanism of the fuzzy system, via integration using the parts over the interval $\langle 0, 100 \rangle$, finds the values of the definite integrals $\int v \cdot \mu_{agg}(v) \, dv$ and $\int \mu_{agg}(v) \, dv$ and, from them, a ratio is obtained:

$$v_u = \int v \cdot \mu_{agg}(v) dv / \int \mu_{agg}(v) dv \qquad (2)$$

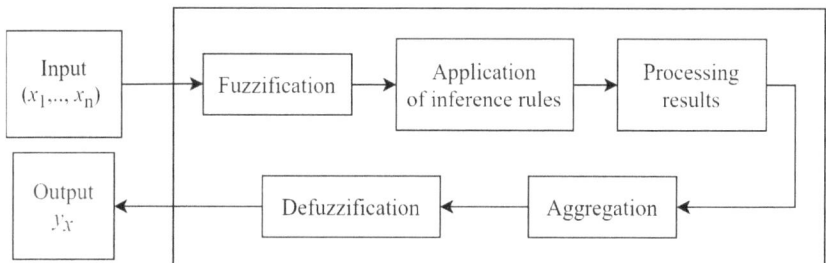

Figure 2. Problem solving process within the framework of the fuzzy approach.

This is a standard defuzzification method in the literature, referred to as the CoG (Center of Gravity) method. The defuzzification of the output consists of finding the centroid of the surface after aggregating the outputs of the rules using the previous aggregation step.

The "defuzzification" block also includes converter OUT (see Figure 1), which converts internal values $v \in \langle 0, 100 \rangle = V$ to external values $y \in \mathcal{R}$.

As part of setting the default parameters for a given task, a knowledgeable expert chooses the intervals $\langle x_{i\_min}, x_{i\_max} \rangle$, $i = 1, \ldots, n$, respectively $\langle y_{min}, y_{max} \rangle$, in which, according to their opinion, the values of external numerical variables can occur. Therefore, for each of these variables, the expert enters the parameter $x_{i\_ref} = x_{i\_max} - x_{i\_min}$, respectively, $y_{ref} = y_{max} - y_{min}$. The following functions play key roles in the conversion of the external values of input variables (x_i) and output variables (y) to internal values, with which the internal mechanism of the fuzzy system works; this happens in the IN converter:

$$k(x_i) = 100 \cdot (x_i - x_{i\_min}) / x_{i\_ref}, x_i \in \langle x_{i\_min}, x_{i\_max} \rangle \qquad (3)$$

respectively,

$$k(y) = 100 \cdot (y - y_{min})/y_{ref}, y \in \langle y_{min}, y_{max} \rangle \tag{4}$$

The inputs to the IN converter are divided into two groups: those that have a positive effect on the result from the point of view of the solved problem, and those whose effect on the result is negative. The first of them are recalculated to internal values using the formula $u_i = k(x_i)$, and the second using the formula $u_i = 100 - k(x_i)$. This is an application of the unary t-norm, which is an involutive negation. This approach makes it possible, in the case of having fewer than four input linguistic variables, to apply, among other things, the strategy of the predominant element in the formulation of the inference rules, whereby the expert intervenes only in necessary cases.

The conversion of the output values y from the OUT converter into internal output values $v \in V$ takes place according to the formula $v = k(y)$; in the opposite direction:

$$y_x = k^{-1}(v_u) = (v_u \cdot y_{ref}/100) + y_{min} \tag{5}$$

4. Data

This article aims to predict the short-term trend and share price of Tesla (TSLA) at the beginning of 2023. Methodologically, the task is solved using the fuzzy approach. The inputs to the model are the variables influencing the value of the output variable—the TSLA share price. Table 2 summarises the values of these selected variables and the output variables of the TSLA share price for the monitored period of 2016–2022. The real values at the beginning of 2023 for all the variables are presented in the last column.

The input variables of the task are the EPS (net earnings per share) indicator, whose high value indicates the attractiveness of the share, the Federal Funds Rates (FFR), which are a key tool for managing US monetary policy, the annual close stock price of the NASDAQ Composite Index (COMP), which is heavily loaded with companies in the technology sector, and the year close stock prices of the output variable TSLA (TSLA). The input data, EPS, FFR, and COMP were chosen based on their proven strong regression relationship with the development of stock prices, which has been demonstrated in a number of published works, e.g., by Boyacioglu and Avci [53] and Alenezy et al. [54].

Table 2. Values of EPS, FFR, COMP, and TSLA in period 2016–2023.

Year	2016	2017	2018	2019	2020	2021	2022
x_EPS ($)	−0.31	−0.79	−0.38	−0.33	0.21	1.63	2.55
x_FFR (%)	0.55	1.33	2.40	1.55	0.09	0.07	4.33
x_COMP ($)	5383	6903	6635	8973	12,888	15,645	10,466
Y_TSLA ($)	14	21	22	28	235	352	123

Source: [55,56].

The data were collected through the Macrotrends and Yahoo! Finance databases [57,58] based on annual closing prices. The measurement units of the variables correspond to their nature—the EPS, COMP, and TSLA are expressed in nominal US dollars and the FFR in percentage rate. The annual closing prices of the COMP and TSLA are the market prices traded at the end of December of the year in question.

From the development of the EPS, FFR, COMP, and TSLA variables, one can see a turning point in 2019/2020. The EPS indicator changed from negative to positive, with its value continuing to grow over time. The FFR rate fell by 94% from 2019 to 2020, followed by a sharp increase in 2022 (the FFR increased 48 times from 2020–2022). The COMP share price increased by almost 43% in 2019/2020, which was the smallest increase over the entire period under review. The change in the TSLA share price during the breakthrough period was enormous—the 739% increase was the highest in the entire period under review; from 2020, as in the case of the COMP share price, the continuous phase of growth ended.

The specifics of the breakthrough period of 2019/2020 are considered in the subsequent fuzzy analysis. The short-term trend and TSLA share price predictions for early 2023 (31 January 2023) are based on a range of input variable values for the period of 2020–2022—see Table 2. The data used and their frequency corresponds to the desired range of data carrying the necessary information. A longer range of data would bias the prediction. The nature of the task (the prediction of the price of a volatile title and high market uncertainty) allows for the share price prediction to be realized only in the short term—a maximum within 1 month from the last detected share price.

5. Construction of the Fuzzy Prediction Model

The formulation of the fuzzy model for the prediction of the next member of the time series of the TSLA share price is specific and based on the fact that we know the number of the previous members of the resulting series of this linguistic variable—see Table 2, respectively, Table 3. From them, it is possible to estimate in which phase of its development (decline, growth, and stagnation) that the price of the share is now. Furthermore, we know the historical series of the basal values of the linguistic variables, upon which the price of the share broadly and vaguely depends. This dependence is mainly reflected in the values of the extreme limits in which we look for the result of the prediction.

Table 3. Conversion of basal input data of Table 2 relevant to the historical period 2020–2022 into dimensionless data according to (3) and (4).

Year	2020	2021	2022
EPS	0	61	100
FFR (%)	100	0	0
COMP ($)	47	100	0
Y_TSLA ($)			100

The opinion of the expert plays an important role in the construction of fuzzy prediction models. When formulating the framework for a model, they must consider their experiences and expectations, intervening in the structure of the model when it comes to the definition of the inference rules and the course of the functions of the linguistic variables, accordingly.

In terms of the specified model, if positive and high EPS values are expected, which therefore make the stock more attractive, the expert predicts a rise in the stock price in the next period. In contrast, a rise in FFR interest rates is expected to have the opposite effect; the reason being the inverse relationship between the required interest income, the amount of which is determined by the interest rate, and the share price. If interest rates are lowered (and nothing else changes), stock prices theoretically rise, and vice versa. Due to the importance of technology companies, the expected price development of the COMP index can also be projected into the development of the TSLA share price in the same direction—a positive correlation is assumed.

6. Results: Fuzzy Prediction for TSLA Share PRICE and Short-Term Trend for Beginning of 2023

This model works with the dimensionless input variables u_{EPS}, u_{FFR}, and u_{COMP} of the U_{EPS}, U_{FFR}, and U_{COMP} universes, respectively, with the dimensionless output variable y of the Y_{TSLA} universe on the interval $\langle 0, 100 \rangle$. The conversion of the basic values of the linguistic variable quantities x_{EPS}, x_{FFR}, and x_{COMP} to an interval is mediated by converting them according to relationship (3), where x_{max} is the highest value of x, and, respectively, x_{min} is the lowest value of x statistically recorded in the period of 2020–2022 (see Table 3). The output linguistic variable Y is the TSLA share price in the year immediately following the historically known period. The transformation of the basal output variable x_{TSLA} of the Y_{TSLA} universe of the output linguistic variable y to the dimensionless value $y \in Y_{TSLA} = \langle 0, 100 \rangle$, where Y_{TSLA} is marked as Y in the graphs, is given by relationship (4), where y_{max} is the highest value of y, and, respectively, y_{min} is the lowest value of y statistically recorded in the period of 2020–2022. The conversion in the opposite direction—the recalculation of the dimensionless output variable to the output linguistic variable x_{TSLA}—is performed using (5).

The progress of Y_{TSLA} (see Table 2) in the years of 2016–2022 shows an upward trend—the TSLA share price increased almost 20 times. However, due to the expected recession in the USA, accompanied by the expected increase in basic interest rates, it is possible to consider a stagnation of the TSLA share price for the year 2023. This consideration is compensated by the expected positive development of the EPS. Nothing definite can be said about the expected development of the price of the COMP index based on the historical data in relation to the future TSLA share price.

The formulation of the prediction fuzzy model for the beginning of 2023 is based on the numerical characteristics of the period 2020–2022 and assumptions about the development of variables that have an impact on the TSLA share price.

From the last column in Table 3, the triple $(100, 0, 0) \in U_{EPS} \times U_{FFR} \times U_{COMP}$ is created. Due to the external and internal uncertainty of the model, points a, b, c, and d of the interval $\langle 0, 100 \rangle$ are evenly distributed, i.e., a = 20, b = 40, c = 60, and d = 80. The course of the membership functions of all the converted variables (input and output) over the set of the universe Y values of the output linguistic variable (Y_TSLA) are identical (see Figure 3).

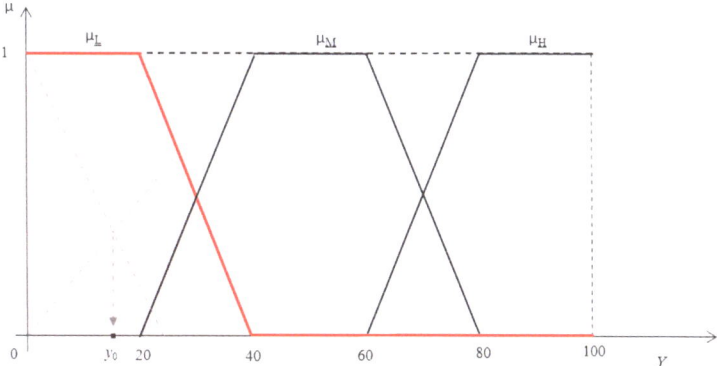

Figure 3. Courses of the membership functions with even distribution of points a = 20, b = 40, c = 60 and d = 80 within the interval values $y \in Y = \langle 0, 100 \rangle$ and the course of the membership function μ_{AGG}. The value y_0 is the dimensionless output of the linguistic variable (Y_TSLA).

The fuzzification table (see Table 4) applies to i = EPS, FFR, COMP. The contents are the values of the membership functions $\mu_{Ai}(u_i)$, where $A \in \{L, M, H\}$ are derived from the aforementioned equations and inequalities (see (1)).

Table 4. Fuzzification table valid for i = EPS, FFR, COMP (see Table 1).

Interval	$u_i < 20$	$20 \leq u_i < 40$	$40 \leq u_i < 60$	$60 \leq u_i < 80$	$u_i \geq 80$
L_i	1	$(40 - u_i)/20$	0	0	0
M_i	0	$(u_i - 20)/20$	1	$(80 - u_i)/20$	0
H_i	0	0	0	$(u_i - 60)/20$	1

Only non-zero elements are considered in the fuzzification table. With its help, set $X = \{(H_{EPS}, 1), (L_{FFR}, 1), (L_{COMP}, 1)\}$ is created from the input data vector $(100, 0, 0) \in U_{EPS} \times U_{FFR} \times U_{COMP}$. The three-element set $LF = \{(H, L, L)\}$ is created from its three elements. The triplet (H, L, L) of the input fuzzy sets from mapping F assigns the output fuzzy set L

based on the strategy of the prevailing element of the rule, whereby $\mu_{AGG}(y) = \mu_{\underline{L}}(y)$ applies (see μ_{AGG} function (in red) in Figure 3).

For the dimensionless variable, y_0 applies (see relationship (2)):

$y_0 = \int y \cdot \mu_{\underline{L}}(y) dy / \int \mu_{\underline{L}}(y) dy = 15.55$, which presents a point in the body of the coloured figure under the curve highlighted in red, where the total weight of the body may be thought to be concentrated.

The internal dimensionless output value $y_0 \in \langle 0, 100 \rangle = V$ is then converted into the external value $y_x \in \mathcal{R}$ by the output converter OUT. Thus, for the predicted basal value x_0 of the x_{Y_TSLA} variable at the beginning of 2023 (31st January), it applies (see (5)):

$$x_0 \sim 123 + 15.55 \cdot (352 - 123)/100 = \$159.$$

The result of the fuzzy model depends significantly on the experience and expectations of the expert. This was reflected in the model by the expert in terms of the definition of the inference rules and the progressions of the membership functions of the linguistic variables. In our case:

- The expert chose the frequency and data range of the input and output variables. The expert's rationale was: the frequency of the annual closing prices of a three-year historically known period sufficiently determines the price shift in the subsequent short-term horizon of 1–12 months; the choice of a longer period for analysis is inadequate, given the landmark development of the input and output variables, both recent and current.
- The expert chose a uniform distribution of points a, b, c, and d on the interval of the values $y \in Y = \langle 0, 100 \rangle$ by which they defined the progressions of the membership functions for all the variables. The expert's rationale was: the courses of the membership functions defined in this way correspond to the state of the highest degree of uncertainty, which corresponds to the situation of uncertainty regarding the future decisions of the FED, the development of the economy, the marketing behaviour of TSLA, and the public engagement of the owner, Elon Musk.
- The expert chose to select the prevailing element rule. The expert's rationale was: the choice of the prevailing element is the choice of a rational decision maker with a neutral attitude to risk.

The resulting fuzzy prediction for the value of the TSLA share price at the beginning of 2023 came in at 159 USD (31st January), with a clear upward trend compared to the last recorded value at the end of December 2022 (123 USD). The real TSLA share price on 31 January 2023 was 173 USD (Yahoo! Finance [56])—see Figure 4.

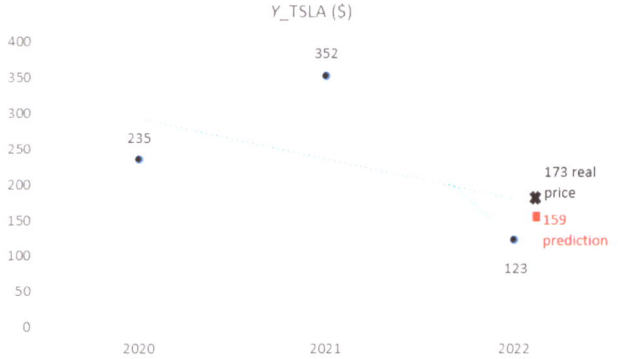

Figure 4. Evolution of annual closing prices of TSLA (solid line) relevant to fuzzy prediction including regression price trend (dotted line), short-term fuzzy prediction (■) versus actual closing price of TSLA on 31 January 2023 (×).

The difference between the growth rate of the actual and fuzzy predicted price is a consequence of the experts' limited information and the imprecise assumptions they made when setting up the model.

7. Discussion: Functioning of Fuzzy Neural Network and Fuzzy System in the Field of Time Series Value Prediction

The fuzzy system presented here differs from the predictive fuzzy neural network in many ways. The fuzzy system is based on Kahneman's allegorical model of the functioning of the human mind, which deals with the cognitive processes taking place on a macro level within two cooperating subsystems, S_1 and S_2, as represented by the conscious and subconscious components of thinking [59]. Just as system S_1 classifies particulars into vaguely defined content of intuitive concepts, which it retrieves based on the input stimulus through a mental process called "free-floating of associations" by cognitive scientists (see Thagard [60]), the fuzzy mechanism of the fuzzy system, through the "fuzzification" phase, "blurs" the vector u = (u_1, u_2,..., u_n) into n-tuples (T_1, T_2,..., T_n) of the fuzzy subsets of the support S, with which it coincides, and classifies "particulars" into the framework of the relevant terms that the vector u "resembles". Just as system S_2, based on a rational analysis of what is presented by S_1, arrives at appropriate conclusions, the fuzzy mechanism of the fuzzy system, during the subsequent four phases of the solving process, according to Figure 2, identifies an adequate resulting reaction y_x to the arising situation described by the vector x = (x_1, x_2,..., x_n).

What a fuzzy neural network has in common with a fuzzy system in the field of time series value prediction is the training set [61], even though the elements are worked with differently in each of them. In our training set, the four time series values of the variables EPS, FFR, COMP, and TSLA, entered in the unshaded part of Table 2, play a crucial role. Here, similar to fuzzy neural networks, each column of values can be viewed as a time-indexed element ((EPS, FFR, and COMP), TSLA) within a seven-element "training set", in which the triplet (EPS, FFR, and COMP) is the input and TSLA is the desired output. This seven-element set is then reduced to a three-element set (columns for the years 2020, 2021, and 2022—see Table 3) for the reasons justified above, from which the value in the last column (Y_TSLA) in Table 2 should be derived.

In a fuzzy neural network with a teacher, the teacher inserts, on the input side of the neural network, the EPS, FFR, and COMP values of the given element ((EPS, FFR, and COMP) TSLA) of the aforementioned three-element training set, with the network reacting to this with some output. Then, according to the instructions of the learning fuzzy backpropagation algorithm, the teacher changes the weights of the synaptic connections in the network, bringing its output closer to the desired TSLA value (bottom row and last column of Table 2). The same is repeated with the next element of the training set, over and over again, until a situation occurs where the fuzzy learning algorithm no longer requires any change in its weights. This ends the learning process. As a result, the network is optimally set and ready to search for the correct TSLA values, even for other triplets of the EPS, FFR, and COMP values that it did not encounter in the training set.

In the fuzzy system presented here, the analogy of a fuzzy neural network with a teacher is the fuzzy mechanism described in Section 3, with the role of the teacher played by the solver/expert. They, like the teacher in the previous case, inserted, on the input side of the fuzzy system, the three values of the EPS, FFR, and COMP of the element ((EPS, FFR, and COMP), TSLA), with the fuzzy system subsequently "giving" them the answer through its output. Here, instead of changing the weights of the synapses in the network, the solver/expert changed parameters a, b, c, and d in the numerical universe of the output linguistic variable (see Figure 3), thereby changing its terms (i.e., the supports and cores of the respective fuzzy sets) in order to bring the output of the fuzzy system closer in line with the desired TSLA value, or changing the setting of the inference rules. However, unlike the fuzzy neural network teacher, the solver/expert with a fuzzy system did not have any learning algorithm at their disposal that advised them how to change parameters a, b, c, and d, or how to reset the inference rules. Their decisions were based on attempts, such as successes, errors, or iterative methods, to help them get as close as possible to the desired TSLA value. They did the same with each element of the training set and went through it so many times that the outputs from the fuzzy system for all three elements of the training set stabilised at values close to the desired TSLA values. As in the case of backpropagation, the convergence of this procedure cannot be proven in general. However, practice shows that it works.

If we compare a fuzzy system and a fuzzy neural network from the point of view of their applicability and effectiveness in the field of solving the problems of predicting time series values, we come to the conclusion that a fuzzy neural network, implementing a semi-parallel computing process, reacts to input data by producing outputs much faster than a fuzzy system implementing a disproportionately more complex sequential computing process [62]. Fuzzy neural network learning is also more user friendly. The fuzzy neural network teacher does not have to think too much and acts completely mechanically. A fuzzy learning algorithm tells them what to do [63]. In contrast, the fuzzy system teacher has no proven learning algorithm at their disposal and must think about what they will do in the learning process. However, if they are sufficiently knowledgeable and experienced, they have a chance to achieve a higher convergence speed of the learning process than a fuzzy neural network.

The limitations of the proposed fuzzy system are derived from the limitations of the fuzzy logic that are mainly based on the limited information available to the expert and their possible ignorance or inexperience, which are circumstances that are interrelated. Specifically, the disadvantages of using fuzzy logic to solve problems include (a) its complexity (while fuzzy logic systems are generally simpler than other AI systems, they can still be quite complex), (b) its lack of accuracy (fuzzy logic systems may not be suitable for applications that require a very high level of accuracy), and (c) the difficulty in refining the system (it can be difficult to debug fuzzy logic systems because it is not always clear how the system has arrived at a particular result)—see [64]. To overcome this to some extent, the model in its future form may include a suitable intelligent agent to assist with expert reasoning.

Among other alternative approaches to stock price prediction that share the commonality of deriving the information base for price prediction from sources other than the historical series of realized prices, there is BERT-LSTM, for example, which extracts informative features regarding the direction of stock price movements from news on Twitter, using a natural language processing (NLP) model called BERT [65]. Selected tweets from Twitter were included for prediction in Velu et al. [66]. In order to obtain attributes for prediction, the study employed an approach to a sentiment analysis that combined psychological labelling and valence rating, representing the strength of emotional expression. Worth mentioning is the analysis by Bhadkamar and Bhattachary [67], who examined the relationship between Elon Musk's tweets and the value of Tesla's stock. A data analysis was used as the primary method to discern patterns within the preprocessed dataset, which had all stop words removed. The combination of methodologies and elements yielded a conclusion that an increase in the number of tweets/interactions corresponded to an increase in Tesla's closing price, and conversely.

These predictive approaches combine different methods; the fuzzy system presented here is a unique model based on fuzzy logic. They share the commonality that the prediction results are unequivocally supported by underlying drivers of the researched subject, which ultimately leads to a higher accuracy in estimation compared to approaches that rely purely on a historical time series analysis. These methods are based on statistical inference and require the correct selection of a model, which incorporates knowledge about the system. The fuzzy approach requires us to choose a predictive algorithm based on its empirical capabilities. The justification of the inference model typically relies on whether we feel if it adequately captures the essence of the system [68].

8. Conclusions

Investors and analysts use a number of models to predict stock market trends and future share prices. Modern approaches make it possible to solve problems in which uncertainty is involved, which is typical for stock markets. Fuzzy logic enables this to some extent, especially in relation to uncertain problems linked to insufficient knowledge. By combining methods that implement learning paradigms with methods based on fuzzy logic, predictive fuzzy neural networks have been created, which were discussed in the introduction.

The purpose of this article was to present the predictive capabilities of a tool for multi-criteria evaluation, a so-called fuzzy system, operating on the basis of a non-fuzzy neural approach, but also one that was capable of implementing a learning paradigm and working with vague concepts. The principle behind the fuzzy system was the transformation of input data by the IN converter, which converted real external input values into internal input values in the interval of 0–100. The internal output value for the given interval was then converted by the output converter OUT into an external output value, which was the result of the process of what happened inside the "black box" of the fuzzy system, i.e., what took place in the process of solving problems within the fuzzy approach, i.e., the algorithm described in steps 1 to 5. The formulation of the fuzzy model for predicting the next member of the time series was based on the knowledge of a number of previous members of the target variable, a historical series of variable values upon which the target variable depended, and on the opinion and experience of the solver/expert. When formulating the model, the solver/expert took into account their expectations by setting inference rules and defining the course of the functions of the linguistic variables.

The task of predicting the short-term trend of Tesla share price development was solved by applying a fuzzy system. An upward trend was identified with an early 2023 price estimate of 159 USD (31st January); the actual closing price of the end of January 2023 was 173 USD, a difference of 14 USD.

The difference between the rate of growth of the actual price and the price predicted by the fuzzy system was a consequence of the limited knowledge of the solver/expert and the assumptions they used when constructing the model. A fuzzy system is an analogy of a fuzzy neural network with a teacher, whereby the role of the teacher is played by the solver/expert. They place influencing variables on the inputs of the fuzzy system, which subsequently "gives" them the answer in the form of output. The learning of the system is achieved by changing parameters a, b, c, and d in the numerical universe of the linguistic variables, thereby changing their terms in order to bring the output of the fuzzy system closer in line with the desired value, or changing the setting of the inference rules. Unlike the fuzzy neural network teacher, the solver/expert with a fuzzy system does not have any learning algorithm available to them to advise them how to change parameters a, b, c, and d, or how to reset the inference rules.

In terms of its applicability and effectiveness in solving prediction problems, a fuzzy neural network reacts to input data by producing outputs faster than a fuzzy system, which implements a more complex sequential calculation process, and is more "user friendly". The fuzzy neural network teacher works completely mechanically and follows a fuzzy learning algorithm. In contrast, the fuzzy system teacher has no proven learning algorithm at their disposal and has to think about what they will do in the learning process. However, if they are sufficiently knowledgeable and experienced, they have a chance to achieve a higher convergence speed of the learning process than a fuzzy neural network.

The above shows that the presented fuzzy system can be an alternative to fuzzy neural networks from the point of view of prediction, and that it has real capabilities to predict short-term development trends.

The perspective of the author's fuzzy system is in translating the generally described fuzzy system into the set of algorithms of the fuzzy process in order to design a "fuzzy calculator", a programming language of a software that will significantly speed up and facilitate the process of multi-criteria evaluation and prediction. The fuzzy calculator will then be tested in order to verify the basic hypotheses and assumptions regarding its practical applicability to the stock market data of publicly traded companies for the purpose of prediction and to the companies´ accounting statements for the purpose of multi-criteria decision making.

Author Contributions: Conceptualization, S.H. and P.Š.; methodology, S.H. and P.Š; formal analysis, S.H.; investigation, R.K.; resources, R.K.; data curation, R.K.; writing—original draft preparation, S.H.; writing—review and editing, S.H. and P.Š.; visualization, R.K.; supervision, S.H.; project administration, S.H.; funding acquisition, S.H. All authors have read and agreed to the published version of the manuscript.

Funding: This research was funded by Grant No. IVSUZO2302, School of Expertness and Valuation, Institute of Technology and Business in Ceske Budejovice.

Data Availability Statement: Macrotrends Federal Funds Rate—62 Year Historical Chart. 2023. Available online: https://www.macrotrends.net/2015/fed-funds-rate-historical-chart. Yahoo!Finance Tesla, Inc. (TSLA). 2023. Available online: https://finance.yahoo.com/quote/TSLA/history/. Macrotrends NASDAQ Composite—45 Year Historical Chart. 2023. Available online: https://www.macrotrends.net/1320/nasdaq-historical-chart. Macrotrends Tesla EPS—Earnings per Share 2010–2022 | TSLA. 2023. Available online: https://www.macrotrends.net/stocks/charts/TSLA/tesla/eps-earnings-per-share-diluted.

Conflicts of Interest: The authors declare no conflict of interest. The funders had no role in the design of the study; in the collection, analyses, or interpretation of data; in the writing of the manuscript; or in the decision to publish the results.

References

1. Stefko, R.; Heckova, J.; Gavurova, B.; Valentiny, T.; Chapcakova, A.; Kascakova, D.R. An analysis of the impact of economic context of selected determinants of cross-border mergers and acquisitions in the EU. *Econ. Res.-Ekon. Istraživanja* **2022**, *35*, 6385–6402. [CrossRef]
2. Ng'ang'a, F.W.; Oleche, M. Modelling and Forecasting of Crude Oil Price Volatility Comparative Analysis of Volatility Models. *J. Financ. Risk Manag.* **2022**, *11*, 154–187. [CrossRef]
3. Chu, J.; Chan, S.; Nadarajah, S.; Osterrieder, J. GARCH modelling of cryptocurrencies. *J. Risk Financ. Manag.* **2017**, *10*, 17. [CrossRef]
4. Debnath, P.; Srivastava, H.M. Optimal Returns in Indian Stock Market during Global Pandemic: A Comparative Study. *J. Risk Financ. Manag.* **2021**, *14*, 592. [CrossRef]
5. Nikou, M.; Mansourfar, G.; Bagherzadeh, J. Stock price prediction using DEEP learning algorithm and its comparison with machine learning algorithms. *Intell. Syst. Account. Financ. Manag.* **2019**, *26*, 164–174. [CrossRef]
6. Vochozka, M.; Horák, J.; Šuleř, P. Equalizing seasonal time series using artificial neural networks in predicting the Euro–Yuan exchange rate. *J. Risk Financ. Manag.* **2019**, *12*, 76. [CrossRef]
7. Castillo, O.; Melin, P. Forecasting of COVID-19 time series for countries in the world based on a hybrid approach combining the fractal dimension and fuzzy logic. *Chaos Solitons Fractals* **2020**, *140*, 110242. [CrossRef]
8. Polishchuk, V.; Kelemen, M.; Gavurová, B.; Varotsos, C.; Andoga, R.; Gera, M.; Christodoulakis, J.; Soušek, R.; Kozuba, J.; Blišťan, P.; et al. A Fuzzy Model of Risk Assessment for Environmental Start-Up Projects in the Air Transport Sector. *Int. J. Environ. Res. Public Health* **2019**, *16*, 3573. [CrossRef]
9. Wang, H.-F.; Tsaur, R.-C. Insight of a fuzzy regression model. *Fuzzy Sets Syst.* **2000**, *112*, 355–369. [CrossRef]
10. Tsaur, R.C. A fuzzy time series-Markov chain model with an application to forecast the exchange rate between the Taiwan and US dollar. *Int. J. Innov. Comput. Inf. Control.* **2012**, *8*, 4931–4942.
11. Souza, P.V.D.C. Fuzzy neural networks and neuro-fuzzy networks: A review the main techniques and applications used in the literature. *Appl. Soft Comput.* **2020**, *92*, 106275. [CrossRef]
12. Wang, G.; Zhang, Y.; Ye, X.; Mou, X. *Machine Learning for Tomographic Imaging*; IOP Publishing: Bristol, UK, 2019. [CrossRef]
13. Khuat, T.T.; Le, M.H. An Application of Artificial Neural Networks and Fuzzy Logic on the Stock Price Prediction Problem. *JOIV: Int. J. Inform. Vis.* **2017**, *1*, 40–49. [CrossRef]
14. Kulkarni, A.; Kulkarni, N. Fuzzy Neural Network for Pattern Classification. *Procedia Comput. Sci.* **2020**, *167*, 2606–2616. [CrossRef]
15. Kumar, G.; Jain, S.; Singh, U.P. Stock Market Forecasting Using Computational Intelligence: A Survey. *Arch. Comput. Methods Eng.* **2020**, *28*, 1069–1101. [CrossRef]
16. Hašková, S.; Šuleř, P.; Krulický, T. Advantages of fuzzy approach compared to probabilistic approach in project evaluation. *Entrep. Sustain. Issues* **2021**, *9*, 446–456. [CrossRef]
17. Cheng, S.-H.; Chen, S.-M.; Jian, W.-S. Fuzzy time series forecasting based on fuzzy logical relationships and similarity measures. *Inf. Sci.* **2016**, *327*, 272–287. [CrossRef]

18. Zhang, R.; Ashuri, B.; Deng, Y. A novel method for forecasting time series based on fuzzy logic and visibility graph. *Adv. Data Anal. Classif.* **2017**, *11*, 759–783. [CrossRef]
19. Moshkin, V.; Kurilo, D.; Yarushkina, N. Integration of Fuzzy Ontologies and Neural Networks in the Detection of Time Series Anomalies. *Mathematics* **2023**, *11*, 1204. [CrossRef]
20. Maciel, L.; Ballini, R. Functional Fuzzy Rule-Based Modeling for Interval-Valued Data: An Empirical Application for Exchange Rates Forecasting. *Comput. Econ.* **2020**, *57*, 743–771. [CrossRef]
21. Gündoğdu, F.K.; Kahraman, C. A novel spherical fuzzy analytic hierarchy process and its renewable energy application. *Soft Comput.* **2019**, *24*, 4607–4621. [CrossRef]
22. Ben Jabeur, S.; Hassine, R.B.; Mefteh-Wali, S. Firm financial performance during the financial crisis: A French case study. *Int. J. Financ. Econ.* **2020**, *26*, 2800–2812. [CrossRef]
23. Song, Y.; Liu, T.; Liang, D.; Li, Y.; Song, X. A Fuzzy Stochastic Model for Carbon Price Prediction Under the Effect of Demand-related Policy in China's Carbon Market. *Ecol. Econ.* **2018**, *157*, 253–265. [CrossRef]
24. Hašková, S. Analysis of Prediction of Current Profit and Interval Fuzzy Profit in the Case of Subsidized Projects. In Proceedings of the 13th International Days of Statistics and Economics, Prague, Czech Republic, 5–7 September 2019. [CrossRef]
25. Taliento, M. Corporate Valuation: Looking Beyond the Forecast Period Through New "Fuzzy Lenses". *IEEE Trans. Eng. Manag.* **2019**, *68*, 467–482. [CrossRef]
26. Lo, F.-Y.; Wong, W.-K.; Geovani, J. Optimal combinations of factors influencing the sustainability of Taiwanese firms. *Int. J. Emerg. Mark.* **2021**, *16*, 909–928. [CrossRef]
27. Shao, J.; Liang, C.; Liu, Y.; Xu, J.; Zhao, S. Relief demand forecasting based on intuitionistic fuzzy case-based reasoning. *Socio-Econ. Plan. Sci.* **2020**, *74*, 100932. [CrossRef]
28. Bayram, M.; Akat, M. Market-neutral trading with fuzzy inference, a new method for the pairs trading strategy. *Eng. Econ.* **2019**, *30*, 411–421. [CrossRef]
29. Wu, H.; Long, H.; Wang, Y.; Wang, Y. Stock index forecasting: A new fuzzy time series forecasting method. *J. Forecast.* **2020**, *40*, 653–666. [CrossRef]
30. Chourmouziadis, K.; Chourmouziadou, D.K.; Chatzoglou, P.D. Embedding Four Medium-Term Technical Indicators to an Intelligent Stock Trading Fuzzy System for Predicting: A Portfolio Management Approach. *Comput. Econ.* **2020**, *57*, 1183–1216. [CrossRef]
31. Mohamed, E.A.; Ahmed, I.E.; Mehdi, R.; Hussain, H. Impact of corporate performance on stock price predictions in the UAE markets: Neuro-fuzzy model. *Intell. Syst. Account. Financ. Manag.* **2021**, *28*, 52–71. [CrossRef]
32. Chang, P.-C.; Liu, C.-H. A TSK type fuzzy rule based system for stock price prediction. *Expert Syst. Appl.* **2008**, *34*, 135–144. [CrossRef]
33. Liu, C.-F.; Yeh, C.-Y.; Lee, S.-J. Application of type-2 neuro-fuzzy modeling in stock price prediction. *Appl. Soft Comput.* **2012**, *12*, 1348–1358. [CrossRef]
34. Xie, C.; Rajan, D.; Chai, Q. An interpretable Neural Fuzzy Hammerstein-Wiener network for stock price prediction. *Inf. Sci.* **2021**, *577*, 324–335. [CrossRef]
35. Nasiri, H.; Ebadzadeh, M.M. MFRFNN: Multi-Functional Recurrent Fuzzy Neural Network for Chaotic Time Series Prediction. *Neurocomputing* **2022**, *507*, 292–310. [CrossRef]
36. Kumar, A.; Sharma, A. Systematic literature review of fuzzy logic based text summarization. *Iran. J. Fuzzy Syst.* **2019**, *16*, 45–59. [CrossRef]
37. Talpur, N.; Abdulkadir, S.J.; Alhussian, H.; Hasan, H.; Aziz, N.; Bamhdi, A. A comprehensive review of deep neuro-fuzzy system architectures and their optimization methods. *Neural Comput. Appl.* **2022**, *34*, 1837–1875. [CrossRef]
38. Kar, S.; Das, S.; Ghosh, P.K. Applications of neuro fuzzy systems: A brief review and future outline. *Appl. Soft Comput.* **2014**, *15*, 243–259. [CrossRef]
39. Agrawal, S.C. Deep learning based non-linear regression for Stock Prediction. In *IOP Conference Series: Materials Science and Engineering*; IOP Publishing: Bristol, UK, 2021; Volume 1116, p. 012189.
40. Barapatre, O.; Tete, E.; Sahu, C.L.; Kumar, D.; Kshatriya, H. Stock price prediction using artificial neural network. *Int. J. Adv. Res. Ideas Innov. Technol.* **2018**, *4*, 916–922.
41. Aldhyani, T.H.H.; Alzahrani, A. Framework for Predicting and Modeling Stock Market Prices Based on Deep Learning Algorithms. *Electronics* **2022**, *11*, 3149. [CrossRef]
42. Alkhatib, K.; Khazaleh, H.; Alkhazaleh, H.A.; Alsoud, A.R.; Abualigah, L. A New Stock Price Forecasting Method Using Active Deep Learning Approach. *J. Open Innov. Technol. Mark. Complex.* **2022**, *8*, 96. [CrossRef]
43. Czakon, W.; Niemand, T.; Gast, J.; Kraus, S.; Frühstück, L. Designing coopetition for radical innovation: An experimental study of managers' preferences for developing self-driving electric cars. *Technol. Forecast. Soc. Chang.* **2020**, *155*, 119992. [CrossRef]
44. Yang, Y.; Zhang, K.; Kannan, P. Identifying Market Structure: A Deep Network Representation Learning of Social Engagement. *J. Mark.* **2021**, *86*, 37–56. [CrossRef]
45. Chen, Y.; Chowdhury, S.D.; Donada, C. Mirroring hypothesis and integrality: Evidence from Tesla Motors. *J. Eng. Technol. Manag.* **2019**, *54*, 41–55. [CrossRef]
46. Kozinets, R.V. YouTube utopianism: Social media profanation and the clicktivism of capitalist critique. *J. Bus. Res.* **2019**, *98*, 65–81. [CrossRef]

47. Zadeh, L.A. Fuzzy sets. *Inf. Control.* **1965**, *8*, 338–353. [CrossRef]
48. Novák, V.; Perfilieva, I.; Mockor, J. *Mathematical Principles of Fuzzy Logic*; Springer Science & Business Media: Berlin/Heidelberg, Germany, 2012; Volume 517.
49. Běhounek, L.; Cintula, P. Fuzzy class theory. *Fuzzy Sets Syst.* **2005**, *154*, 34–55. [CrossRef]
50. Hašková, S.; Fiala, P. A fuzzy approach for the estimation of foreign investment risk based on values of rating indices. *Risk Manag.* **2019**, *21*, 183–199. [CrossRef]
51. Bai, Y.; Wang, D. Fundamentals of Fuzzy Logic Control—Fuzzy Sets, Fuzzy Rules and Defuzzifications. In *Advances in Industrial Control*; Springer: London, UK, 2006; pp. 17–36. [CrossRef]
52. Shen, Q.; Chouchoulas, A. A rough-fuzzy approach for generating classification rules. *Pattern Recognit.* **2002**, *35*, 2425–2438. [CrossRef]
53. Boyacioglu, M.A.; Avci, D. An Adaptive Network-Based Fuzzy Inference System (ANFIS) for the prediction of stock market return: The case of the Istanbul Stock Exchange. *Expert Syst. Appl.* **2010**, *37*, 7908–7912. [CrossRef]
54. Alenezy, A.H.; Ismail, M.T.; Al Wadi, S.; Tahir, M.; Hamadneh, N.N.; Jaber, J.J.; Khan, W.A. Forecasting Stock Market Volatility Using Hybrid of Adaptive Network of Fuzzy Inference System and Wavelet Functions. *J. Math.* **2021**, *2021*, 9954341. [CrossRef]
55. Macrotrends Federal Funds Rate—62 Year Historical Chart. 2023. Available online: https://www.macrotrends.net/2015/fed-funds-rate-historical-chart (accessed on 10 December 2022).
56. Yahoo!Finance Tesla, Inc. (TSLA). 2023. Available online: https://finance.yahoo.com/quote/TSLA/history/ (accessed on 5 February 2023).
57. Macrotrends NASDAQ Composite—45 Year Historical Chart. 2023. Available online: https://www.macrotrends.net/1320/nasdaq-historical-chart (accessed on 10 December 2022).
58. Macrotrends Tesla EPS—Earnings per Share 2010–2022 | TSLA. 2023. Available online: https://www.macrotrends.net/stocks/charts/TSLA/tesla/eps-earnings-per-share-diluted (accessed on 10 December 2022).
59. Kannengiesser, U.; Gero, J.S. Empirical evidence for Kahneman's System 1 and System 2 thinking in design. *Hum. Behav. Design.* **2019**, 89–100. [CrossRef]
60. Thagard, P. *Mind: Introduction to Cognitive Science*; MIT Press: Cambridge, MA, USA, 2005.
61. Vimal, K.; Vinodh, S. Application of artificial neural network for fuzzy logic based leanness assessment. *J. Manuf. Technol. Manag.* **2013**, *24*, 274–292. [CrossRef]
62. Nguyen, H.T.; Sugeno, M. (Eds.) *Fuzzy Systems: Modeling and Control*; Springer Science & Business Media: Berlin/Heidelberg, Germany, 2012; Volume 2.
63. Shihabudheen, K.; Pillai, G. Recent advances in neuro-fuzzy system: A survey. *Knowl.-Based Syst.* **2018**, *152*, 136–162. [CrossRef]
64. Mittal, K.; Jain, A.; Vaisla, K.S.; Castillo, O.; Kacprzyk, J. A comprehensive review on type 2 fuzzy logic applications: Past, present and future. *Eng. Appl. Artif. Intell.* **2020**, *95*, 103916. [CrossRef]
65. Dong, Y.; Yan, D.; Almudaifer, A.I.; Yan, S.; Jiang, Z.; Zhou, Y. Belt: A pipeline for stock price prediction using news. In Proceedings of the 2020 IEEE International Conference on Big Data, Online, 10–13 December 2020; pp. 1137–1146.
66. Velu, S.R.; Ravi, V.; Tabianan, K. Multi-Lexicon Classification and Valence-Based Sentiment Analysis as Features for Deep Neural Stock Price Prediction. *Sci* **2023**, *5*, 8. [CrossRef]
67. Bhadkamar, A.; Bhattacharya, S. Tesla Inc. Stock Prediction using Sentiment Analysis. *Australas. Account. Bus. Financ. J.* **2022**, *16*, 52–66. [CrossRef]
68. Ij, H. Statistics versus machine learning. *Nat. Methods* **2018**, *15*, 233.

Disclaimer/Publisher's Note: The statements, opinions and data contained in all publications are solely those of the individual author(s) and contributor(s) and not of MDPI and/or the editor(s). MDPI and/or the editor(s) disclaim responsibility for any injury to people or property resulting from any ideas, methods, instructions or products referred to in the content.

Article

Simulations and Bisimulations between Weighted Finite Automata Based on Time-Varying Models over Real Numbers

Predrag S. Stanimirović [1,2], Miroslav Ćirić [1], Spyridon D. Mourtas [2,3], Pavle Brzaković [4] and Darjan Karabašević [4,5,*]

1. Faculty of Sciences and Mathematics, University of Niš, Višegradska 33, 18000 Niš, Serbia; pecko@pmf.ni.ac.rs (P.S.S.); miroslav.ciric@pmf.edu.rs (M.Ć.)
2. Laboratory "Hybrid Methods of Modelling and Optimization in Complex Systems", Siberian Federal University, Prosp. Svobodny 79, Krasnoyarsk 660041, Russia; spirmour@econ.uoa.gr
3. Department of Economics, Division of Mathematics-Informatics and Statistics-Econometrics, National and Kapodistrian University of Athens, Sofokleous 1 Street, 10559 Athens, Greece
4. Faculty of Applied Management, Economics and Finance, University Business Academy in Novi Sad, Jevrejska 24, 11000 Belgrade, Serbia; pavle.brzakovic@mef.edu.rs
5. College of Global Business, Korea University, Sejong 30019, Republic of Korea
* Correspondence: darjan.karabasevic@mef.edu.rs

Citation: Stanimirović, P.S.; Ćirić, M.; Mourtas, S.D.; Brzaković, P.; Karabašević, D. Simulations and Bisimulations between Weighted Finite Automata Based on Time-Varying Models over Real Numbers. *Mathematics* **2024**, *12*, 2110. https://doi.org/10.3390/math12132110

Academic Editor: Francisco Rodrigues Lima Junior

Received: 5 June 2024
Revised: 27 June 2024
Accepted: 3 July 2024
Published: 5 July 2024

Copyright: © 2024 by the authors. Licensee MDPI, Basel, Switzerland. This article is an open access article distributed under the terms and conditions of the Creative Commons Attribution (CC BY) license (https://creativecommons.org/licenses/by/4.0/).

Abstract: The zeroing neural network (ZNN) is an important kind of continuous-time recurrent neural network (RNN). Meanwhile, the existence of forward and backward simulations and bisimulations for weighted finite automata (WFA) over the field of real numbers has been widely investigated. Two types of quantitative simulations and two types of bisimulations between WFA are determined as solutions to particular systems of matrix and vector inequalities over the field of real numbers $\mathbb{R}$. The approach used in this research is unique and based on the application of a ZNN dynamical evolution in solving underlying matrix and vector inequalities. This research is aimed at the development and analysis of four novel ZNN dynamical systems for addressing the systems of matrix and/or vector inequalities involved in simulations and bisimulations between WFA. The problem considered in this paper requires solving a system of two vector inequalities and a couple of matrix inequalities. Using positive slack matrices, required matrix and vector inequations are transformed into corresponding equations and then the derived system of matrix and vector equations is transformed into a system of linear equations utilizing vectorization and the Kronecker product. The solution to the ZNN dynamics is defined using the pseudoinverse solution of the generated linear system. A detailed convergence analysis of the proposed ZNN dynamics is presented. Numerical examples are performed under different initial state matrices. A comparison between the ZNN and linear programming (LP) approach is presented.

Keywords: weighted finite automata; Zhang neural network; forward simulation; backward simulation; pseudoinverse

MSC: 65F20; 68T05; 68Q70

1. Preliminaries on Weighted Finite Automata and Zeroing Neural Networks

Simulations between WFA ensure its containment, while bisimulations ensure the equivalence of WFA. As a result of the transition from various boolean to quantitative systems, both simulations and bisimulations become quantitative. Corresponding models are based on the use of matrices whose entries supply a quantitative measurement of the relationship between states of underlying systems.

Hereafter, $\mathbb{R}$ denotes the field of real numbers, and $\mathbb{N}$ denotes the set of natural numbers without zero, while the set of all positive real numbers is denoted by $\mathbb{R}_+$. Additionally, $X = \{x_1, \ldots, x_r\}$ is a non-empty finite set with k elements, where $k \in \mathbb{N}$, called an *alphabet*, while $X^+ = \{x_1 x_2 \ldots x_s \mid s \in \mathbb{N}, x_1, x_2, \ldots, x_s \in X\}$ is the set of all finite sequences of elements of X, which are called *words* over the alphabet X, and $X^* = X^+ \cup \{\varepsilon\}$, where $\varepsilon \notin X^+$ is a symbol that denotes the *empty word* of length 0. With respect to the conventional concatenation operation on words (sequences), X^+ forms a semigroup, while X^* is a structure representing a monoid with the identity element ε.

A *weighted finite automaton* over the field of real numbers $\mathbb{R}$ and the alphabet X is defined as a quadruple $\mathcal{A} = \left(m, \sigma^A, \{M_x^A\}_{x \in X}, \tau^A\right)$, where $m \in \mathbb{N}$ denotes the *dimension* of $\mathcal{A}$; $\sigma^A \in \mathbb{R}^{1 \times m}$, $\tau^A \in \mathbb{R}^{m \times 1}$ are the *initial vector* and *terminal vector*, respectively, and $\{M_x^A\}_{x \in X} \subset \mathbb{R}^{m \times m}$ is a collection of *transition matrices*. The initial vector σ^A is treated as a row vector, while the terminal vector τ^A is treated as a column vector. The behavior of a weighted finite automaton is expressed as the product σ^A, representing the initial weights, matrices $\{M_x^A\}_{x \in X}$ representing the weights of the transitions induced by input letters, and the column vector τ^A representing the terminal weights.

The collection $\{M_x^A\}_{x \in X}$ is extended up to a collection $\{M_u^A\}_{u \in X^*} \subset \mathbb{R}^{m \times m}$ of *compound transition matrices* expressed as

$$M_u^A = \begin{cases} I_m, & u = \varepsilon, \\ M_{x_1}^A M_{x_2}^A \cdots M_{x_s}^A, & u = x_1 x_2 \cdots x_s \in X^+, \end{cases} \quad (1)$$

where I_m denotes the $m \times m$ identity matrix. The matrices M_u^A, $u \in X^*$, defined in (1), are known as the *compound transition matrices* of $\mathcal{A}$. The multiplication of transition matrices carry numerical values over $\mathbb{R}$, known as weights. A function $f : X^* \to \mathbb{R}$ is called a *word function*. In particular, each weighted finite automaton $\mathcal{A} = \left(m, \sigma^A, \{M_x^A\}_{x \in X}, \tau^A\right)$ gives rise to a word function $[\![\mathcal{A}]\!] : X^* \to \mathbb{R}$ defined as follows:

$$[\![\mathcal{A}]\!](u) = \begin{cases} \sigma^A M_u^A \tau^A = \sigma^A M_{x_1}^A M_{x_2}^A \cdots M_{x_s}^A \tau^A, & u = x_1 x_2 \ldots x_s \in X^+, \\ \sigma^A M_\varepsilon^A \tau^A = \sigma^A \tau^A, & u = \varepsilon. \end{cases} \quad (2)$$

The word function $[\![\mathcal{A}]\!]$ defined in (2) is called the *behavior* of $\mathcal{A}$, or a *word function computed by* $\mathcal{A}$. The behavior of an automaton is a mapping that relates a weight to words over a semiring.

Consider the weighted finite automata (WFA) $\mathcal{A} = \left(m, \sigma^A, \{M_x^A\}_{x \in X}, \tau^A\right)$ and $\mathcal{B} = \left(n, \sigma^B, \{M_x^B\}_{x \in X}, \tau^B\right)$ over the field of real numbers $\mathbb{R}$ and X. The following notations are used:

$[\![\mathcal{A}]\!] = [\![\mathcal{B}]\!] \iff [\![\mathcal{A}]\!](u) = [\![\mathcal{B}]\!](u)$, for every $u \in X^*$;

$[\![\mathcal{A}]\!] \leqslant [\![\mathcal{B}]\!] \iff [\![\mathcal{A}]\!](u) \leqslant [\![\mathcal{B}]\!](u)$, for every $u \in X^*$.

WFA $\mathcal{A}$ and $\mathcal{B}$ over $\mathbb{R}$ and the alphabet X are said to be *equivalent* if $[\![\mathcal{A}]\!] = [\![\mathcal{B}]\!]$. On the other hand, if $[\![\mathcal{A}]\!] \leqslant [\![\mathcal{B}]\!]$, then $\mathcal{A}$ is said to be *contained in* $\mathcal{B}$. The problem of determining whether WFA are equivalent is called the *equivalence problem*, and the problem of determining whether one of two WFA is contained in another is called the *containment problem*. A solution to the equivalence problem decides whether two WFA compute the same word function. On the other hand, a solution to the containment problem determines whether the word function computed by one WFA is less than or equal to the word function corresponding to the other WFA

A matrix (resp. vector) is said to be a *positive matrix* (resp. *positive vector*) if all its entries are positive real numbers, and a weighted finite automaton $\mathcal{A}$ is said to be a *positive automaton* if its initial and terminal vectors, as well as all its transition matrices, are positive.

Weighted automata have been applied to describe quantitative properties in various systems, as well as to represent probabilistic models, image compression, speech recognition, and finite representations of formal languages. Context–free grammars are used in the development of programming languages as well as in artificial intelligence.

The theoretical foundations of current investigations involve two types of simulations and two types of bisimulations defined in [1], in the general context of WFA over a semiring. The approach we use consists of defining quantitative simulations and bisimulations as matrices that are solutions to certain systems of matrix inequations. Such an approach was introduced in [2], where quantitative simulations and bisimulations between fuzzy finite automata were introduced and their basic properties were examined. Algorithms for testing their existence were developed in [3]. The same algorithms compute the greatest simulations and bisimulations in cases when they exist. Then, the same approach was applied to the study of bisimulations and simulations for non-deterministic automata [4], WFA over an additively idempotent semiring [5], and max-plus automata [6], as well as for WFA over an arbitrary semiring [1,7], which encompass all the previous ones. It turns out that an almost identical methodology can also be applied to social networks [8]. In [9], it was proven that two probabilistic finite automata are equivalent if and only if there is a bisimulation between them, where the bisimulation is defined as a classical binary relation between the vector spaces corresponding to those automata.

In the present paper, we investigate forward and backward simulations and bisimulations for WFA over the field of real numbers. It is worth noting that there are some very important specifics in this case. For most WFA types, the *problem of equivalence* (determining whether two automata compute the same word function) and the *minimization problem* (determining an automaton with the minimal number of states equivalent to a given automaton) are computationally hard. In these cases, bisimulations have two very important roles. The first role is to provide an efficient procedure for witnessing the existence of the equivalence of two automata, and the second one is to provide an efficient way to construct an automaton equivalent to a given one, with a not necessarily minimal but reasonably smaller number of states. However, it is not the case with WFA over the field of real numbers, for which there are efficient algorithms for testing the equivalence and performing minimization. Despite this observation, the importance of bisimulations for these automata is not diminished. Bisimulations are still needed as a means of determining the measure of similarity between the states of different automata, which algorithms for testing the equivalence are unable to do. In the context of weighted automata over the field of real numbers, such measures have already been studied in [10] by means of bisimulation seminorms and pseudometrics, and in [11] by means of linear bisimulations; in our upcoming research, we will deal with the relationships between bisimulation seminorms, linear bisimulations, and our concepts of bisimulations.

Following the definitions of simulations and bisimulations over various algebraic structures, an analogous approach has been used in defining simulations and bisimulations for WFA over the field of real numbers. The problem of simulations and bisimulations for WFA over the field of real numbers reduces to the system of two vector inequations and a number of matrix inequations. There is a notable lack of numerical methods for solving simulation and bisimulationproblems. Urabe and Hasuo proposed the idea of reducing the problem of testing the existence of simulations to the problem of linear programming (LP) and implemented it in [7] (Section 5). Seen more generally, the research described in this paper shows that the ZNN design is usable in solving systems of matrix and vector inequations in linear algebra. Our goal is to show that the zeroing neural network (ZNN) dynamics are an effective tool to decide on the containment or equivalence between WFA. A comparison between the ZNN and LP approach is presented.

On the other hand, the application of dynamical systems is a robust tool for solving various matrix algebra problems, primarily owing to the global exponential convergence, parallel distributed essence, convenience of hardware implementation, suitability for online computations involving TV objects, and possibility of providing convergence in a finite time frame [12,13]. First, ZNN models have been used to solve the TV matrix inversion problem [14]. Standard and finite-time convergent ZNN dynamical systems aimed at solving time-varying (TV) linear matrix equations have been widely investigated [12,15–18]. The applications of ZNN design, mainly focusing on robot manipulator path tracking, motion planning, and chaotic systems, were surveyed in [19]. ZNN dynamical systems for solving TV linear matrix–vector inequalities (TVLMVI) and TV linear matrix inequalities (TVLMI) have been broadly investigated [12,15,20–27]. Moreover, various ZNN models for solving TVLMI have been applied, mainly in obstacle avoidance for redundant robots and robot manipulator control [12,28,29]. Typically, TVLMVI and TVLMI of type "$\leq$" are solved by utilizing an additional matrix or vector of appropriate dimensions with non-negative entries. A TV matrix inequality of the Stein form $A(t)X(t)B(t) + X(t) \leq C(t)$ was considered in [21]. A TVLMVI problem of the general form $A(t)x(t) \leq b(t)$ was considered in [24,26,27]. Two ZNN models for solving systems of two TVLMVI were developed in [15]. In [22], the authors proposed ZNNs for solving TV nonlinear inequalities. Finite-time dynamics for solving general TVLMVI $A(t)X(t)B(t) \leq C(t)$ were proposed in [25]. A comparison between ZNN and gradient-based networks for solving $A(t)x(t) \leq b(t)$ was investigated in [23]. The computational time for solving TV equations increases due to the large number of calculations of TV requirements [30].

The problem under consideration is more complex because it requires us to solve systems of linear matrix and vector inequations. The structure of ZNN models developed in the current research is based on composite models with a prescribed number of error functions in matrix form and two in vector form. The ZNN dynamics aim to force the convergence of the involved error functions to zero over the considered time interval [13]. But the ZNN model in this paper aims to solve several matrix–vector equations that are inconsistent in the general case. Our strategy is to utilize ZNN neurodynamics to generate simulations between two WFA with weights over real numbers. In this way, our objective involves the topic of numerical linear algebra.

This research is aimed at the development and analysis of four novel ZNN models for addressing the systems of matrix and vector inequalities involved in simulations between WFA. The problem considered in this paper is specific and complex, and it requires solving a system of two vector inequalities and a couple of matrix inequalities. Using positive slack matrices, matrix and vector inequalities are transformed into corresponding equalities. In this case, it is useful to utilize the development of ZNN dynamics based on several inequalities and Zhang error functions. ZNN algorithms established upon a few error functions have been investigated in several studies, such as [31–34]. Our motivation for the application of ZNN arises from a verified fact that it is a powerful tool for solving various matrix algebra models, possessing global exponential convergence and a parallel distributed structure [12,13]. Therefore, it is interesting to construct the ZNN evolution for such a problem and study its behavior. A detailed convergence analysis is considered. Numerical examples are performed with different initial state matrices.

The main results are emphasized as follows.

(1) Two types of quantitative simulations and two types of bisimulations between WFA are determined as solutions to particular systems of several matrix and two vector inequations over $\mathbb{R}$.
(2) The approach used to solve the problem of simulations and bisimulations in this research is unique and based on the application of the ZNN dynamical evolution in solving underlying matrix and vector inequations.
(3) A detailed convergence analysis of the proposed ZNN dynamics is presented.
(4) Numerical examples are performed under different initial state matrices, and a comparison between the ZNN and LP approach is presented.

The overall organization of the sections is as follows. Preliminaries on WFA and ZNN are presented in Section 1. Global results are highlighted in the same section. Two types of simulations and four types of bisimulations proposed in [1] in the general context of WFA over a semiring are generalized in the context of WFA over the field of real numbers in Section 2. ZNN designs for simulations and bisimulations of WFA over real numbers are presented in Section 3. Section 4 is aimed at testing the developed ZNN dynamical systems and making comparisons with the LP solver. Concluding remarks are given in Section 5.

2. Simulations and Bisimulations of WFA over Real Numbers

As a continuation of the research presented in [1], here we correspondingly introduce definitions of two types of simulations and two types of bisimulations in the context of WFA over $\mathbb{R}$. For this purpose, consider two WFA $\mathcal{A} = (m, \sigma^A, \{M_x^A\}_{x \in X}, \tau^A)$ and $\mathcal{B} = (n, \sigma^B, \{M_x^B\}_{x \in X}, \tau^B)$ over the field of real numbers $\mathbb{R}$ and the alphabet X. A matrix $U \in \mathbb{R}^{m \times n}$ is called a *forward simulation* between $\mathcal{A}$ and $\mathcal{B}$ if it satisfies the following conditions with respect to U:

$$
\begin{aligned}
&\text{(fs-1)} \quad \sigma^A \leqslant \sigma^B U^\top \\
&\text{(fs-2)} \quad U^\top M_x^A \leqslant M_x^B U^\top \quad (\forall x \in X) \\
&\text{(fs-3)} \quad U^\top \tau^A \leqslant \tau^B,
\end{aligned}
\tag{3}
$$

and it is termed as *backward simulation* between $\mathcal{A}$ and $\mathcal{B}$ if it fulfills

$$
\begin{aligned}
&\text{(bs-1)} \quad \tau^A \leqslant U \tau^B \\
&\text{(bs-2)} \quad M_x^A U \leqslant U M_x^B \quad (\forall x \in X) \\
&\text{(bs-3)} \quad \sigma^A U \leqslant \sigma^B.
\end{aligned}
\tag{4}
$$

Our intention is to apply the notion of transposed automaton from [35] to reverse the transitions' flow direction. If both U and $U^\top$ are forward simulations between $\mathcal{A}$ and $\mathcal{B}$ and vice versa, i.e., if they fulfil

$$
\begin{aligned}
&\text{(fb-1)} \quad \sigma^A \leqslant \sigma^B U^\top, \quad \sigma^B \leqslant \sigma^A U \\
&\text{(fb-2)} \quad U^\top M_x^A \leqslant M_x^B U^\top, \quad U M_x^B \leqslant M_x^A U \quad (\forall x \in X) \\
&\text{(fb-3)} \quad U^\top \tau^A \leqslant \tau^B, \quad U \tau^B \leqslant \tau^A
\end{aligned}
\tag{5}
$$

then U is termed as a *forward bisimulation* between $\mathcal{A}$ and $\mathcal{B}$, and if both U and $U^\top$ are backward simulations between $\mathcal{A}$ and $\mathcal{B}$ and vice versa, i.e., if they satisfy

$$
\begin{aligned}
&\text{(bb-1)} \quad \tau^A \leqslant U \tau^B, \quad \tau^B \leqslant U^\top \tau^A \\
&\text{(bb-2)} \quad M_x^A U \leqslant U M_x^B, \quad M_x^B U^\top \leqslant U^\top M_x^A \quad (\forall x \in X) \\
&\text{(bb-3)} \quad \sigma^A U \leqslant \sigma^B, \quad \sigma^B U^\top \leqslant \sigma^A
\end{aligned}
\tag{6}
$$

then U is known as a *backward bisimulation* between $\mathcal{A}$ and $\mathcal{B}$.

It is important to note that, for any $\omega \in \{\text{fs}, \text{bs}, \text{fb}, \text{bb}\}$, the conditions ($\omega$-1), ($\omega$-2), and ($\omega$-3) can be treated a system of matrix inequations with the unknown matrix U, and simulations or bisimulations of type ω are precisely solutions to this system. This is extremely important because simulations between weighted automata over the field of real numbers are searched for by solving the corresponding systems of matrix inequalities.

Another important note is that the main role of simulations is to witness containment between automata $\mathcal{A}$ and $\mathcal{B}$, while the main role of bisimulations is to witness equivalence between $\mathcal{A}$ and $\mathcal{B}$. However, forward and backward simulations and bisimulations are defined by matrix inequations. On that note, in order to prove that simulations achieve containment and bisimulations achieve equivalence, we need the inequations to be preserved by multiplying, on either side, by the transition matrices, as well as by the initial and terminal vectors. Multiplication by matrices and vectors containing negative entries can violate inequalities, and, therefore, in order for simulations and bisimulations defined by systems of inequations to make full sense, we consider these types of bisimulations and simulations only between positive automata.

Theorem 1 is a modified version of [1] (Theorem 1).

Theorem 1. *The following statements are valid for positive WFA $\mathcal{A}$ and $\mathcal{B}$ over $\mathbb{R}$:*

(a) *For $\omega \in \{\text{fs}, \text{bs}\}$, if there is a simulation of type ω between $\mathcal{A}$ and $\mathcal{B}$, then $[\![\mathcal{A}]\!] \leqslant [\![\mathcal{B}]\!]$.*

(b) *For $\omega \in \{\text{fb}, \text{bb}\}$, if there is a bisimulation of type ω between $\mathcal{A}$ and $\mathcal{B}$, then $[\![\mathcal{A}]\!] = [\![\mathcal{B}]\!]$.*

The modification is reflected in the following. A slightly different version of Theorem 1 was proved in [1] [Theorem 1] for WFA over a positive semiring. Theorem 1 could also be formulated for $\mathcal{A}$ and $\mathcal{B}$ as WFA over the positive semiring $\mathbb{R}_+$ of nonnegative real numbers, but such a formulation would mean that the simulations and bisimulations between $\mathcal{A}$ and $\mathcal{B}$ should also be over the semiring $\mathbb{R}_+$, that is, they should be positive matrices, which is not necessary. Namely, for positive WFA over an arbitrary ordered semiring (not necessarily positive), the proof of [1] (Theorem 1) also holds for simulations and bisimulations that contain negative entries, and Theorem 1 is formulated to allow for such simulations and bisimulations as well.

As this article is primarily concerned with solving systems of matrix inequalities, nothing important will change if we consider the more general case and allow the transition matrices, as well as the initial and terminal vectors, to have negative entries, which is performed below. On the other hand, in some applications of simulations and bisimulations, for example in the dimensionality reduction for WFA, there is a need to find positive solutions of the considered systems of matrix inequalities. For this reason, we consider systems with an additional condition requiring the positivity of the solution. It should be noted that the proposed procedures for solving the systems remain valid even in the case when this condition is omitted, and in the same way, in that case we obtain solutions that do not have to be positive.

3. ZNN Designs for Simulations and Bisimulations of WFA over Real Numbers

This section defines and analyzes four novel ZNN models for addressing the systems of inequations (3)–(6). For the remainder of this section, let $\mathcal{A} = \left(m, \sigma^A, \left\{M_{x_i}^A\right\}_{x_i \in X}, \tau^A\right)$ and $\mathcal{B} = \left(n, \sigma^B, \left\{M_{x_i}^B\right\}_{x_i \in X}, \tau^B\right)$ be two WFA over $\mathbb{R}$, where $M_{x_i}^A \in \mathbb{R}^{m \times m}$, $\sigma^A \in \mathbb{R}^{1 \times m}$, $\tau^A \in \mathbb{R}^{m \times 1}$ and $M_{x_i}^B \in \mathbb{R}^{n \times n}$, $\sigma^B \in \mathbb{R}^{1 \times n}$, $\tau^B \in \mathbb{R}^{n \times 1}$ with $i = 1, \ldots r$.

Also, it is crucial to mention that the process of building a ZNN model usually involves two primary steps. The error matrix equation's (EME) function, $E(t)$, must be initially declared. Secondly, the dynamic system represented by the continuous differential equation of the general form

$$\dot{E}(t) = -\lambda E(t), \tag{7}$$

needs to be employed. The dynamical evolution (7) relates the time derivative $\dot{E}(t)$ to $E(t)$ in proportion to the positive real coefficient λ. The convergence rate of the dynamical system (7) is altered by manipulating the parameter $\lambda \in \mathbb{R}^+$. More precisely, with increasing values of λ, any ZNN model converges even faster [13,36,37]. The primary goal of the dynamics (7) is to force $E(t)$ to approach 0 as $t \to \infty$. The continuous learning principle that emerges from the EME's construction in Equation (7) is used to manage this goal. EME is, therefore, considered as a tracking indication in the context of the ZNN model's learning.

Special attention should be paid to a few notations that are used in the remainder of this work. The $p \times 1$ matrices with all ones and all zeros as entries are indicated by $\mathbf{1}_p$ and $\mathbf{0}_p$, whereas the $p \times r$ matrices with all ones and all zeros as entries are indicated by $\mathbf{1}_{p,r}$ and $\mathbf{0}_{p,r}$. Furthermore, the $p \times p$ identity matrix is indicated by I_p, whereas $\text{vec}()$, $\otimes$, $\odot$, $()^{\odot}$, $()^\dagger$, and $\|\|_F$ stand for the vectorization process, the Kronecker product, the Hadamard (or elementwise) product, the Hadamard exponential, pseudoinversion, and the matrix Frobenius norm, respectively. Finally, $\text{rand}(m,n)$ denotes an $m \times n$ matrix whose entries consist of random numbers.

3.1. The ZNN-fs Model

In line with (3), the following group of inequations must be satisfied:

$$\begin{cases} U^{\mathrm{T}}(t)\tau^A - \tau^B \leqslant \mathbf{0}_n, \\ \sigma^A - \sigma^B U^{\mathrm{T}}(t) \leqslant \mathbf{0}_m^{\mathrm{T}}, \\ U^{\mathrm{T}}(t)M_{x_i}^A - M_{x_i}^B U^{\mathrm{T}}(t) \leqslant \mathbf{0}_{n,m}, \ i=1,\ldots,r, \\ U(t) \geqslant \mathbf{0}_{m,n}, \end{cases} \quad (8)$$

with respect to an unknown matrix $U(t) \in \mathbb{R}^{m \times n}$. Utilizing the vectorization in conjunction with the Kronecker product, the system (8) is reformulated into the vector inequations form

$$\begin{cases} \left((\tau^A)^{\mathrm{T}} \otimes I_n\right)\mathrm{vec}(U^{\mathrm{T}}(t)) - \tau^B \leqslant \mathbf{0}_n, \\ -\left(I_m \otimes \sigma^B\right)\mathrm{vec}(U^{\mathrm{T}}(t)) + (\sigma^A)^{\mathrm{T}} \leqslant \mathbf{0}_m, \\ \left((M_{x_i}^A)^{\mathrm{T}} \otimes I_n - I_m \otimes M_{x_i}^B\right)\mathrm{vec}(U^{\mathrm{T}}(t)) \leqslant \mathbf{0}_{mn}, \ i=1,\ldots,r, \\ -\mathrm{vec}(U(t)) \leqslant \mathbf{0}_{mn}. \end{cases} \quad (9)$$

To calculate $U(t)$ more efficiently, (9) must be simplified. Thus, the vectorization-related Lemma 1 derived from [38] is given.

Lemma 1. *The vectorization* $\mathrm{vec}(W^{\mathrm{T}}) \in \mathbb{R}^{mn}$ *of the transpose* W^{T} *of* $W \in \mathbb{R}^{m \times n}$ *is defined by*

$$\mathrm{vec}(W^{\mathrm{T}}) = P\,\mathrm{vec}(W), \quad (10)$$

where $P \in \mathbb{R}^{mn \times mn}$ *is a constant permutation matrix that depends on the number of columns n and number of rows m in W.*

The algorithmic procedure for generating the permutation matrix P in (10) is presented in the following Algorithm 1.

Algorithm 1 The permutation matrix P formation.

Input: The number of rows m and columns n of a matrix $W \in \mathbb{R}^{m \times n}$.
1: **procedure** PERM_MAT(m,n)
2:    Put $g = \mathrm{eye}(mn)$ and $W = \mathrm{reshape}(1:mn, n, m)$
3:    **return** $P = g(:,\mathrm{reshape}(W^{\mathrm{T}}, 1, mn))$
4: **end procedure**
Output: P

Using the permutation matrix P for generating $\mathrm{vec}(U^{\mathrm{T}}(t))$, inequations (9) can be rewritten in the form

$$\begin{cases} \left((\tau^A)^{\mathrm{T}} \otimes I_n\right)P\,\mathrm{vec}(U(t)) - \tau^B \leqslant \mathbf{0}_n, \\ -\left(I_m \otimes \sigma^B\right)P\,\mathrm{vec}(U(t)) + (\sigma^A)^{\mathrm{T}} \leqslant \mathbf{0}_m, \\ \left((M_{x_i}^A)^{\mathrm{T}} \otimes I_n - I_m \otimes M_{x_i}^B\right)P\,\mathrm{vec}(U(t)) \leqslant \mathbf{0}_{mn}, \ i=1,\ldots,r, \\ -\mathrm{vec}(U(t)) \leqslant \mathbf{0}_{mn}, \end{cases} \quad (11)$$

wherein the last constraint imposes non-negativity on the solution. The corresponding block matrix form of (11) is given by

$$L_{fs}\,\mathrm{vec}(U(t)) - \mathbf{b}_{fs} \leqslant \mathbf{0}_z, \quad (12)$$

such that $z = (r+1)mn + m + n$ and

$$L_{fs} = \begin{bmatrix} ((\tau^A)^T \otimes I_n)P \\ -(I_m \otimes \sigma^B)P \\ W_{fs} \\ -I_{mn} \end{bmatrix} \in \mathbb{R}^{z \times mn}, \quad \mathbf{b}_{fs} = \begin{bmatrix} \tau^B \\ -(\sigma^A)^T \\ \mathbf{0}_{(r+1)mn} \end{bmatrix} \in \mathbb{R}^z, \quad W_{fs} = \begin{bmatrix} ((M^A_{x_1})^T \otimes I_n - I_m \otimes M^B_{x_1})P \\ ((M^A_{x_2})^T \otimes I_n - I_m \otimes M^B_{x_2})P \\ \dots \\ ((M^A_{x_r})^T \otimes I_n - I_m \otimes M^B_{x_r})P \end{bmatrix} \in \mathbb{R}^{rmn \times mn}. \tag{13}$$

Then, considering the vector of slack variables $K(t) = \begin{bmatrix} k_1(t) \\ \dots \\ k_z(t) \end{bmatrix} \in \mathbb{R}^z$, the inequation (12) can be converted into the corresponding equation

$$L_{fs}\,\text{vec}(U(t)) - \mathbf{b}_{fs} + K^{\odot 2}(t) = \mathbf{0}_z, \tag{14}$$

in which $K^{\odot 2}(t) = \begin{bmatrix} k_1^2(t) \\ \dots \\ k_z^2(t) \end{bmatrix}$ is the time-varying term with secured non-negative entries.

Thereafter, the ZNN approach considers the following EME, which is based on (12), to simultaneously satisfy all the inequations in (8):

$$E_{fs}(t) = L_{fs}\,\text{vec}(U(t)) - \mathbf{b}_{fs} + K^{\odot 2}(t), \tag{15}$$

where $U(t)$ and $K(t)$ are the unknown matrices that need to be found. The ZNN design (7) exploits the first time derivative of (15)

$$\dot{E}_{fs}(t) = L_{fs}\text{vec}(\dot{U}(t)) + 2(I_z \odot K(t))\dot{K}(t). \tag{16}$$

Combining Equations (15) and (16) with the generic ZNN design (7), we obtain

$$L_{fs}\text{vec}(\dot{U}(t)) + 2(I_z \odot K(t))\dot{K}(t) = -\lambda E_{fs}(t). \tag{17}$$

As a result, setting

$$H_{fs} = \begin{bmatrix} L_{fs} & 2(I_z \odot K(t)) \end{bmatrix} \in \mathbb{R}^{z \times (mn+z)}, \quad \dot{\mathbf{x}}(t) = \begin{bmatrix} \text{vec}(\dot{U}(t)) \\ \dot{K}(t) \end{bmatrix} \in \mathbb{R}^{mn+z}, \quad \mathbf{x}(t) = \begin{bmatrix} \text{vec}(U(t)) \\ K(t) \end{bmatrix} \in \mathbb{R}^{mn+z},$$

the next system of linear equations with respect to $\dot{\mathbf{x}}$ is obtained:

$$H_{fs}\dot{\mathbf{x}} = -\lambda E_{fs}(t). \tag{18}$$

The ZNN dynamics are applicable in solving (18) if the mass matrix H_{fs} is invertible. To avoid this restriction, it is appropriate to use the pseudoinverse (best approximate) solution

$$\dot{\mathbf{x}} = H_{fs}^{\dagger}\left(-\lambda E_{fs}(t)\right). \tag{19}$$

An appropriate ode MATLAB R2022a solver can be used to handle the ZNN dynamics (19), additionally referred to as the ZNN-fs model. The ZNN-fs model's convergence and stability investigation is shown in Theorem 2.

Theorem 2. Let $\mathcal{A} = \left(m, \sigma^A, \{M^A_{x_i}\}_{x_i \in X}, \tau^A\right)$ and $\mathcal{B} = \left(n, \sigma^B, \{M^B_{x_i}\}_{x_i \in X}, \tau^B\right)$ be the WFA over $\mathbb{R}$ and the alphabet $X = \{x_1, \dots, x_r\}$, where $M^A_{x_i} \in \mathbb{R}^{m \times m}$, $\sigma^A \in \mathbb{R}^{1 \times m}$, $\tau^A \in \mathbb{R}^{m \times 1}$ and $M^B_{x_i} \in \mathbb{R}^{n \times n}$, $\sigma^B \in \mathbb{R}^{1 \times n}$, $\tau^B \in \mathbb{R}^{n \times 1}$ with $i = 1, \dots, r$. The dynamics (17) in linewith the ZNN method (7) lead to the theoretical solution (TSOL), determined by $\mathbf{x}_S(t) = \begin{bmatrix} \text{vec}(U_S(t))^T & K_S^T(t) \end{bmatrix}^T$, which is stable according to Lyapunov.

Proof. Let

$$\begin{cases} U_{\mathcal{S}}^{\text{T}}(t)\tau^A - \tau^B \leqslant \mathbf{0}_n, \\ \sigma^A - \sigma^B U_{\mathcal{S}}^{\text{T}}(t) \leqslant \mathbf{0}_m^{\text{T}}, \\ U_{\mathcal{S}}^{\text{T}}(t)M_{x_i}^A - M_{x_i}^B U_{\mathcal{S}}^{\text{T}}(t) \leqslant \mathbf{0}_{n,m}, \ i = 1, \ldots, r, \\ U_{\mathcal{S}}(t) \geqslant \mathbf{0}_{m,n}. \end{cases} \quad (20)$$

Using vectorization, Kronecker product, and the permutation matrix P for constructing $\text{vec}(U^{\text{T}}(t))$, defined by Algorithm 1, the system (20) is reformulated as

$$\begin{cases} \left((\tau^A)^{\text{T}} \otimes I_n\right) P \text{vec}(U_{\mathcal{S}}(t)) - \tau^B \leqslant \mathbf{0}_n, \\ -\left(I_m \otimes \sigma^B\right) P \text{vec}(U_{\mathcal{S}}(t)) + (\sigma^A)^{\text{T}} \leqslant \mathbf{0}_m, \\ \left((M_{x_i}^A)^{\text{T}} \otimes I_n - I_m \otimes M_{x_i}^B\right) P \text{vec}(U_{\mathcal{S}}(t)) \leqslant \mathbf{0}_{mn}, \ i = 1, \ldots, r, \\ -\text{vec}(U_{\mathcal{S}}(t)) \leqslant \mathbf{0}_{mn}. \end{cases} \quad (21)$$

The equivalent form of (21) is

$$L_{fs} \text{vec}(U_{\mathcal{S}}(t)) - \mathbf{b}_{fs} \leqslant \mathbf{0}_z \quad (22)$$

where L_{fs} and $\mathbf{b}_{fs}$ are declared in Equation (13). Then, considering the slack variable $K_{\mathcal{S}}(t) \in \mathbb{R}^z$, the inequation (22) can be converted into the equation

$$L_{fs} \text{vec}(U_{\mathcal{S}}(t)) - \mathbf{b}_{fs} + K_{\mathcal{S}}^{\odot 2}(t)(t) = \mathbf{0}_z,$$

in which $K_{\mathcal{S}}^{\odot 2}(t)$ is always a non-negative time-varying term.

The substitution

$$\mathbf{x}_{\mathcal{O}}(t) := -\mathbf{x}(t) + \mathbf{x}_{\mathcal{S}}(t) = \begin{bmatrix} -\text{vec}(U(t)) + \text{vec}(U_{\mathcal{S}}(t)) \\ -K(t) + K_{\mathcal{S}}(t) \end{bmatrix} := \begin{bmatrix} \text{vec}(U_{\mathcal{O}}(t)) \\ K_{\mathcal{O}}(t) \end{bmatrix}$$

gives

$$\mathbf{x}(t) = \mathbf{x}_{\mathcal{S}}(t) - \mathbf{x}_{\mathcal{O}}(t) = \begin{bmatrix} \text{vec}(U_{\mathcal{S}}(t)) - \text{vec}(U_{\mathcal{O}}(t)) \\ K_{\mathcal{S}}(t) - K_{\mathcal{O}}(t) \end{bmatrix}.$$

The 1st derivative of $\mathbf{x}(t)$ is equal to

$$\dot{\mathbf{x}}(t) = \dot{\mathbf{x}}_{\mathcal{S}}(t) - \dot{\mathbf{x}}_{\mathcal{O}}(t) = \begin{bmatrix} \text{vec}(\dot{U}_{\mathcal{S}}(t)) - \text{vec}(\dot{U}_{\mathcal{O}}(t)) \\ \dot{K}_{\mathcal{S}}(t) - \dot{K}_{\mathcal{O}}(t) \end{bmatrix}.$$

As a result, after substituting (14) for $\mathbf{x}(t) = \mathbf{x}_{\mathcal{S}}(t) - \mathbf{x}_{\mathcal{O}}(t)$, the following holds

$$E_{\mathcal{S}}(t) = L_{fs}(\text{vec}(U_{\mathcal{S}}(t)) - \text{vec}(U_{\mathcal{O}}(t))) - \mathbf{b}_{fs} + (K_{\mathcal{S}}(t) - K_{\mathcal{O}}(t))^{\odot 2},$$

or

$$E_{\mathcal{S}}(t) = [L_{fs} \quad (I_z \odot (K_{\mathcal{S}}(t) - K_{\mathcal{O}}(t)))](\mathbf{x}_{\mathcal{S}}(t) - \mathbf{x}_{\mathcal{O}}(t)) - \mathbf{b}_{fs},$$

where L_{fs} and $\mathbf{b}_{fs}$ are declared in (13). Then, the following results follow from (7):

$$\dot{E}_{\mathcal{S}}(t) = L_{fs} \text{vec}(\dot{U}(t) + \ddot{U}(t)) + 2(I_z \odot (K_{\mathcal{S}}(t) - K_{\mathcal{O}}(t)))(\dot{K}_{\mathcal{S}}(t) - \dot{K}_{\mathcal{O}}(t)) = -\lambda E_{\mathcal{S}}(t),$$

or equivalently

$$\dot{E}_{\mathcal{S}}(t) = [L_{fs} \quad 2(I_z \odot (K_{\mathcal{S}}(t) - K_{\mathcal{O}}(t)))](\dot{\mathbf{x}}_{\mathcal{S}}(t) - \dot{\mathbf{x}}_{\mathcal{O}}(t)) = -\lambda E_{\mathcal{S}}(t). \quad (23)$$

Next, for confirming the convergence, we choose the plausible Lyapunov function

$$\mathcal{Z}(t) = \frac{1}{2}\|E_{\mathcal{S}}(t)\|_{\text{F}}^2 = \frac{1}{2}\text{tr}\Big(E_{\mathcal{S}}(t)(E_{\mathcal{S}}(t))^{\text{T}}\Big).$$

The following is confirmed for $\mathcal{Z}(t)$:

$$\dot{\mathcal{Z}}(t) = \frac{2\text{tr}\left((E_{\mathcal{S}}(t))^{\text{T}}\dot{E}_{\mathcal{S}}(t)\right)}{2} = \text{tr}\left((E_{\mathcal{S}}(t))^{\text{T}}\dot{E}_{\mathcal{S}}(t)\right) = -\lambda \text{tr}\left((E_{\mathcal{S}}(t))^{\text{T}}E_{\mathcal{S}}(t)\right). \quad (24)$$

Because of (24), the following is valid:

$$\dot{\mathcal{Z}}(t)\begin{cases} <0, E_{\mathcal{S}}(t) \neq 0, \\ =0, E_{\mathcal{S}}(t) = 0, \end{cases}$$

$$\Leftrightarrow \dot{\mathcal{Z}}(t)\begin{cases} <0, \left[L_{fs} \quad (I_z \odot (K_{\mathcal{S}}(t) - K_{\mathcal{O}}(t)))\right](\mathbf{x}_{\mathcal{S}}(t) - \mathbf{x}_{\mathcal{O}}(t)) - \mathbf{b}_{fs} \neq 0, \\ =0, \left[L_{fs} \quad (I_z \odot (K_{\mathcal{S}}(t) - K_{\mathcal{O}}(t)))\right](\mathbf{x}_{\mathcal{S}}(t) - \mathbf{x}_{\mathcal{O}}(t)) - \mathbf{b}_{fs} = 0, \end{cases}$$

$$\Leftrightarrow \dot{\mathcal{Z}}(t)\begin{cases} <0, \left[L_{fs} \quad (I_z \odot (K_{\mathcal{S}}(t) - K_{\mathcal{O}}(t)))\right]\begin{bmatrix} \text{vec}(U_{\mathcal{S}}(t)) - \text{vec}(U_{\mathcal{O}}(t)) \\ K_{\mathcal{S}}(t) - K_{\mathcal{O}}(t) \end{bmatrix} - \mathbf{b}_{fs} \neq 0, \\ =0, \left[L_{fs} \quad (I_z \odot (K_{\mathcal{S}}(t) - K_{\mathcal{O}}(t)))\right]\begin{bmatrix} \text{vec}(U_{\mathcal{S}}(t)) - \text{vec}(U_{\mathcal{O}}(t)) \\ K_{\mathcal{S}}(t) - K_{\mathcal{O}}(t) \end{bmatrix} - \mathbf{b}_{fs} = 0, \end{cases}$$

$$\Leftrightarrow \dot{\mathcal{Z}}(t)\begin{cases} <0, \begin{bmatrix} \text{vec}(U_{\mathcal{O}}(t)) \\ K_{\mathcal{O}}(t) \end{bmatrix} \neq 0, \\ =0, \begin{bmatrix} \text{vec}(U_{\mathcal{O}}(t)) \\ K_{\mathcal{O}}(t) \end{bmatrix} = 0. \end{cases}$$

$$\Leftrightarrow \dot{\mathcal{Z}}(t)\begin{cases} <0, \mathbf{x}_{\mathcal{O}}(t) \neq 0, \\ =0, \mathbf{x}_{\mathcal{O}}(t) = 0. \end{cases}$$

With $\mathbf{x}_{\mathcal{O}}(t)$ being the equilibrium point of the system (23), we have

$$\forall \mathbf{x}_{\mathcal{O}}(t) \neq 0, \quad \dot{\mathcal{Z}}(t) \leq 0.$$

It appears that the equilibrium state

$$\mathbf{x}_{\mathcal{O}}(t) = -\mathbf{x}(t) + \mathbf{x}_{\mathcal{S}}(t) = \begin{bmatrix} -\text{vec}(U(t)) + \text{vec}(U_{\mathcal{S}}(t)) \\ -K(t) + K_{\mathcal{S}}(t) \end{bmatrix} = 0$$

is stable in accordance with Lyapunov theory. Afterwards, when $t \to \infty$, the following holds:

$$\mathbf{x}(t) = \begin{bmatrix} \text{vec}(U(t)) \\ K(t) \end{bmatrix} \to \mathbf{x}_{\mathcal{S}}(t) = \begin{bmatrix} \text{vec}(U_{\mathcal{S}}(t)) \\ K_{\mathcal{S}}(t) \end{bmatrix},$$

which finalizes the proof. □

Theorem 3. Let $\mathcal{A} = \left(m, \sigma^A, \left\{M^A_{x_i}\right\}_{x_i \in X}, \tau^A\right)$ and $\mathcal{B} = \left(n, \sigma^B, \left\{M^B_{x_i}\right\}_{x_i \in X}, \tau^B\right)$ be the WFA over $\mathbb{R}$ and $X = \{x_1, \ldots, x_r\}$, where $M^A_{x_i} \in \mathbb{R}^{m \times m}$, $\sigma^A \in \mathbb{R}^{1 \times m}$, $\tau^A \in \mathbb{R}^{m \times 1}$ and $M^B_{x_i} \in \mathbb{R}^{n \times n}$, $\sigma^B \in \mathbb{R}^{1 \times n}$, $\tau^B \in \mathbb{R}^{n \times 1}$ with $i = 1, \ldots, r$. Beginning from any initial point $\mathbf{x}(0)$, the ZNN-fs model of (19) converges exponentially to $\mathbf{x}^*(t)$, which refers to the TSOL of (3).

Proof. Firstly, the system of (8) is considered to find the solution $\mathbf{x}(t) = [\text{vec}(U(t))^{\text{T}}, K^{\text{T}}(t)]^{\text{T}}$ that is affiliated to the time-varying backward-forward bisimulation between $\mathcal{A}$ and $\mathcal{B}$ of (3). Secondly, the system of (8) is reformulated into the system of (9) utilizing vectorization and the Kronecker product and, then, into the system of (12) utilizing the operational permutation matrix P for $\text{vec}(U^{\text{T}}(t))$. Thirdly, considering the slack variable $K(t)$, the inequality constraint of the system of (12) is converted into an equality constraint in the system of (14). Fourthly, the EME of (15) is constructed, in keeping with the ZNN technique and the system of (14), to generate the solution $\mathbf{x}(t)$ that is affiliated with the system of (3).

Fifthly, the model of (17) is yielded in accordance to the ZNN technique of (7) for zeroing (15). According to Theorem 2, the EME of (15) converges to zero as $t \to \infty$. Consequently, the solution of (19) converges to $\mathbf{x}^*(t) = \left[\mathrm{vec}(U^*(t))^{\mathrm{T}},\ (K^*(t))^{\mathrm{T}}\right]^{\mathrm{T}}$ as $t \to \infty$. Furthermore, it is obvious that (19) is (17) in a different form because of the derivation process. After that, the proof is accomplished. □

3.2. The ZNN-bs Model

In line with (4), the following group of inequations must be satisfied:

$$\begin{cases} \tau^A - U(t)\tau^B \leqslant \mathbf{0}_m, \\ \sigma^A U(t) - \sigma^B \leqslant \mathbf{0}_n^{\mathrm{T}}, \\ M_{x_i}^A U(t) - U(t) M_{x_i}^B \leqslant \mathbf{0}_{m,n},\ i = 1,\ldots,r, \\ U(t) \geqslant \mathbf{0}_{m,n}, \end{cases} \quad (25)$$

where $U(t) \in \mathbb{R}^{m \times n}$ denotes the unknown matrix to be found. Utilizing vectorization and the Kronecker product, the system of inequations (25) is rewritten in the equivalent form

$$\begin{cases} -\left((\tau^B)^{\mathrm{T}} \otimes I_m\right) \mathrm{vec}(U(t)) + \tau^A \leqslant \mathbf{0}_m, \\ (I_n \otimes \sigma^A) \mathrm{vec}(U(t)) - (\sigma^B)^{\mathrm{T}} \leqslant \mathbf{0}_n, \\ \left(I_n \otimes M_{x_i}^A - (M_{x_i}^B)^{\mathrm{T}} \otimes I_m\right) \mathrm{vec}(U(t)) \leqslant \mathbf{0}_{mn},\ i = 1,\ldots,r, \\ -\mathrm{vec}(U(t)) \leqslant \mathbf{0}_{mn}, \end{cases}$$

and its corresponding matrix form is

$$L_{bs} \mathrm{vec}(U(t)) - \mathbf{b}_{bs} \leqslant \mathbf{0}_z, \quad (26)$$

where

$$L_{bs} = \begin{bmatrix} -(\tau^B)^{\mathrm{T}} \otimes I_m \\ I_n \otimes \sigma^A \\ W_{bs} \\ -I_{mn} \end{bmatrix} \in \mathbb{R}^{z \times mn},\ \mathbf{b}_{bs} = \begin{bmatrix} -\tau^A \\ (\sigma^B)^{\mathrm{T}} \\ \mathbf{0}_{(r+1)mn} \end{bmatrix} \in \mathbb{R}^z,\ W_{bs} = \begin{bmatrix} I_n \otimes M_{x_1}^A - (M_{x_1}^B)^{\mathrm{T}} \otimes I_m \\ I_n \otimes M_{x_2}^A - (M_{x_2}^B)^{\mathrm{T}} \otimes I_m \\ \cdots \\ I_n \otimes M_{x_r}^A - (M_{x_r}^B)^{\mathrm{T}} \otimes I_m \end{bmatrix} \in \mathbb{R}^{rmn \times mn}.$$

Then, considering the slack variable $K(t) \in \mathbb{R}^z$, the inequation (26) can be converted into the equation

$$L_{bs} \mathrm{vec}(U(t)) - \mathbf{b}_{bs} + K^{\odot 2}(t) = \mathbf{0}_z, \quad (27)$$

where $K^{\odot 2}(t)$ is always a non-negative time-varying term.

Thereafter, the ZNN approach considers the following EME, which is based on (27), for simultaneously satisfying all the inequations in (25):

$$E_{bs}(t) = L_{bs} \mathrm{vec}(U(t)) - \mathbf{b}_{bs} + K^{\odot 2}(t), \quad (28)$$

where $U(t)$ and $K(t)$ are the unknown matrices to be found. The first time derivative of (28) is

$$\dot{E}_{bs}(t) = L_{bs} \mathrm{vec}(\dot{U}(t)) + 2(I_z \odot K(t))\dot{K}(t). \quad (29)$$

Then, combining Equations (28) and (29) with the ZNN design (7), we obtain

$$L_{bs} \mathrm{vec}(\dot{U}(t)) + 2(I_z \odot K(t))\dot{K}(t) = -\lambda E_{bs}(t). \quad (30)$$

As a result, setting

$$H_{bs} = \begin{bmatrix} L_{bs} & 2(I_z \odot K(t)) \end{bmatrix} \in \mathbb{R}^{z \times (mn+z)},\ \dot{\mathbf{x}}(t) = \begin{bmatrix} \mathrm{vec}(\dot{U}(t)) \\ \dot{K}(t) \end{bmatrix} \in \mathbb{R}^{mn+z},\ \mathbf{x}(t) = \begin{bmatrix} \mathrm{vec}(U(t)) \\ K(t) \end{bmatrix} \in \mathbb{R}^{mn+z},$$

the next model is obtained:
$$H_{ls}\dot{x} = -\lambda E_{ls}(t). \tag{31}$$

Since the ZNN dynamics in solving (31) requires invertibility of the mass matrix H_{ls}, it is practical to use the best approximate solution to (31), which leads to

$$\dot{x} = H_{ls}^{\dagger}(-\lambda E_{ls}(t)). \tag{32}$$

An appropriate ode MATLAB solver can be used to handle the ZNN model of (32), additionally referred to as the ZNN-bs flow. The ZNN-bs model's convergence and stability investigation is shown in Theorem 4.

Theorem 4. *Let* $\mathcal{A} = \left(m, \sigma^A, \left\{M_{x_i}^A\right\}_{x_i \in X}, \tau^A\right)$ *and* $\mathcal{B} = \left(n, \sigma^B, \left\{M_{x_i}^B\right\}_{x_i \in X}, \tau^B\right)$ *be WFA over* $\mathbb{R}$, *where* $M_{x_i}^A \in \mathbb{R}^{m \times m}, \sigma^A \in \mathbb{R}^{1 \times m}, \tau^A \in \mathbb{R}^{m \times 1}$ *and* $M_{x_i}^B \in \mathbb{R}^{n \times n}, \sigma^B \in \mathbb{R}^{1 \times n}, \tau^B \in \mathbb{R}^{n \times 1}$ *with* $i = 1, \ldots, r$. *The dynamics (30) in line with the ZNN method of (7) lead to the TSOL, shown by* $\mathbf{x}_S(t) = \left[\text{vec}(U_S(t))^T \ K_S^T(t)\right]^T$, *which is stable according to Lyapunov.*

Proof. The proof is omitted since it is similar to the proof of Theorem 2. □

Theorem 5. *Let* $\mathcal{A} = \left(m, \sigma^A, \left\{M_{x_i}^A\right\}_{x_i \in X}, \tau^A\right)$ *and* $\mathcal{B} = \left(n, \sigma^B, \left\{M_{x_i}^B\right\}_{x_i \in X}, \tau^B\right)$ *be WFA over* $\mathbb{R}$, *where* $M_{x_i}^A \in \mathbb{R}^{m \times m}, \sigma^A \in \mathbb{R}^{1 \times m}, \tau^A \in \mathbb{R}^{m \times 1}$ *and* $M_{x_i}^B \in \mathbb{R}^{n \times n}, \sigma^B \in \mathbb{R}^{1 \times n}, \tau^B \in \mathbb{R}^{n \times 1}$ *with* $i = 1, \ldots, r$. *Beginning from any initial point* $\mathbf{x}(0)$, *the ZNN-bs design (32) converges exponentially to* $\mathbf{x}^*(t)$, *which refers to the TSOL of (4).*

Proof. The proof is omitted since it is similar to the proof of Theorem 3. □

3.3. The ZNN-fb Model

In line with (5), the following group of inequations must be satisfied:

$$\begin{cases} U^T(t)\tau^A - \tau^B \leqslant \mathbf{0}_n, \\ U(t)\tau^B - \tau^A \leqslant \mathbf{0}_m, \\ \sigma^A - \sigma^B U^T(t) \leqslant \mathbf{0}_m^T, \\ \sigma^B - \sigma^A U(t) \leqslant \mathbf{0}_n^T, \\ U^T(t)M_{x_i}^A - M_{x_i}^B U^T(t) \leqslant \mathbf{0}_{n,m}, \ i = 1, \ldots, r, \\ U(t)M_{x_i}^B - M_{x_i}^A U(t) \leqslant \mathbf{0}_{m,n}, \ i = 1, \ldots, r, \\ U(t) \geqslant \mathbf{0}_{m,n}, \end{cases} \tag{33}$$

where $U(t) \in \mathbb{R}^{m \times n}$ implies the unknown matrix to be generated. Utilizing vectorization in combination with the Kronecker product, the system of (33) is reformulated as

$$\begin{cases} \left((\tau^A)^T \otimes I_n\right)\text{vec}(U^T(t)) - \tau^B \leqslant \mathbf{0}_n, \\ \left((\tau^B)^T \otimes I_m\right)\text{vec}(U(t)) - \tau^A \leqslant \mathbf{0}_m, \\ -\left(I_m \otimes \sigma^B\right)\text{vec}(U^T(t)) + (\sigma^A)^T \leqslant \mathbf{0}_m, \\ -\left(I_n \otimes \sigma^A\right)\text{vec}(U(t)) + (\sigma^B)^T \leqslant \mathbf{0}_n, \\ \left((M_{x_i}^A)^T \otimes I_n - I_m \otimes M_{x_i}^B\right)\text{vec}(U^T(t)) \leqslant \mathbf{0}_{mn}, \ i = 1, \ldots, r, \\ \left((M_{x_i}^B)^T \otimes I_m - I_n \otimes M_{x_i}^A\right)\text{vec}(U(t)) \leqslant \mathbf{0}_{mn}, \ i = 1, \ldots, r, \\ -\text{vec}(U(t)) \leqslant \mathbf{0}_{mn}. \end{cases} \tag{34}$$

Using the permutation matrix P for $\text{vec}(U^T(t))$, (34) is rewritten as

$$\begin{cases} ((\tau^A)^T \otimes I_n)P\text{vec}(U(t)) - \tau^B \leqslant \mathbf{0}_n, \\ ((\tau^B)^T \otimes I_m)\text{vec}(U(t)) - \tau^A \leqslant \mathbf{0}_m, \\ -(I_m \otimes \sigma^B)P\text{vec}(U(t)) + (\sigma^A)^T \leqslant \mathbf{0}_m, \\ -(I_n \otimes \sigma^A)\text{vec}(U(t)) + (\sigma^B)^T \leqslant \mathbf{0}_n, \\ \left((M^A_{x_i})^T \otimes I_n - I_m \otimes M^B_{x_i}\right)P\text{vec}(U(t)) \leqslant \mathbf{0}_{mn}, \ i=1,\ldots,r, \\ \left((M^B_{x_i})^T \otimes I_m - I_n \otimes M^A_{x_i}\right)\text{vec}(U(t)) \leqslant \mathbf{0}_{mn}, \ i=1,\ldots,r, \\ -\text{vec}(U(t)) \leqslant \mathbf{0}_{mn}, \end{cases}$$

and its corresponding matrix form is

$$L_{fb}\text{vec}(U(t)) - \mathbf{b}_{fb} \leqslant \mathbf{0}_y, \tag{35}$$

where $y = (2r+1)mn + 2m + 2n$ and

$$L_{fb} = \begin{bmatrix} ((\tau^A)^T \otimes I_n)P \\ (\tau^B)^T \otimes I_m \\ -(I_m \otimes \sigma^B)P \\ -I_n \otimes \sigma^A \\ W_{fb} \\ -I_{mn} \end{bmatrix} \in \mathbb{R}^{y \times mn}, \ \mathbf{b}_{fb} = \begin{bmatrix} \tau^B \\ \tau^A \\ -(\sigma^A)^T \\ -(\sigma^B)^T \\ \mathbf{0}_{(2r+1)mn} \end{bmatrix} \in \mathbb{R}^y, \ W_{fb} = \begin{bmatrix} ((M^A_{x_1})^T \otimes I_n - I_m \otimes M^B_{x_1})P \\ (M^B_{x_1})^T \otimes I_m - I_n \otimes M^A_{x_1} \\ ((M^A_{x_2})^T \otimes I_n - I_m \otimes M^B_{x_2})P \\ (M^B_{x_2})^T \otimes I_m - I_n \otimes M^A_{x_2} \\ \ldots \\ ((M^A_{x_r})^T \otimes I_n - I_m \otimes M^B_{x_r})P \\ (M^B_{x_r})^T \otimes I_m - I_n \otimes M^A_{x_r} \end{bmatrix} \in \mathbb{R}^{2rmn \times mn}.$$

Then, considering the slack variables vector $K(t) \in \mathbb{R}^y$, the inequation (35) is converted into the equation

$$L_{fb}\text{vec}(U(t)) - \mathbf{b}_{fb} + K^{\odot 2}(t) = \mathbf{0}_y.$$

Thereafter, the ZNN approach considers the following EME, which is based on (35), for simultaneously satisfying all the inequations in (33):

$$E_{fb}(t) = L_{fb}\text{vec}(U(t)) - \mathbf{b}_{fb} + K^{\odot 2}(t), \tag{36}$$

where $U(t)$ and $K(t)$ are the unknown matrices to be found. The first time derivative of (36) is equal to

$$\dot{E}_{fb}(t) = L_{fb}\text{vec}(\dot{U}(t)) + 2(I_y \odot K(t))\dot{K}(t). \tag{37}$$

Then, combining Equations (36) and (37) with the ZNN design (7), we obtain the following:

$$L_{fb}\text{vec}(\dot{U}(t)) + 2(I_y \odot K(t))\dot{K}(t) = -\lambda E_{fb}(t). \tag{38}$$

As a result, setting

$$H_{fb} = \begin{bmatrix} L_{fb} & 2(I_y \odot K(t)) \end{bmatrix} \in \mathbb{R}^{y \times (mn+y)}, \ \dot{\mathbf{x}}(t) = \begin{bmatrix} \text{vec}(\dot{U}(t)) \\ \dot{K}(t) \end{bmatrix} \in \mathbb{R}^{mn+y}, \ \mathbf{x}(t) = \begin{bmatrix} \text{vec}(U(t)) \\ K(t) \end{bmatrix} \in \mathbb{R}^{mn+y},$$

(38) is transformed into the model

$$H_{fb}\dot{\mathbf{x}} = -\lambda E_{fb}(t)$$

whose pseudoinverse solution is equal to

$$\dot{\mathbf{x}} = H_{fb}^\dagger(-\lambda E_{fb}(t)). \tag{39}$$

An appropriate ode MATLAB solver can be used to handle the ZNN model (39), additionally referred to as the ZNN-fb model. The ZNN-fb model's convergence and stability investigation is shown in the next theorem.

Theorem 6. *Let* $\mathcal{A} = \left(m, \sigma^A, \left\{M_{x_i}^A\right\}_{x_i \in X'}, \tau^A\right)$ *and* $\mathcal{B} = \left(n, \sigma^B, \left\{M_{x_i}^B\right\}_{x_i \in X'}, \tau^B\right)$ *be the WFA over* $\mathbb{R}$, *where* $M_{x_i}^A \in \mathbb{R}^{m \times m}$, $\sigma^A \in \mathbb{R}^{1 \times m}$, $\tau^A \in \mathbb{R}^{m \times 1}$ *and* $M_{x_i}^B \in \mathbb{R}^{n \times n}$, $\sigma^B \in \mathbb{R}^{1 \times n}$, $\tau^B \in \mathbb{R}^{n \times 1}$ *with* $i = 1, \ldots, r$. *The dynamics of (38) in line with the ZNN method of (7) lead to the TSOL, shown by* $\mathbf{x}_\mathcal{S}(t) = \left[\mathrm{vec}(U_\mathcal{S}(t))^\mathrm{T} \quad K_\mathcal{S}^\mathrm{T}(t)\right]^\mathrm{T}$, *which is stable according to Lyapunov*.

Proof. The proof is omitted since it is similar to the proof of Theorem 2. □

Theorem 7. *Let* $\mathcal{A} = \left(m, \sigma^A, \left\{M_{x_i}^A\right\}_{x_i \in X'}, \tau^A\right)$ *and* $\mathcal{B} = \left(n, \sigma^B, \left\{M_{x_i}^B\right\}_{x_i \in X'}, \tau^B\right)$ *be the WFA over* $\mathbb{R}$, *where* $M_{x_i}^A \in \mathbb{R}^{m \times m}$, $\sigma^A \in \mathbb{R}^{1 \times m}$, $\tau^A \in \mathbb{R}^{m \times 1}$ *and* $M_{x_i}^B \in \mathbb{R}^{n \times n}$, $\sigma^B \in \mathbb{R}^{1 \times n}$, $\tau^B \in \mathbb{R}^{n \times 1}$ *with* $i = 1, \ldots, r$. *Beginning from any initial point* $\mathbf{x}(0)$, *the ZNN-bs model of (39) converges exponentially to* $\mathbf{x}^*(t)$, *which refers to the TSOL of (5)*.

Proof. The proof is similar to the proof of Theorem 3. □

3.4. The ZNN-bb Model

In line with (6), the following group of inequations must be satisfied:

$$\begin{cases} \tau^A - U(t)\tau^B \leqslant \mathbf{0}_m, \\ \tau^B - U^\mathrm{T}(t)\tau^A \leqslant \mathbf{0}_n, \\ \sigma^A U(t) - \sigma^B \leqslant \mathbf{0}_n^\mathrm{T}, \\ \sigma^B U^\mathrm{T}(t) - \sigma^A \leqslant \mathbf{0}_m^\mathrm{T}, \\ M_{x_i}^A U(t) - U(t) M_{x_i}^B \leqslant \mathbf{0}_{m,n}, \ i = 1, \ldots, r, \\ M_{x_i}^B U^\mathrm{T}(t) - U^\mathrm{T}(t) M_{x_i}^A \leqslant \mathbf{0}_{n,m}, \ i = 1, \ldots, r, \\ U(t) \geqslant \mathbf{0}_{m,n}, \end{cases} \tag{40}$$

where $U(t) \in \mathbb{R}^{m \times n}$ stands for the unknown matrix. The system of (40) is reformulated as follows:

$$\begin{cases} -\left((\tau^B)^\mathrm{T} \otimes I_m\right)\mathrm{vec}(U(t)) + \tau^A \leqslant \mathbf{0}_m, \\ -\left((\tau^A)^\mathrm{T} \otimes I_n\right)\mathrm{vec}(U^\mathrm{T}(t)) + \tau^B \leqslant \mathbf{0}_n, \\ \left(I_n \otimes \sigma^A\right)\mathrm{vec}(U(t)) - (\sigma^B)^\mathrm{T} \leqslant \mathbf{0}_n, \\ \left(I_m \otimes \sigma^B\right)\mathrm{vec}(U^\mathrm{T}(t)) - (\sigma^A)^\mathrm{T} \leqslant \mathbf{0}_m, \\ \left(I_n \otimes M_{x_i}^A - (M_{x_i}^B)^\mathrm{T} \otimes I_m\right)\mathrm{vec}(U(t)) \leqslant \mathbf{0}_{mn}, \ i = 1, \ldots, r, \\ \left(I_m \otimes M_{x_i}^B - (M_{x_i}^A)^\mathrm{T} \otimes I_n\right)\mathrm{vec}(U^\mathrm{T}(t)) \leqslant \mathbf{0}_{mn}, \ i = 1, \ldots, r, \\ -\mathrm{vec}(U(t)) \leqslant \mathbf{0}_{mn}. \end{cases} \tag{41}$$

Using the permutation matrix P for generating $\mathrm{vec}(U^\mathrm{T}(t))$, (41) is rewritten as

$$\begin{cases} -\left((\tau^B)^\mathrm{T} \otimes I_m\right)\mathrm{vec}(U(t)) + \tau^A \leqslant \mathbf{0}_m, \\ -\left((\tau^A)^\mathrm{T} \otimes I_n\right)P\,\mathrm{vec}(U(t)) + \tau^B \leqslant \mathbf{0}_n, \\ \left(I_n \otimes \sigma^A\right)\mathrm{vec}(U(t)) - (\sigma^B)^\mathrm{T} \leqslant \mathbf{0}_n, \\ \left(I_m \otimes \sigma^B\right)P\,\mathrm{vec}(U(t)) - (\sigma^A)^\mathrm{T} \leqslant \mathbf{0}_m, \\ \left(I_n \otimes M_{x_i}^A - (M_{x_i}^B)^\mathrm{T} \otimes I_m\right)\mathrm{vec}(U(t)) \leqslant \mathbf{0}_{mn}, \ i = 1, \ldots, r, \\ \left(I_m \otimes M_{x_i}^B - (M_{x_i}^A)^\mathrm{T} \otimes I_n\right)P\,\mathrm{vec}(U(t)) \leqslant \mathbf{0}_{mn}, \ i = 1, \ldots, r, \\ -\mathrm{vec}(U(t)) \leqslant \mathbf{0}_{mn}, \end{cases}$$

and its corresponding matrix form is the following:

$$L_{lb}\text{vec}(U(t)) - \mathbf{b}_{lb} \leqslant \mathbf{0}_y, \tag{42}$$

where

$$L_{lb} = \begin{bmatrix} -(\tau^B)^{\text{T}} \otimes I_m \\ -((\tau^A)^{\text{T}} \otimes I_n)P \\ I_n \otimes \sigma^A \\ (I_m \otimes \sigma^B)P \\ W_{lb} \\ -I_{mn} \end{bmatrix} \in \mathbb{R}^{y \times mn}, \quad \mathbf{b}_{lb} = \begin{bmatrix} -\tau^A \\ -\tau^B \\ (\sigma^B)^{\text{T}} \\ (\sigma^A)^{\text{T}} \\ \mathbf{0}_{(2r+1)mn} \end{bmatrix} \in \mathbb{R}^y,$$

$$W_{lb} = \begin{bmatrix} I_n \otimes M^A_{x_1} - (M^B_{x_1})^{\text{T}} \otimes I_m \\ (I_m \otimes M^B_{x_1} - (M^A_{x_1})^{\text{T}} \otimes I_n)P \\ I_n \otimes M^A_{x_2} - (M^B_{x_2})^{\text{T}} \otimes I_m \\ (I_m \otimes M^B_{x_2} - (M^A_{x_2})^{\text{T}} \otimes I_n)P \\ \cdots \\ I_n \otimes M^A_{x_r} - (M^B_{x_r})^{\text{T}} \otimes I_m \\ (I_m \otimes M^B_{x_r} - (M^A_{x_r})^{\text{T}} \otimes I_n)P \end{bmatrix} \in \mathbb{R}^{2rmn \times mn}.$$

Then, considering the slack variable $K(t) \in \mathbb{R}^y$, the inequation (42) can be converted into the equation

$$L_{lb}\text{vec}(U(t)) - \mathbf{b}_{lb} + K^{\odot 2}(t) = \mathbf{0}_y,$$

in which $K^{\odot 2}(t)$ is always a non-negative time-varying term.

Thereafter, the ZNN approach considers the following EME, which is based on (42), for simultaneously satisfying all the equations in (40):

$$E_{lb}(t) = L_{lb}\text{vec}(U(t)) - \mathbf{b}_{lb} + K^{\odot 2}(t), \tag{43}$$

where $U(t)$ and $K(t)$ are the unknown matrices to be found. The first time derivative of (43) is given as

$$\dot{E}_{lb}(t) = L_{lb}\text{vec}(\dot{U}(t)) + 2(I_y \odot K(t))\dot{K}(t). \tag{44}$$

Then, combining Equations (43) and (44) with the ZNN design of (7), we can obtain

$$L_{lb}\text{vec}(\dot{U}(t)) + 2(I_y \odot K(t))\dot{K}(t) = -\lambda E_{lb}(t). \tag{45}$$

As a result, setting

$$H_{lb} = [L_{lb} \quad 2(I_y \odot K(t))] \in \mathbb{R}^{y \times (mn+y)}, \quad \dot{\mathbf{x}}(t) = \begin{bmatrix} \text{vec}(\dot{U}(t)) \\ \dot{K}(t) \end{bmatrix} \in \mathbb{R}^{mn+y}, \quad \mathbf{x}(t) = \begin{bmatrix} \text{vec}(U(t)) \\ K(t) \end{bmatrix} \in \mathbb{R}^{mn+y},$$

the next model is obtained:

$$H_{lb}\dot{\mathbf{x}} = -\lambda E_{lb}(t),$$

or an equivalent:

$$\dot{\mathbf{x}} = H_{lb}^{\dagger}(-\lambda E_{lb}(t)). \tag{46}$$

An appropriate ode MATLAB solver can be used to handle the ZNN model of (46), additionally referred to as the ZNN-bb model. The ZNN-bb model's convergence and stability investigation is shown in the next theorem.

Theorem 8. Let $\mathcal{A} = \left(m, \sigma^A, \left\{M^A_{x_i}\right\}_{x_i \in X}, \tau^A\right)$ and $\mathcal{B} = \left(n, \sigma^B, \left\{M^B_{x_i}\right\}_{x_i \in X}, \tau^B\right)$ be the WFA over $\mathbb{R}$, where $M^A_{x_i} \in \mathbb{R}^{m \times m}$, $\sigma^A \in \mathbb{R}^{1 \times m}$, $\tau^A \in \mathbb{R}^{m \times 1}$ and $M^B_{x_i} \in \mathbb{R}^{n \times n}$, $\sigma^B \in \mathbb{R}^{1 \times n}$, $\tau^B \in \mathbb{R}^{n \times 1}$ with $i = 1, \ldots, r$. The dynamics of (45) in line with the ZNN method of (7) lead to the TSOL, shown by $\mathbf{x}_S(t) = \begin{bmatrix} \text{vec}(U_S(t))^{\text{T}} & K^{\text{T}}_S(t) \end{bmatrix}^{\text{T}}$, which is stable according to Lyapunov.

Proof. The proof is omitted since it is similar to the proof of Theorem 2. □

Theorem 9. *Let* $\mathcal{A} = \left(m, \sigma^A, \{M^A_{x_i}\}_{x_i \in X}, \tau^A\right)$ *and* $\mathcal{B} = \left(n, \sigma^B, \{M^B_{x_i}\}_{x_i \in X}, \tau^B\right)$ *be the WFA over* $\mathbb{R}$, *where* $M^A_{x_i} \in \mathbb{R}^{m \times m}$, $\sigma^A \in \mathbb{R}^{1 \times m}$, $\tau^A \in \mathbb{R}^{m \times 1}$ *and* $M^B_{x_i} \in \mathbb{R}^{n \times n}$, $\sigma^B \in \mathbb{R}^{1 \times n}$, $\tau^B \in \mathbb{R}^{n \times 1}$ *with* $i = 1, \ldots, r$. *Beginning from any initial point* $\mathbf{x}(0)$, *the ZNN-bb model of* (46) *converges exponentially to* $\mathbf{x}^*(t)$, *which refers to the TSOL of* (6).

Proof. The proof is similar to the proof of Theorem 3. □

4. ZNN Experiments

The performances of the ZNN-fs model of (19), the ZNN-bs model of (32), the ZNN-fb model of (39), and the ZNN-bb model of (46) are examined in each of the five numerical experiments presented in this section. Keep in mind that during the computation in all experiments, the MATLAB `ode45` solver was applied with time span of $[0, 10]$ under a relative and absolute tolerance of 10^{-12} and 10^{-8}, respectively. Additionally, we contrast the output of the ZNN models with the results of the MATLAB function `linprog` (with the default settings). Following the model proposed in [7], the zero initial point is used.

Example 1. *Let us choose* $m = 2$, $n = 3$, $r = 2$, *and* $X = \{x_1, x_2\}$, *and consider WFA over* $\mathbb{R}$ *defined by* $\mathcal{A} = \left(m, \sigma^A, \{M^A_{x_i}\}_{x_i \in X}, \tau^A\right)$ *and* $\mathcal{B} = \left(n, \sigma^B, \{M^B_{x_i}\}_{x_i \in X}, \tau^B\right)$. *Clearly,* $M^A_{x_i} \in \mathbb{R}^{m \times m}$, $\sigma^A \in \mathbb{R}^{1 \times m}$, $\tau^A \in \mathbb{R}^{m \times 1}$ *and* $M^B_{x_i} \in \mathbb{R}^{n \times n}$, $\sigma^B \in \mathbb{R}^{1 \times n}$, $\tau^B \in \mathbb{R}^{n \times 1}$. *Consider* $\mathcal{A} = \left(m, \sigma^A, \{M^A_{x_i}, = 1, 2\}, \tau^A\right)$ *defined by*

$$\sigma^A = \begin{bmatrix} -7 & -8 \end{bmatrix}, \quad \tau^A = \begin{bmatrix} -13 & -13 \end{bmatrix}^T,$$

$$M^A_{x_1} = \begin{bmatrix} -6 & 9 \\ -13 & -14 \end{bmatrix}, \quad M^A_{x_2} = \begin{bmatrix} 13 & -1 \\ -15 & -9 \end{bmatrix}$$

and $\mathcal{B} = \left(n, \sigma^B, \{M^B_{x_i}, i = 1, 2\}, \tau^B\right)$ *defined by*

$$\sigma^B = \begin{bmatrix} 5 & -14 & 9 \end{bmatrix}, \quad \tau^B = \begin{bmatrix} -3 & 1 & 1 \end{bmatrix}^T,$$

$$M^B_{x_1} = \begin{bmatrix} -4 & -2 & 2 \\ -13 & 17 & 9 \\ -15 & -14 & 17 \end{bmatrix}, \quad M^B_{x_2} = \begin{bmatrix} -2 & 12 & 7 \\ -1 & 16 & -3 \\ -2 & -5 & 7 \end{bmatrix}.$$

Furthermore, the design parameter of ZNN is set to $\lambda = 10$, *and the following initial conditions (ICs) are used:*

- *IC1:* $\mathbf{x}(0) = \mathbf{1}_{23}$,
- *IC2:* $\mathbf{x}(0) = -\mathbf{1}_{23}$,
- *IC3:* $\mathbf{x}(0) = \text{rand}(23, 1)$.

The results of the ZNN-fs model are presented in Figure 1.

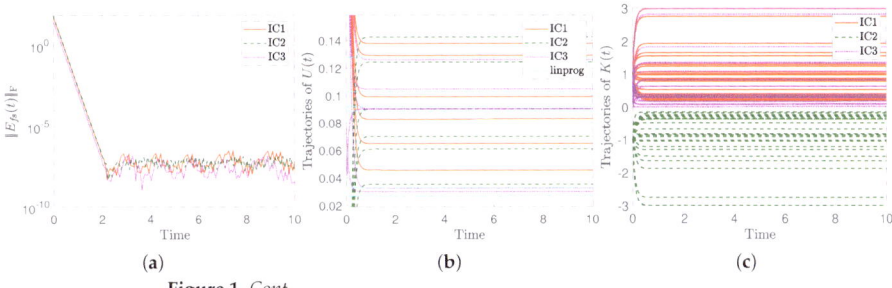

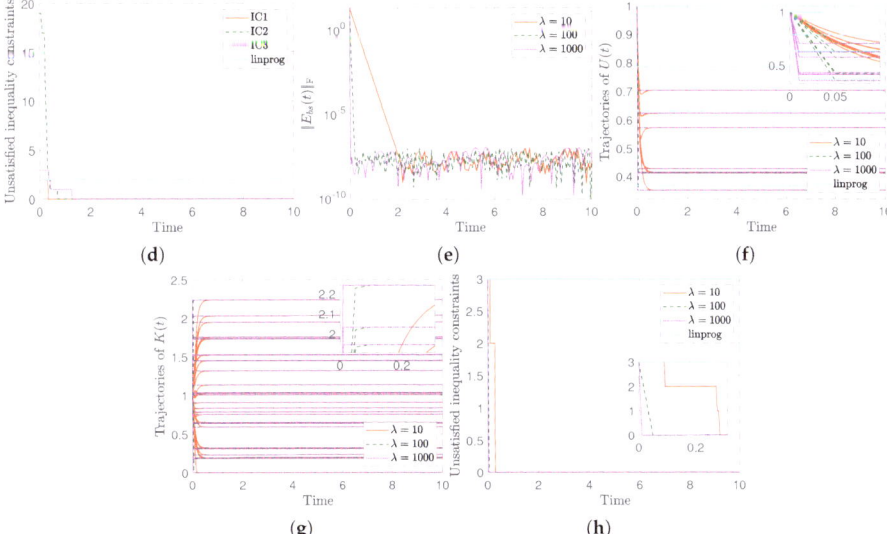

Figure 1. Errors and trajectories in Examples 1 and 2. (**a**) Example 1: EME errors. (**b**) Example 1: Trajectories of $U(t)$. (**c**) Example 1: Trajectories of $K(t)$. (**d**) Example 1: Number of unsatisfied constraints. (**e**) Example 2: EME errors. (**f**) Example 2: Trajectories of $U(t)$. (**g**) Example 2: Trajectories of $K(t)$. (**h**) Example 2: Number of unsatisfied constraints.

Example 2. Let $m = 4$, $n = 2$, $r = 2$, and $X = \{x_1, x_2\}$, and consider WFA $\mathcal{A} = \left(m, \sigma^A, \left\{M_{x_i}^A\right\}_{x_i \in X}, \tau^A\right)$ and $\mathcal{B} = \left(n, \sigma^B, \left\{M_{x_i}^B\right\}_{x_i \in X}, \tau^B\right)$. Clearly, $M_{x_i}^A \in \mathbb{R}^{m \times m}$, $\sigma^A \in \mathbb{R}^{1 \times m}$, $\tau^A \in \mathbb{R}^{m \times 1}$ and $M_{x_i}^B \in \mathbb{R}^{n \times n}$, $\sigma^B \in \mathbb{R}^{1 \times n}$, $\tau^B \in \mathbb{R}^{n \times 1}$. Consider $\mathcal{A} = \left(m, \sigma^A, \{M_{x_i}^A, = 1, 2\}, \tau^A\right)$ defined by

$$\sigma^A = \begin{bmatrix} -1 & 1 & -2 & 1 \end{bmatrix}, \quad \tau^A = \begin{bmatrix} 1 & 1 & 1 & 1 \end{bmatrix}^T,$$

$$M_{x_1}^A = \begin{bmatrix} 2 & -1 & 3 & -1 \\ 1 & -2 & 1 & -2 \\ 3 & -1 & 2 & -1 \\ 3 & -1 & 2 & -1 \end{bmatrix}, \quad M_{x_2}^A = \begin{bmatrix} 1 & 4 & -2 & 4 \\ 2 & -1 & 2 & -1 \\ -2 & 4 & 1 & 4 \\ -2 & 4 & 1 & 4 \end{bmatrix}$$

and $\mathcal{B} = \left(n, \sigma^B, \{M^B_{x_i}, i=1,2\}, \tau^B\right)$ defined by

$$\sigma^B = \begin{bmatrix} 1 & 2 \end{bmatrix}, \quad \tau^B = \begin{bmatrix} 1 & 1 \end{bmatrix}^T,$$
$$M^B_{x_1} = \begin{bmatrix} 6 & 4 \\ -4 & 4 \end{bmatrix}, \quad M^B_{x_2} = \begin{bmatrix} 4 & 6 \\ 6 & 4 \end{bmatrix}.$$

Also, the design parameters of ZNN are $\lambda = 10$, $\lambda = 100$ and $\lambda = 100$, whereas the IC is set to $\mathbf{x}(0) = \mathbf{1}_{30}$. The results of the ZNN-bs model are presented in Figure 1.

Example 3. Let $m = n = k = 10$, $r = 2$, and $X = \{x_1, x_2\}$, and consider WFA $\mathcal{A} = \left(m, \sigma^A, \{M^A_{x_i}\}_{x_i \in X}, \tau^A\right)$ and $\mathcal{B} = \left(n, \sigma^B, \{M^B_{x_i}\}_{x_i \in X}, \tau^B\right)$ over $\mathbb{R}$. Clearly, $M^A_{x_i} \in \mathbb{R}^{m \times m}$, $\sigma^A \in \mathbb{R}^{1 \times m}$, $\tau^A \in \mathbb{R}^{m \times 1}$ and $M^B_{x_i} \in \mathbb{R}^{n \times n}$, $\sigma^B \in \mathbb{R}^{1 \times n}$, $\tau^B \in \mathbb{R}^{n \times 1}$. Consider $\mathcal{A} = \left(m, \sigma^A, \{M^A_{x_i}, = 1, 2\}, \tau^A\right)$ defined by

$$\sigma^A = \mathbf{1}_k^T, \quad \tau^A = \mathbf{1}_k, \quad M^A_{x_1} = I_k, \quad M^A_{x_2} = I_k$$

and $\mathcal{B} = \left(n, \sigma^B, \{M^B_{x_i}, i=1,2\}, \tau^B\right)$ defined by

$$\sigma^B = 5 \cdot \mathbf{1}_k^T, \quad \tau^B = 5 \cdot \mathbf{1}_k, \quad M^B_{x_1} = 5 \cdot I_k, \quad M^B_{x_2} = 5 \cdot I_k.$$

Furthermore, the design parameter of ZNN is set to $\lambda = 10, 100, 1000$, and the following ICs are used:

- IC1: $\mathbf{x}(0) = \mathbf{1}_{420}$,
- IC2: $\mathbf{x}(0) = -\mathbf{1}_{420}$,
- IC3: $\mathbf{x}(0) = \mathrm{rand}(420, 1)$.

The results of the ZNN-fs model are presented in Figure 2.

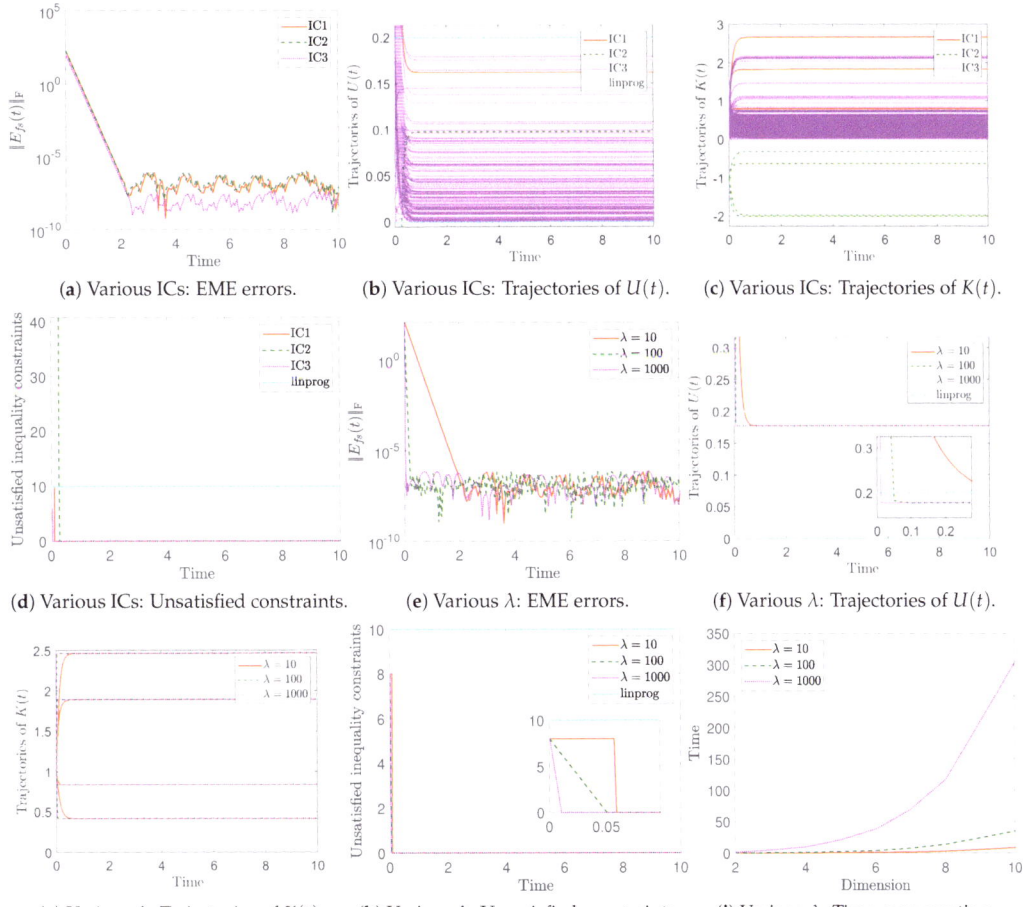

Figure 2. Errors and trajectories in Example 3.

Example 4. *Let $m = n = k = 10$, $r = 2$, and $X = \{x_1, x_2\}$, and consider WFA $\mathcal{A} = \left(m, \sigma^A, \{M_{x_i}^A\}_{x_i \in X}, \tau^A\right)$ and $\mathcal{B} = \left(n, \sigma^B, \{M_{x_i}^B\}_{x_i \in X}, \tau^B\right)$ over $\mathbb{R}$. Clearly, $M_{x_i}^A \in \mathbb{R}^{m \times m}$, $\sigma^A \in \mathbb{R}^{1 \times m}$, $\tau^A \in \mathbb{R}^{m \times 1}$ and $M_{x_i}^B \in \mathbb{R}^{n \times n}$, $\sigma^B \in \mathbb{R}^{1 \times n}$, $\tau^B \in \mathbb{R}^{n \times 1}$. Consider $\mathcal{A} = \left(m, \sigma^A, \{M_{x_i}^A, = 1, 2\}, \tau^A\right)$ defined by*

$$\sigma^A = \mathbf{1}_k^T, \quad \tau^A = \mathbf{1}_k, \quad M_{x_1}^A = I_k, \quad M_{x_2}^A = I_k$$

and $\mathcal{B} = \left(n, \sigma^B, \{M_{x_i}^B, i = 1, 2\}, \tau^B\right)$ defined by

$$\sigma^B = 2 \cdot \mathbf{1}_k^T, \quad \tau^B = 2 \cdot \mathbf{1}_k, \quad M_{x_1}^B = 2 \cdot I_k, \quad M_{x_2}^B = 2 \cdot I_k.$$

Furthermore, the design parameter of ZNN is set to $\lambda = 10, 100, 1000$, and the following ICs are used:

- *IC1: $\mathbf{x}(0) = \mathbf{1}_{420}$,*
- *IC2: $\mathbf{x}(0) = -\mathbf{1}_{420}$,*
- *IC3: $\mathbf{x}(0) = \text{rand}(420, 1)$.*

The results of the ZNN-bs model are presented in Figure 3.

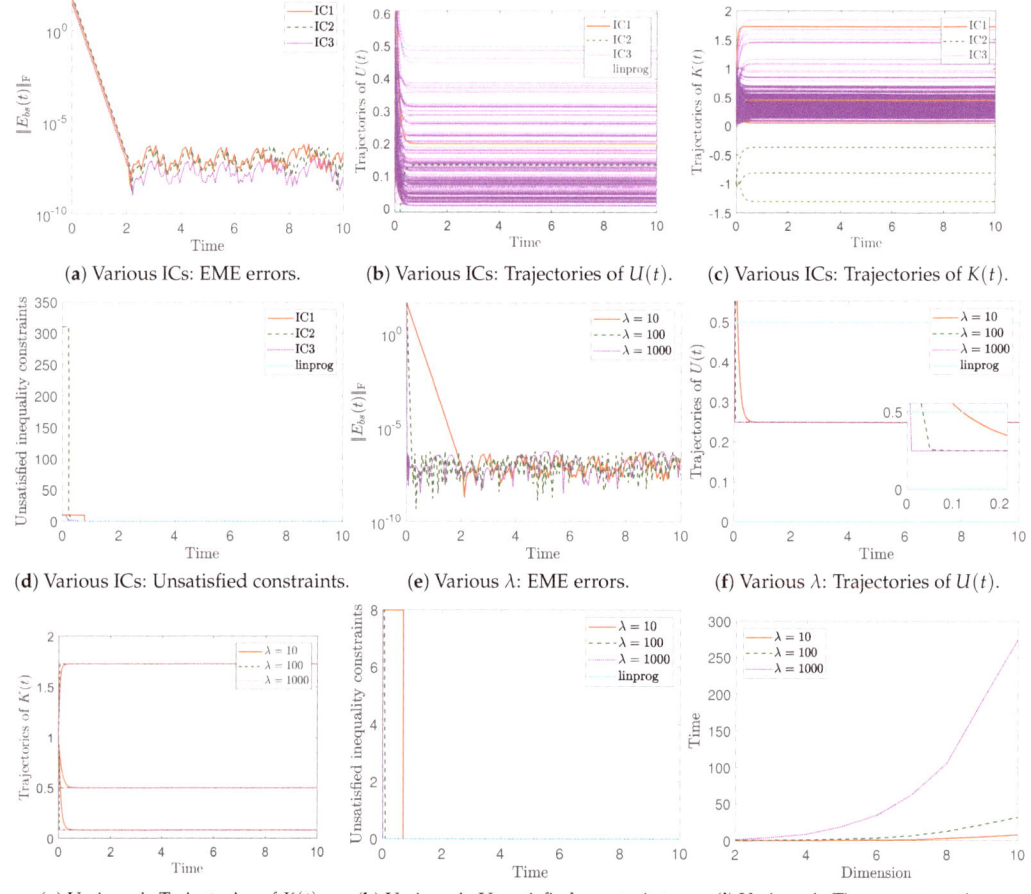

Figure 3. Errors and trajectories in Example 4.

Example 5. *Let $m = k + 1$, $n = k$ with $k = 10$, $r = 2$, and $X = \{x_1, x_2\}$, and consider WFA $\mathcal{A} = \left(m, \sigma^A, \{M^A_{x_i}\}_{x_i \in X}, \tau^A\right)$ and $\mathcal{B} = \left(n, \sigma^B, \{M^B_{x_i}\}_{x_i \in X}, \tau^B\right)$ over $\mathbb{R}$. Clearly, $M^A_{x_i} \in \mathbb{R}^{m \times m}$, $\sigma^A \in \mathbb{R}^{1 \times m}$, $\tau^A \in \mathbb{R}^{m \times 1}$, and $M^B_{x_i} \in \mathbb{R}^{n \times n}$, $\sigma^B \in \mathbb{R}^{1 \times n}$, $\tau^B \in \mathbb{R}^{n \times 1}$. Consider $\mathcal{A} = \left(m, \sigma^A, \{M^A_{x_i}, = 1, 2\}, \tau^A\right)$ defined by*

$$\sigma^A = -\mathbf{1}^T_{k+1}, \quad \tau^A = \begin{bmatrix} -\mathbf{1}_k \\ 1 \end{bmatrix}, \quad M^A_{x_1} = \begin{bmatrix} -\mathbf{1}_{k,k+1} \\ \mathbf{1}^T_{k+1} \end{bmatrix}, \quad M^A_{x_2} = 2 \cdot \begin{bmatrix} -\mathbf{1}_{k,k+1} \\ \mathbf{1}^T_{k+1} \end{bmatrix}$$

and $\mathcal{B} = \left(n, \sigma^B, \{M^B_{x_i}, i = 1, 2\}, \tau^B\right)$ defined by

$$\sigma^B = -\mathbf{1}^T_k, \quad \tau^B = -\mathbf{1}_k, \quad M^B_{x_1} = -\mathbf{1}_{k,k}, \quad M^B_{x_2} = -2 \cdot \mathbf{1}_{k,k}.$$

Furthermore, the design parameter of ZNN is set to $\lambda = 10, 100, 1000$, and the following ICs are used:

- *IC1:* $\mathbf{x}(0) = \mathbf{1}_{702}$,
- *IC2:* $\mathbf{x}(0) = -\mathbf{1}_{702}$,
- *IC3:* $\mathbf{x}(0) = \text{rand}(702, 1)$.

The results of the ZNN-fb model are presented in Figure 4.

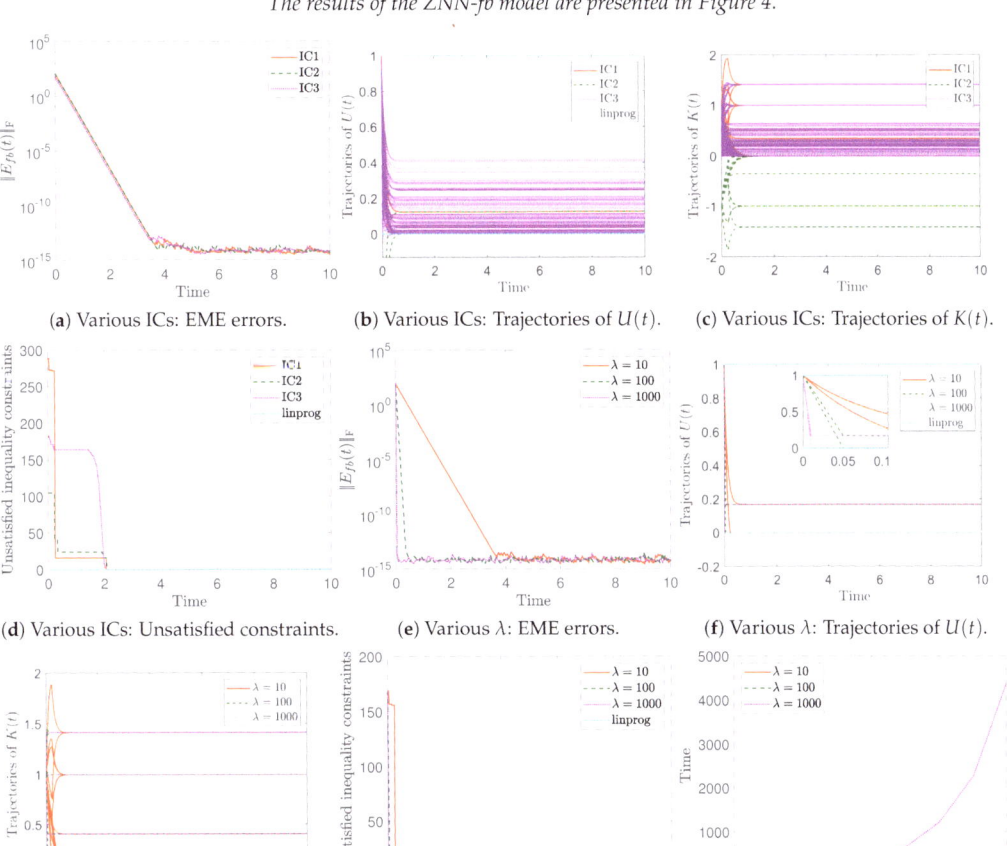

Figure 4. Errors and trajectories in Example 5.

Example 6. Let $m = k+1$, $n = k$ with $k = 10$, $r = 2$, and $X = \{x_1, x_2\}$, and consider WFA $\mathcal{A} = \left(m, \sigma^A, \{M_{x_i}^A\}_{x_i \in X}, \tau^A\right)$ and $\mathcal{B} = \left(n, \sigma^B, \{M_{x_i}^B\}_{x_i \in X}, \tau^B\right)$ over $\mathbb{R}$. Clearly, $M_{x_i}^A \in \mathbb{R}^{m \times m}$, $\sigma^A \in \mathbb{R}^{1 \times m}$, $\tau^A \in \mathbb{R}^{m \times 1}$, and $M_{x_i}^B \in \mathbb{R}^{n \times n}$, $\sigma^B \in \mathbb{R}^{1 \times n}$, $\tau^B \in \mathbb{R}^{n \times 1}$. Consider $\mathcal{A} = \left(m, \sigma^A, \{M_{x_i}^A, = 1, 2\}, \tau^A\right)$ defined by

$$\sigma^A = [2 \cdot \mathbf{1}_k^T \quad 1], \tau^A = \begin{bmatrix} -2 \cdot \mathbf{1}_k \\ -1 \end{bmatrix}, M_{x_1}^A = [-\mathbf{1}_{k+1,k} \quad \mathbf{1}_{k+1}], M_{x_2}^A = 2 \cdot [-\mathbf{1}_{k+1,k} \quad \mathbf{1}_{k+1}]$$

and $\mathcal{B} = \left(n, \sigma^B, \{M_{x_i}^B, i = 1, 2\}, \tau^B\right)$ defined by

$$\sigma^B = 2 \cdot \mathbf{1}_k^T, \quad \tau^B = -2 \cdot \mathbf{1}_k, \quad M_{x_1}^B = -\mathbf{1}_{k,k}, \quad M_{x_2}^B = -2 \cdot \mathbf{1}_{k,k}.$$

Furthermore, the design parameter of ZNN is set to $\lambda = 10, 100, 1000$, and the following ICs are used:

- IC1: $\mathbf{x}(0) = \mathbf{1}_{702}$,
- IC2: $\mathbf{x}(0) = -\mathbf{1}_{702}$,
- IC3: $\mathbf{x}(0) = \text{rand}(702, 1)$.

The results of the ZNN-bb model are presented in Figure 5.

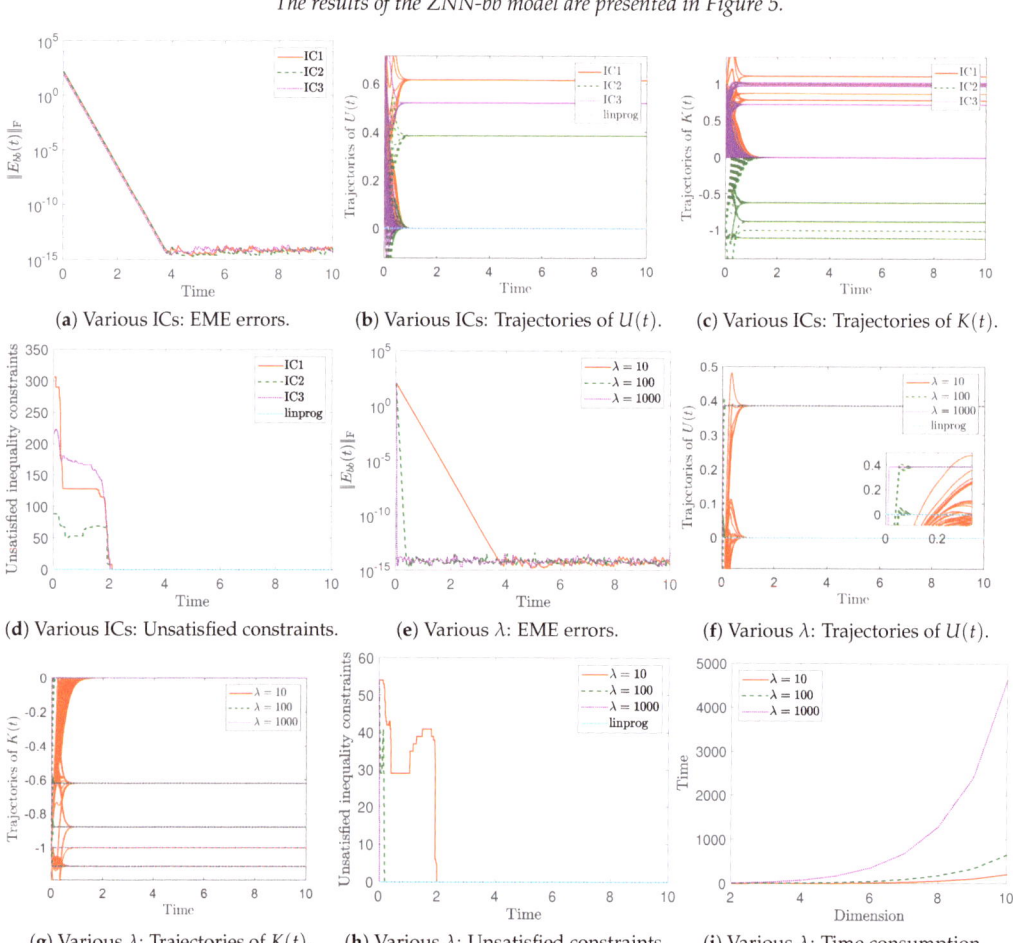

Figure 5. Errors and trajectories in Example 6.

Results Discussion

This part discusses the findings from the four numerical examples that look at how effectively the ZNN models perform.

More precisely, in Example 1, we obtain the next outcomes for the ZNN-fs model by IC1, IC2, and IC3 for $\lambda = 10$. Figure 1e shows the ZNN-fs model's EMEs. All instances start from a huge error price at $t = 0$, and all EMEs conclude in the interval $[10^{-8}, 10^{-7}]$ with a negligible error price at $t = 2$. Put another way, the ZNN-fs model validates Theorem 3 by converging to a value close to zero for three distinct ICs. The trajectories of $U(t)$ and $K(t)$, i.e., the model's solutions, are shown in Figures 1f,g, respectively. These results indicate that $U(t)$ and $K(t)$ do not have similar trajectories via IC1, IC2, and IC3, but their convergence speeds are similar. Therefore, the ZNN-fs model appears to give different solutions for a range of ICs, and its solutions' convergence pattern is proven to be matched up with the convergence pattern of the linked EMEs. Moreover, given that the ZNN-fs model must satisfy $z = 23$ in number inequality constraints, Figure 1h illustrates the number of inequality constraints that remain unsatisfied during the ZNN learning process. This number equals 0 when all of the inequality constraints are satisfied. In this example, this number becomes 0 at $t = 0.5$ for IC1, at $t = 0.8$ for IC2, and at $t = 1.3$ for IC3. Therefore, for a variety of ICs, the ZNN-fs model seems to have varying convergence speeds when it comes to satisfying the inequality constraints. Comparing the ZNN-fs model to the linprog, we see in Figure 1f that the linprog yields different $U(t)$ trajectories than ZNN. Furthermore, we see in Figure 1h that 2 of the 23 inequality constraints are not satisfied by the linprog solution. As a result, the ZNN-fs model outperforms the linprog in this particular example.

In Example 2, under $\lambda = 10, 100, 1000$, the next outcomes for the ZNN-bs model are obtained. Figure 1a shows the ZNN-bs model's EMEs. All instances in this figure start with a large error price at $t = 0$ and conclude at $[10^{-10}, 10^{-7}]$ at $t = 0.02$ for $\lambda = 1000$, at $t = 0.2$ for $\lambda = 100$, and at $t = 2$ for $\lambda = 10$, with a negligible error price. Put another way, the ZNN approach's convergence features are confirmed by the ZNN-bs model's EME, which is dependent on λ, and the ZNN-bs model validates Theorem 5 by converging to a value close to zero. The trajectories of $U(t)$ and $K(t)$, i.e., the model's solutions, are shown in Figure 1b,d, respectively. These results indicate that the convergence speed of the trajectories of $U(t)$ and $K(t)$ is much faster via $\lambda = 1000$ than via $\lambda = 100$, while the convergence speed of the trajectories of $U(t)$ and $K(t)$ is much faster via $\lambda = 100$ than via $\lambda = 10$. Also, it is observable that $U(t)$ and $K(t)$ have similar trajectories with each other via $\lambda = 10, 100, 1000$, respectively. So, the ZNN-bs model appears to give the same $U(t)$ and $K(t)$ solutions for a range of λ values, and its solutions' convergence pattern is proven to be matched up with the convergence pattern of the linked EMEs. Moreover, given that the ZNN-bs model must satisfy $z = 30$ in the number inequality constraints, Figure 1d illustrates the number of inequality constraints that remain unsatisfied during the ZNN learning process. This number becomes 0 at $t = 0.3$ for $\lambda = 10$, at $t = 0.05$ for $\lambda = 100$, and at $t = 0.005$ for $\lambda = 1000$. Therefore, for higher values of λ, the ZNN-bs model seems to have faster convergence speeds when it comes to satisfying the inequality constraints. Comparing the ZNN-bs model to the linprog, we see in Figure 1b that the linprog yields different $U(t)$ trajectories than ZNN. Furthermore, we see in Figure 1d that 3 of the 30 inequality constraints are not satisfied by the linprog solution. As a result, the ZNN-bs model outperforms the linprog in this example.

In Examples 3–6, we obtain the next outcomes for the ZNN-fs, ZNN-bs, ZNN-fb, and ZNN-bb models by IC1, IC2, and IC3 for $\lambda = 10$. Figures 2a, 3a, 4a and 5a show the ZNN model's EMEs. All instances start from a huge error price at $t = 0$, and all EMEs conclude in the interval $[10^{-15}, 10^{-13}]$ with a negligible error price at $t = 3.6$. Put another way, the ZNN-fs, ZNN-bs, ZNN-fb, and ZNN-bb models validate Theorems 3, 5, 7 and 9, respectively, by converging to a value close to zero for two distinct ICs. The trajectories of $U(t)$, generated by the ZNN-fs, ZNN-bs, ZNN-fb, and ZNN-bb models, are shown in Figures 2b, 3b, 4b and 5b, and the trajectories of $K(t)$ are shown in Figures 2c, 3c, 4c and 5c, respectively. For each case, these results indicate that $U(t)$ and $K(t)$ do not have similar trajectories via IC1, IC2, and IC3, but their convergence speeds are similar. Therefore, all ZNN models appear to give different solutions for a range of ICs, and their solutions' convergence pattern is proven to be matched up with the convergence pattern of the linked EMEs. Moreover, given that the ZNN-fs and ZNN-bs models must satisfy $z = 320$ in the number inequality constraints and the ZNN-fb and ZNN-bb models must satisfy $y = 592$ in the number inequality constraints, Figures 2d, 3d, 4d and 5d illustrate the number of inequality constraints that remain unsatisfied during the ZNN learning process. This number becomes 0 at around $t = 2$ for all ICs. Therefore, for a variety of ICs, the ZNN models seem to have varying convergence rates when it comes to satisfying the inequality constraints.

Additionally, the next outcomes for the ZNN-fs, ZNN-bs, ZNN-fb, and ZNN-bb models are obtained in Examples 3–6 under $\lambda = 10, 100, 1000$. Figures 2e, 3e, 4e and 5e show the ZNN models' EMEs. All instances in these figures start with a large error price at $t = 0$ and conclude at $[10^{-16}, 10^{-14}]$ at $t = 0.04$ for $\lambda = 1000$, at $t = 0.4$ for $\lambda = 100$, and at $t = 3.8$ for $\lambda = 10$, with a negligible error price. Put another way, the ZNN approach's convergence features are confirmed by the ZNN models' EME, which is dependent on λ, and the ZNN-fs, ZNN-bs, ZNN-fb, and ZNN-bb models validate Theorems 3, 5, 7 and 9, respectively, by convergence to a value close to zero. The trajectories of $U(t)$ generated by the ZNN-fs, ZNN-bs, ZNN-fb, and ZNN-bb models are shown in Figures 2f, 3f, 4f and 5f, and the trajectories of $K(t)$ are shown in Figures 2g, 3g, 4g and 5g, respectively. These results indicate that the convergence speed of the trajectories of $U(t)$ and $K(t)$ is much faster via $\lambda = 1000$ than via $\lambda = 100$, while the convergence speed of the trajectories of $U(t)$ and $K(t)$ is much faster via $\lambda = 100$ than via $\lambda = 10$. Also, it is observable that $U(t)$ and $K(t)$ have similar trajectories via $\lambda = 10, 100$, and 1000, respectively. So, for each case, the ZNN model appears to give the same $U(t)$ and $K(t)$ solutions for a range of λ values, and its solutions' convergence pattern is proven to be matched up with the convergence pattern of the linked EMEs. Moreover, given that the ZNN-fs and ZNN-bs models must satisfy $z = 320$ in the number of inequality constraints and the ZNN-fb and ZNN-bb models must satisfy $y = 592$ in the number of inequality constraints, Figures 2h, 3h, 4h and 5h illustrate the number of inequality constraints that remain unsatisfied during the ZNN learning process. This number becomes 0 at $t = 1.9$ for $\lambda = 10$, at $t = 0.3$ for $\lambda = 100$, and at $t = 0.03$ for $\lambda = 1000$. Therefore, for higher values of λ, the ZNN models seem to have faster convergence speeds when it comes to satisfying the inequality constraints.

Comparing the ZNN models in Examples 3–6 to the `linprog`, we see in Figures 2b,f, 3b,f, 4b,f and 5b,f that the `linprog` yields different $U(t)$ trajectories than ZNN. It is important to note that the zero solution is produced by the `linprog` in Example 6 and that 10 of the 320 inequality constraints are not satisfied by the `linprog` solution in Example 3. Thus, the ZNN model performs similarly to the `linprog` in Examples 4–6, while the ZNN model outperforms the `linprog` in Example 3. Furthermore, Figures 2i, 3i, 4i and 5i show the time consumption of the ZNN-fs, ZNN-bs, ZNN-fb, and ZNN-bb models in Examples 3–6, respectively, using the MATLAB R2022a environment on an Intel® Core™ i5-6600K CPU 3.50 GHz, 16 GB RAM, running on Windows 10 64 bit Operating System. In these figures, as the dimensions of the matrices (i.e., the value of k) rise, we find that the ZNN models' time consumption increases considerably more via $\lambda = 1000$ than via $\lambda = 100$, and that the ZNN models' time consumption increases considerably more via $\lambda = 100$ than via $\lambda = 10$. Therefore, for higher values of λ, the ZNN models seem to have a higher time consumption.

When everything is considered, the ZNN-fs, ZNN-bs, ZNN-fb, and ZNN-bb models perform admirably in finding the solution of Equations (3)–(6), respectively. Upon comparing the ZNN models to the `linprog`, it is discovered that each ZNN model exhibits comparable or superior performance to the `linprog`. Additionally, all ZNN model performances are affected by the value of λ, and their solutions are affected by the value of the ICs. Keep in mind that the values of λ and the ICs in the experiments of this section were chosen at random. As a corollary, the approximation to the TSOL, $x^*(t)$, in the ZNN-fs, ZNN-bs, ZNN-fb, and ZNN-bb models, is achieved faster via $\lambda = 1000$ than via $\lambda = 100$ and $\lambda = 10$, while the time consumption is higher via $\lambda = 1000$ than via $\lambda = 100$ and $\lambda = 10$.

5. Concluding Remarks

Practically, this research is focused on solving the equivalence problem (determining whether two automata determine the same word function) or solving the containment problem (determining whether the word function of one WFA is bounded from above by the word function of another). Our intention was to unify two important topics, namely, the zeroing neural network (ZNN) and the existence of forward and backward simulations and bisimulations for weighted finite automata (WFA) over the field of real numbers $\mathbb{R}$. Two types of quantitative simulations and two types of bisimulations were defined as solutions to particular systems of matrix and vector inequations over $\mathbb{R}$. This research was aimed at the development and analysis of two novel ZNN models, termed as ZNN-bs and ZNN-fs, for addressing the systems of matrix and vector inequations involved in simulations as well as at bisimulations between WFA and two novel ZNN models, termed ZNN-fb and ZNN-bb, for addressing the systems of matrix and vector inequations involved in bisimulations between WFA. The problem considered in this paper requires solving a system of two vector inequalities and a couple of matrix inequalities. Using positive slack matrices, the required matrix and vector inequations were transformed into corresponding equations which are solvable by the proposed ZNN dynamical systems. A detailed convergence analysis was considered. Numerical examples were performed with different initial state matrices. A comparison with a known LP approach proposed in [7] was presented, and the better performance of the ZNN design was confirmed. The models solved in current research utilized the development of ZNN dynamics based on several inequations and Zhang error functions. The derived models can be viewed as extensions from equations to inequations of ZNN algorithms established upon a few error functions. Such models have been investigated in several papers, such as [31–34].

Seen more generally, the research described in this paper shows that the ZNN design is usable in solving systems of matrix and vector inequations in linear algebra. Further research can be aimed at solving the minimization problems (determining an automaton with the minimal number of states equivalent to a given automaton).

Simulations and bisimulations have already been studied through solving systems of matrix inequalities in the context of fuzzy finite automata [2,3], nondeterministic automata [4], WFA over an additively idempotent semiring [5], and max-plus automata [6]. The methodology used there, based on the concept of residuation, is fundamentally different from the methodology applied in this article to WFA over the field of real numbers. Perhaps some general ideas of this article could be applied to solving systems of matrix inequalities in the context of fuzzy finite automata, for example, the use of neuro-fuzzy systems (fuzzy neural networks), which could be the topic of our future research. On the other hand, the proposed methodology could be more directly applied to some special WFA over a field of real numbers, such as WFA over a semiring of nonnegative real numbers and probabilistic automata. This will also be one of the topics of our future research.

Author Contributions: Conceptualization, M.Ć., P.S.S. and S.D.M.; methodology, P.S.S. and S.D.M.; software, S.D.M.; validation, M.Ć., P.S.S. and D.K.; formal analysis, P.S.S., M.Ć., P.B. and D.K.; investigation, P.S.S., S.D.M. and D.K.; resources, S.D.M.; data curation, S.D.M. and D.K.; writing—original draft preparation, M.Ć. and P.S.S.; writing—review and editing, P.S.S., S.D.M. and M.Ć.; visualization, S.D.M.; supervision, M.Ć. and P.S.S.; project administration, P.B., P.S.S. and D.K.; funding acquisition, P.B. and D.K. All authors have read and agreed to the published version of the manuscript.

Funding: This work was supported by the Ministry of Science and Higher Education of the Russian Federation (Grant No. 075-15-2022-1121).

Data Availability Statement: The data that support the findings of this study are available on request to the authors.

Acknowledgments: Predrag Stanimirović and Miroslav Ćirić acknowledge the support by the Science Fund of the Republic of Serbia, Grant No. 7750185, Quantitative Automata Models: Fundamental Problems and Applications—QUAM. Predrag Stanimirović and Miroslav Ćirić are also supported by the Ministry of Science, Technological Development and Innovation, Republic of Serbia, Contract No. 451-03-65/2024-03/200124.

Conflicts of Interest: The authors declare no conflicts of interest.

References

1. Ćirić, M.; Ignjatović, J.; Stanimirović, P.S. Bisimulations for weighted finite automata over semirings. *Res. Sq.* **2022** . [CrossRef]
2. Ćirić, M.; Ignjatović, J.; Damljanović, N.; Bašić, M. Bisimulations for fuzzy automata. *Fuzzy Sets Syst.* **2012**, *186*, 100–139. [CrossRef]
3. Ćirić, M.; Ignjatović, J.; Jančić, I.; Damljanović, N. Computation of the greatest simulations and bisimulations between fuzzy automata. *Fuzzy Sets Syst.* **2012**, *208*, 22–42. [CrossRef]
4. Ćirić, M.; Ignjatović, J.; Bašić, M.; Jančić, I. Nondeterministic automata: Equivalence, bisimulations, and uniform relations. *Inf. Sci.* **2014**, *261*, 185–218. [CrossRef]

5. Damljanović, N.; Ćirić, M.; Ignjatović, J. Bisimulations for weighted automata over an additively idempotent semiring. *Theor. Comput. Sci.* **2014**, *534*, 86–100. [CrossRef]
6. Ćirić, M.; Micić, I.; Matejić, J.; Stamenković, A. Simulations and bisimulations for max-plus automata. *Discret. Event Dyn. Syst.* **2024**, *34*, 269–295. [CrossRef]
7. Urabe, N.; Hasuo, I. Quantitative simulations by matrices. *Inf. Comput.* **2017**, *252*, 110–137. [CrossRef]
8. Stanković, I.; Ćirić, M.; Ignjatović, J. Bisimulations for weighted networks with weights in a quantale. *Filomat* **2023**, *37*, 3335–3355.
9. Doyen, L.; Henzinger, T.A.; Raskin, J.F. Equivalence of labeled Markov chains. *Int. J. Found. Comput. Sci.* **2008**, *19*, 549–563. [CrossRef]
10. Balle, B.; Gourdeau, P.; Panangaden, P. Bisimulation metrics and norms for real weighted automata. *Inf. Comput.* **2022**, *282*, 104649. [CrossRef]
11. Boreale, M. Weighted bisimulation in linear algebraic form. In *CONCUR 2009—Concurrency Theory, 20th International Conference, Bologna, Italy, 1–4 September 2009*; Bravetti, M., Zavattaro, G., Eds.; LNCS; Springer: Berlin/Heidelberg, Germany, 2009; Volume 5710, pp. 163–177.
12. Xiao, L.; Jia, L. *Zeroing Neural Networks: Finite-time Convergence Design, Analysis and Applications*; John Wiley & Sons, Inc.: Hoboken, NJ, USA, 2022.
13. Zhang, Y.; Yi, C. *Zhang Neural Networks and Neural-Dynamic Method*; Nova Science Publishers, Inc.: New York, NY, USA, 2011.
14. Zhang, Y.; Ge, S.S. Design and analysis of a general recurrent neural network model for time-varying matrix inversion. *IEEE Trans. Neural Netw.* **2005**, *16*, 1477–1490. [CrossRef] [PubMed]
15. Wu, W.; Zheng, B. Two new zhang neural networks for solving time-varying linear equations and inequalities systems. *IEEE Trans. Neural Netw. Learn. Syst.* **2023**, *34*, 4957–4965. [CrossRef] [PubMed]
16. Xiao, L.; Tan, H.; Dai, J.; Jia, L.; Tang, W. High-order error function designs to compute time-varying linear matrix equations. *Inf. Sci.* **2021**, *576*, 173–186. [CrossRef]
17. Li, X.; Yu, J.; Li, S.; Ni, L. A nonlinear and noise-tolerant ZNN model solving for time-varying linear matrix equation. *Neurocomputing* **2018**, *317*, 70–78. [CrossRef]
18. Dai, J.; Li, Y.; Xiao, L.; Jia, L. Zeroing neural network for time-varying linear equations with application to dynamic positioning. *IEEE Trans. Ind. Inform.* **2022**, *18*, 1552–1561. [CrossRef]
19. Wang, T.; Zhang, Z.; Huang, Y.; Liao, B.; Li, S. Applications of Zeroing Neural Networks: A Survey. *IEEE Access* **2024**, *12*, 51346–51363. [CrossRef]
20. Guo, D.; Yan, L.; Zhang, Y. Zhang Neural Networks for online solution of time-varying linear inequalities. In *Artificial Neural Networks*; Rosa, J.L.G., Ed.; IntechOpen: Rijeka, Crooatia, 2016. [CrossRef]
21. Sun, J.; Wang, S.; Wang, K. Zhang neural networks for a set of linear matrix inequalities with time-varying coefficient matrix. *Inf. Process. Lett.* **2016**, *116*, 603–610. [CrossRef]
22. Xiao, L.; Zhang, Y. Two new types of Zhang Neural Networks solving systems of time-varying nonlinear inequalities. *IEEE Trans. Circuits Syst. Regul. Pap.* **2012**, *59*, 2363–2373. [CrossRef]
23. Xiao, L.; Zhang, Y. Zhang Neural Network versus Gradient Neural Network for solving time-varying linear inequalities. *IEEE Trans. Neural Netw.* **2011**, *22*, 1676–1684. [CrossRef]
24. Xiao, L.; Zhang, Y. Different Zhang functions resulting in different ZNN models demonstrated via time-varying linear matrix–vector inequalities solving. *Neurocomputing* **2013**, *121*, 140–149. [CrossRef]
25. Zeng, Y.; Xiao, L.; Li, K.; Li, J.; Li, K.; Jian, Z. Design and analysis of three nonlinearly activated ZNN models for solving time-varying linear matrix inequalities in finite time. *Neurocomputing* **2020**, *390*, 78–87. [CrossRef]
26. Guo, D.; Zhang, Y. A new variant of the Zhang neural network for solving online time-varying linear inequalities. *Proc. R. Soc.* **2012**, *2255*, 2271. [CrossRef]
27. Zheng, B.; Yue, C.; Wang, Q.; Li, C.; Zhang, Z.; Yu, J.; Liu, P. A new super-predefined-time convergence and noise-tolerant RNN for solving time-variant linear matrix–vector inequality in noisy environment and its application to robot arm. *Neural Comput. Appl.* **2024**, *36*, 4811–4827. [CrossRef]
28. Zhang, Y.; Yang, M.; Huang, H.; Xiao, M.; Hu, H. New discrete solution model for solving future different level linear inequality and equality with robot manipulator control. *IEEE Trans. Ind. Inform.* **2019**, *15*, 1975–1984. [CrossRef]
29. Guo, D.; Zhang, Y. A new inequality-based obstacle-avoidance MVN scheme and its application to redundant robot manipulators. *IEEE Trans. Syst. Man Cybern. Part (Appl. Rev.)* **2012**, *42*, 1326–1340. [CrossRef]
30. Kong, Y.; Jiang, Y.; Lou, J. Terminal computing for Sylvester equations solving with application to intelligent control of redundant manipulators. *Neurocomputing* **2019**, *335*, 119–130. [CrossRef]
31. Katsikis, V.; Mourtas, S.; Stanimirović, P.; Zhang, Y. Solving complex-valued time-varying linear matrix equations via QR decomposition with applications to robotic motion tracking and on angle-of-arrival localization. *IEEE Trans. Neural Netw. Learn. Syst.* **2021**, *33*, 3415–3424. [CrossRef] [PubMed]
32. Li, X.; Lin, C.L.; Simos, T.; Mourtas, S.; Katsikis, V. Computation of time-varying 2,3- and 2,4-inverses through zeroing neural networks. *Mathematics* **2022**, *10*, 4759. [CrossRef]
33. Simos, T.; Katsikis, V.; Mourtas, S.; Stanimirović, P. Unique non-negative definite solution of the time-varying algebraic Riccati equations with applications to stabilization of LTV system. *Math. Comput. Simul.* **2022**, *202*, 164–180. [CrossRef]

34. Stanimirović, P.; Mourtas, S.; Mosić, D.; Katsikis, V.; Cao, X.; Li, S. Zeroing Neural Network approaches for computing time-varying minimal rank outer inverse. *Appl. Math. Comput.* **2024**, *465*. [CrossRef]
35. Buchholz, P. Bisimulation relations for weighted automata. *Theor. Comput. Sci.* **2008**, *393*, 109–123. [CrossRef]
36. Hua, C.; Cao, X.; Xu, Q.; Liao, B.; Li, S. Dynamic neural network models for time-varying problem solving: A survey on model structures. *IEEE Access* **2023**, *11*, 65991–66008. [CrossRef]
37. Jin, L.; Li, S.; Liao, B.; Zhang, Z. Zeroing neural networks: A survey. *Neurocomputing* **2017**, *267*, 597–604. [CrossRef]
38. Graham, A. *Kronecker Products and Matrix Calculus with Applications*; Courier Dover Publications: Mineola, NY, USA, 2018.

Disclaimer/Publisher's Note: The statements, opinions and data contained in all publications are solely those of the individual author(s) and contributor(s) and not of MDPI and/or the editor(s). MDPI and/or the editor(s) disclaim responsibility for any injury to people or property resulting from any ideas, methods, instructions or products referred to in the content.

Article

Integrating Fuzzy C-Means Clustering and Explainable AI for Robust Galaxy Classification

Gabriel Marín Díaz [1,*], Raquel Gómez Medina [2] and José Alberto Aijón Jiménez [2]

1. Faculty of Statistics, Complutense University, Puerta de Hierro, 28040 Madrid, Spain
2. Science and Aerospace Department, Universidad Europea de Madrid, Villaviciosa de Odón, 28670 Madrid, Spain; raquel.gomez@universidadeuropea.es (R.G.M.); josealberto.aijon@universidadeuropea.es (J.A.A.J.)
* Correspondence: gmarin03@ucm.es

Abstract: The classification of galaxies has significantly advanced using machine learning techniques, offering deeper insights into the universe. This study focuses on the typology of galaxies using data from the Galaxy Zoo project, where classifications are based on the opinions of non-expert volunteers, introducing a degree of uncertainty. The objective of this study is to integrate Fuzzy C-Means (FCM) clustering with explainability methods to achieve a precise and interpretable model for galaxy classification. We applied FCM to manage this uncertainty and group galaxies based on their morphological characteristics. Additionally, we used explainability methods, specifically SHAP (SHapley Additive exPlanations) values and LIME (Local Interpretable Model-Agnostic Explanations), to interpret and explain the key factors influencing the classification. The results show that using FCM allows for accurate classification while managing data uncertainty, with high precision values that meet the expectations of the study. Additionally, SHAP values and LIME provide a clear understanding of the most influential features in each cluster. This method enhances our classification and understanding of galaxies and is extendable to environmental studies on Earth, offering tools for environmental management and protection. The presented methodology highlights the importance of integrating FCM and XAI techniques to address complex problems with uncertain data.

Keywords: Fuzzy C-Means; explainable AI; XAI; SHAP values; LIME; citizen science; astronomy; machine learning (ML)

MSC: 85A35; 62H30

Citation: Marín Díaz, G.; Gómez Medina, R.; Aijón Jiménez, J.A. Integrating Fuzzy C-Means Clustering and Explainable AI for Robust Galaxy Classification. *Mathematics* **2024**, *12*, 2797. https://doi.org/10.3390/math12182797

Academic Editor: Francisco Rodrigues Lima Junior

Received: 15 August 2024
Revised: 5 September 2024
Accepted: 9 September 2024
Published: 10 September 2024

Copyright: © 2024 by the authors. Licensee MDPI, Basel, Switzerland. This article is an open access article distributed under the terms and conditions of the Creative Commons Attribution (CC BY) license (https://creativecommons.org/licenses/by/4.0/).

1. Introduction

Galaxy Zoo is a citizen science project that has revolutionized the way we classify galaxies [1]. Through the collaboration of volunteers worldwide, we have been able to analyze and classify millions of galaxy images obtained by telescopes like the Sloan Digital Sky Survey (SDSS) [2]. This approach popularizes science, allowing individuals without specialized training to contribute significantly while addressing the immense task of processing astronomical data that would otherwise be impossible for professional astronomers to manage alone.

The drive behind this type of galaxy classification is the need to better understand the universe we live in. Galaxies are the fundamental building blocks of the cosmos, and studying their shapes, structures, and distributions provides critical insights into the formation and evolution of the universe. However, classifying galaxies is not a trivial task; it requires detailed and careful analysis, traditionally dependent on human visual perception.

Human perception of the real world and what we visualize in cosmic images becomes a highly effective resource when properly channeled. In Galaxy Zoo, thousands of volunteers observe and classify galaxies according to various criteria, such as shape, presence of bars, orientation, and more. Each galaxy is evaluated by multiple individuals, introducing a degree of uncertainty into the data, as not all observers perceive the same characteristics in the same way.

In this context, artificial intelligence (AI) and machine learning (ML) play a significant role. We use advanced AI techniques to process these human classifications and convert them into data that are actionable and understandable by computers. One of the approaches we employ is Fuzzy C-Means (FCM) clustering [3], a technique that allows us to handle the inherent uncertainty in data aggregated from multiple observers. FCM helps us group galaxies into clusters based on their morphological characteristics, reflecting the diversity and variability in human perceptions.

Furthermore, to achieve greater interpretability of AI models, we implement explainability methods such as SHAP (SHapley Additive exPlanations) values [4]. These values enable us to better understand the key factors influencing galaxy classification, providing a clear and comprehensible explanation of how and why certain decisions are made by the model.

Furthermore, to achieve greater interpretability of AI models, we implement explainability methods such as LIME (Local Interpretable Model-agnostic Explanations) [5]. LIME enables us to understand the key factors influencing specific predictions by approximating the model locally around the prediction of interest, providing a clear and comprehensible explanation of how and why certain decisions are made by the model at a local level.

Human–machine collaboration in models of this type is key to obtaining explainability. While human intelligence allows us to offer interpretations and identify fundamental features in images, AI can support the classification of hundreds of thousands of images, making the process more efficient and scalable. This interaction is important for the future development of AI ethics, which combines human expertise with predictive modeling.

The novelty of this work lies in providing a methodology that unifies the processing of information collected through the opinions of individuals with diverse profiles, effectively mitigating the noise introduced by any single person's opinion. Essentially, this approach leverages a group-based assessment to average the weight of each identified variable. Additionally, the use of FCM aids in obtaining a clear classification, enabling the analysis of trends and deviations relative to the most popular selections. Finally, the application of Explainable AI (XAI) techniques allows for a comprehensive understanding of the model's classification process, utilizing SHAP values for global predictions and LIME for local predictions.

The primary purpose of this work is to demonstrate how human perceptions can be channeled and processed through AI techniques to generate actionable insights. By doing this, we enhance our understanding of the cosmos and illustrate how these methodologies can be applied to other fields, including sustainability and environmental management on Earth. This interdisciplinary approach highlights the importance of combining human perception with the power of automated data analysis to tackle complex problems and generate deep and practical insights.

In the remainder of this paper, we will develop and implement the Fuzzy C-Means (FCM) and Explainable AI (XAI) model, following the structure outlined below: In Section 2, we will review the current state of studies on galaxy classification using machine learning (ML) techniques, focusing specifically on Fuzzy C-Means. We will delve into studies that apply a combination of clustering metrics and explainability techniques. Section 3 will describe the Fuzzy C-Means and Explainable AI (FCM-XAI) methodological framework, which enhances both the accuracy and transparency of the classification models. In Section 4, we will apply the methodological framework, FCM-XAI, to galaxy classification based on the data provided by Galaxy Zoo. The analysis will demonstrate how the developed methodology can be used in various contexts that integrate human perception with AI. Finally, Sections 5 and 6 will present the discussions, conclusions, and future work.

2. Related Work

In the related work review, we will follow a structured approach to explore the relevant work in this domain. First, we will examine the studies that have utilized Fuzzy C-Means (FCM) as a clustering tool, providing a broad overview of its applications across various fields. Next, we will focus on the studies that incorporate Explainable Artificial Intelligence (XAI) techniques, emphasizing the remarkable increase in the use of XAI in recent years. Following this, we will explore the integration of XAI with FCM and discuss the related works that have combined these two methodologies. Subsequently, we will analyze the use of machine learning and deep learning techniques in the context of galaxy classification, identifying trends and gaps in this field. Finally, it will be observed that the methodology combining FCM with XAI has not been previously employed in the specific context of galaxy identification, highlighting the novelty of our proposed approach.

2.1. Fuzzy C-Means and XAI

Fuzzy C-Means (FCM), introduced by Bezdek in 1981 [3], is a versatile and widely used technique that assigns data points to clusters with varying degrees of membership. Unlike traditional hard clustering methods, which rigidly allocate each data point to a single cluster, FCM allows for more flexibility by enabling each point to belong to multiple clusters, with specific membership values assigned. This approach offers a more realistic representation of data, particularly in situations where overlapping clusters and uncertainty need to be managed.

FCM has found significant application in various fields due to its capability to handle complex, uncertain data. In medical imaging, for instance, FCM is extensively used for segmenting medical images, helping to identify and delineate regions of interest such as tumors, tissues, and organs [6]. An example is its role in brain MRI segmentation, where it distinguishes between different brain tissues, such as gray matter, white matter, and cerebrospinal fluid, aiding in the diagnosis of neurological conditions [7]. In remote sensing and environmental monitoring, FCM is applied to classify land cover types using satellite imagery to segment features such as forests, water bodies, and urban areas [8]. Similarly, in image processing and pattern recognition, FCM helps improve tasks like image compression, enhancement, and feature extraction [9]. It plays a key role in face recognition systems by grouping facial features [10]. FCM is also used in market segmentation, where companies use it to identify different groups of customers based on their behavior, preferences, and demographic characteristics. This segmentation allows companies to create targeted marketing strategies, increasing customer satisfaction and engagement [11,12].

Lastly, in bioinformatics, FCM is applied to gene expression data analysis, clustering genes with similar expression patterns. This clustering contributes to understanding gene functions and interactions, further advancing genetic research and personalized medicine [13].

Figure 1 shows a preliminary review of research related to Fuzzy C-Means (FCM) in the Web of Science Core Collection, identifying a total of 1282 publications.

As data storage, management, and information processing speed continue to multiply, the successes achieved by AI in predictive models, along with occasional issues arising from the misuse of AI, such as biased data or noise, have made it imperative to approach AI projects with a focus on the interpretability and explainability of machine learning algorithms, with special emphasis on the ethical processes involved in decision-making. It is evident that research on interpretability has evolved significantly in recent years, with a notable increase in the most recent periods, as illustrated in Figure 2 and Table 1.

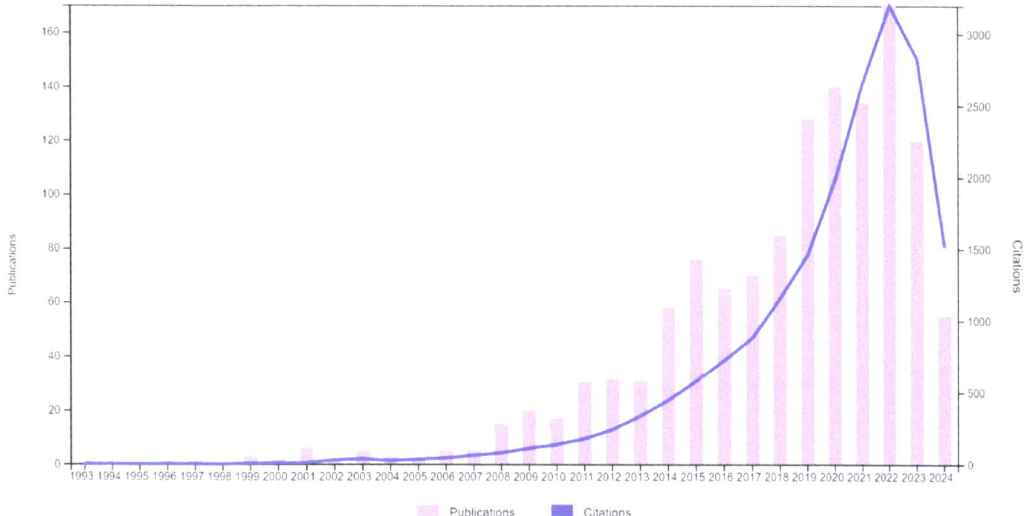

Figure 1. Publications (1282) and citations. TS = (FUZZY C-MEANS CLUSTERING).

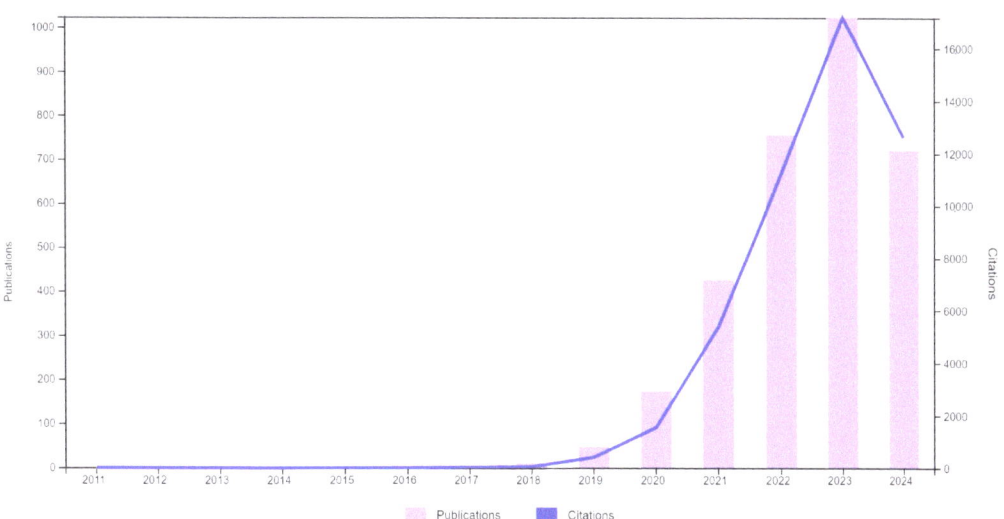

Figure 2. Publications (3178) and citations. TS = ("XAI" OR "EXPLAINABLE ARTIFICIAL INTELLIGENCE").

Table 1. Publications, XAI or Explainable Artificial Intelligence.

Publication Years	Record Count	% of 3178
2024	722	22.72
2023	1023	32.19
2022	757	23.82
2021	427	13.44
2020	175	5.51
2019–2014	74	2.32

To further refine the search, we included the use of Explainable AI (XAI) techniques in conjunction with FCM. Figure 3 provides a graphical representation, followed by a detailed breakdown of the referenced publications.

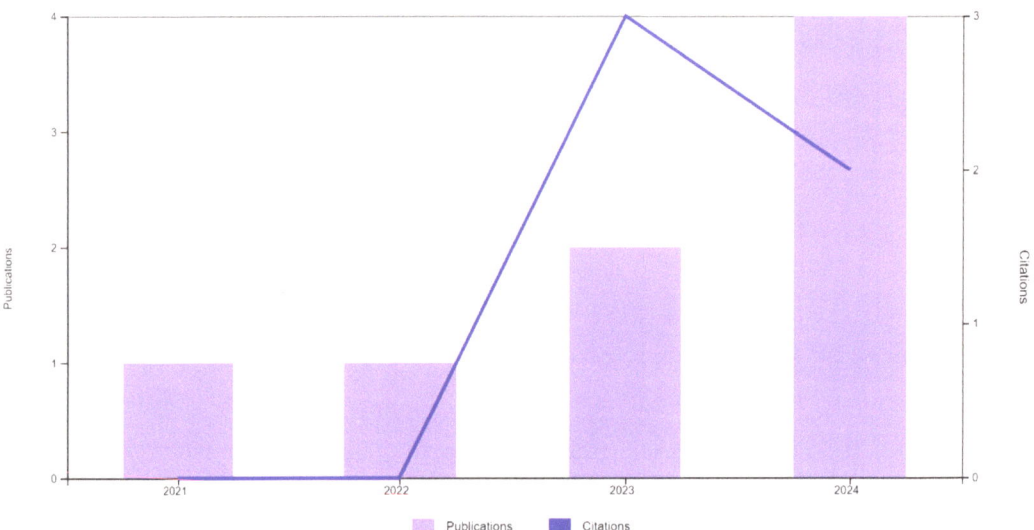

Figure 3. Publications (8) and citations. TS = ("FUZZY C-MEANS") AND TS = ("XAI" OR "EXPLAINABLE ARTIFICIAL INTELLIGENCE").

In the study by V. V. Saradhi et al. [14], the techniques used are based on non-agnostic robust regression models for fuzzy models, providing a global approach to interpretability. I. Ghosh et al. [15] applied agnostic techniques based on feature importance, offering a general view of the most influential variables in the prediction. Additionally, the interpretability methods used are global rather than focused on local explanations. On the other hand, Kmita et al. [16] employ non-agnostic techniques using Semi-Supervised Fuzzy C-Means. While this approach combines labeled and unlabeled data, it lacks an explainable component that connects the model's decisions with human-provided data. Sevas et al. [17] implement agnostic techniques using feature importance, providing a global view and explanation of the predictions. The study by Sirapangi et al. [18] uses global interpretation techniques such as Deep SHAP, offering detailed global explanations of the model. Arabikhan et al. [19] employ non-agnostic techniques based on fuzzy networks, allowing for a mathematical global interpretation of the model. In the work of Priya et al. [20], non-agnostic techniques like Long Short-Term Memory combined with polynomial kernels are used, focusing on temporal prediction. Finally, Akpan et al. [21] use artificial neural networks (ANNs) and non-agnostic techniques for classification and prediction but lack global and local explainable dimensions.

These studies predominantly employ global interpretability techniques such as feature importance, deep SHAP, and fuzzy networks, with no direct human interaction for data collection. In contrast, our study uniquely integrates human opinions through Galaxy Zoo to guide the fuzzification process, combining both global and local interpretability methods (SHAP and LIME) for a more comprehensive understanding of galaxy classifications using FCM. This human–AI hybrid approach offers a novel perspective not explored in the studies mentioned.

2.2. Galaxy Classification

The number of studies related to galaxy classification using machine learning algorithms is depicted in Figure 4. A total of 53 studies have been conducted over the past 10 years. However, as evidenced, none of these studies incorporate algorithm interpretability.

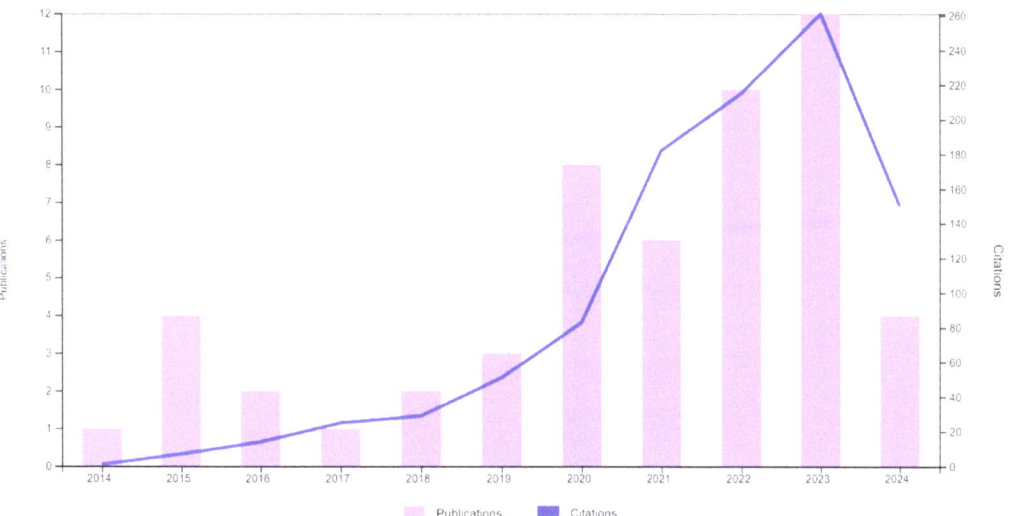

Figure 4. Publications (53) and citations. TS = ("GALAXY CLASSIFICATION") AND TS = ("MACHINE LEARNING" OR "DEEP LEARNING").

Among the studies presented in Figure 4, those from the past two years have been selected, each focusing on different methods for galaxy classification. However, there are key distinctions when compared to our work.

Y. Wu et al. [22] utilize convolutional neural networks (CNNs) for galaxy spectral classification, focusing on deep learning without incorporating interpretability methods, unlike our approach, which uses Fuzzy C-Means (FCM) clustering combined with explainability techniques like SHAP and LIME. Similarly, S. Ndung'u et al. [23] address the morphological classification of radio galaxies but do not integrate explainability or human-driven data. Ma et al. [24] employ hierarchical data learning for galaxy image classification, while our study emphasizes the combination of FCM and XAI to handle uncertainty in user classifications, providing greater transparency. Stoppa et al. [25] use CNNs for galaxy classification without focusing on explainability, whereas our model adds interpretability layers to the classification process. Schneider et al. [26] apply pretraining for galaxy classification, but their approach is purely algorithmic, lacking the human–machine collaboration seen in our study. Lastly, Senel [27] explores hyperparameter optimization for galaxy classification, whereas we prioritize the integration of XAI with FCM to make the classification process understandable, highlighting the unique contributions of human observations in our methodology.

Overall, while the studies focus on improving classification accuracy, they do not address the interpretability and human-driven aspects that our work integrates.

2.3. Conclusion

The review of related work highlights the novelty and distinctiveness of the approach taken in this study. Previous research utilizing Fuzzy C-Means (FCM) clustering and Explainable AI (XAI) techniques has been applied in various fields, such as medical imaging, environmental monitoring, and financial forecasting. However, these studies typically focus on algorithmic precision and theoretical models, with little emphasis on integrating human-driven data into the clustering process. Moreover, while some research applies XAI, it is often limited to global interpretability methods, such as feature importance, without addressing the local interpretability needed for individual predictions.

In the domain of galaxy classification, most studies rely heavily on deep learning techniques, focusing on machine-based image analysis. These approaches, while accurate, lack the interpretability that XAI methods provide. They also fail to incorporate the uncertainty introduced by human classification, which is key for a dataset like Galaxy Zoo, where non-expert volunteers provide classifications based on visual assessments.

Through the integration of FCM and XAI, this work bridges the gap between human decision-making and AI, providing a transparent and ethically informed framework for galaxy classification that can be extended to other domains.

3. Methodology

The methodology used in this study is based on the Knowledge Discovery in Databases (KDD) process [28], which is foundational for structuring data analysis and mining tasks. The KDD process involves several key stages, as depicted in Figure 5:

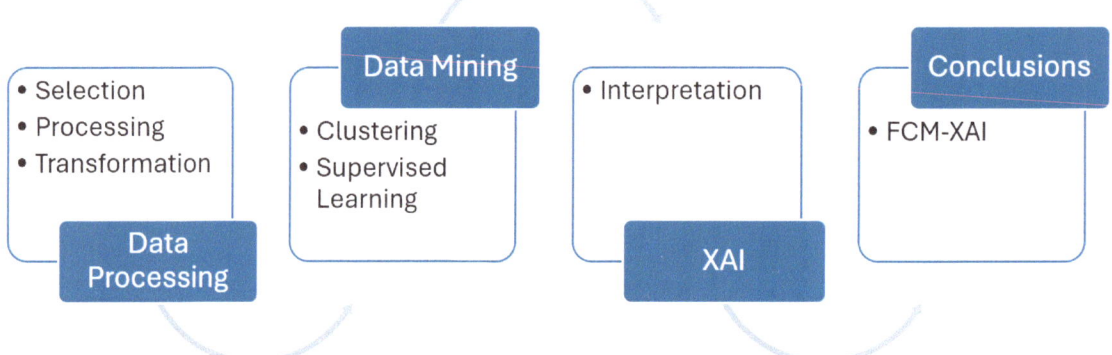

Figure 5. Methodology.

The data collected in this study originates from the Galaxy Zoo project [1]. After selecting this data source, we proceed to develop each of the stages that constitute the model:

3.1. Data Processing

In the field of galaxy classification, the objectives should be clearly defined, including the identification of the key features that influence the classification of galaxies based on their morphological features. Once the objectives have been defined, the next step involves selecting relevant data by focusing on important features from the dataset, such as galaxy shape, the presence of spiral structures, and bulge prominence, which are likely to impact galaxy classification and align with the study's goals. An Exploratory Data Analysis (EDA) is then conducted, which includes analyzing the distribution of galaxy shapes and structures, as well as studying the correlations between features to identify relationships and redundancies. Lastly, a standard scaler is applied to normalize the data, ensuring that each feature has a mean of 0 and a standard deviation of 1 for proper processing.

3.2. Data Mining

3.2.1. Fuzzy C-Means (FCM)

Fuzzy C-Means (FCM) is adept at handling scenarios where data points exhibit characteristics of multiple clusters [3]. This is particularly pertinent in the context of galaxy classification, where morphological features may not be distinctly categorized (Table 3).

FCM allows galaxies to possess varying degrees of membership across multiple clusters, thereby offering a more nuanced classification approach.

In practical applications, including astronomical datasets, entities frequently display a blend of characteristics from distinct categories. FCM's capability to manage overlapping memberships makes it an appropriate choice for such intricate datasets.

The objective of FCM clustering is to group galaxies into clusters based on their morphological characteristics. This method allows for overlapping clusters, meaning each galaxy can belong to multiple clusters with varying degrees of membership.

1. Initialization:

To begin the clustering process, the number of clusters (c) must first be set, representing the desired number of groups into which the data will be partitioned. The next step involves initializing the cluster centers (v_j), where initial cluster centers are randomly selected as starting points for the algorithm to iterate upon.

$$v_j^{(0)}, for\ j = 1, 2, \ldots, c$$

2. Membership Calculation:

Degree of Membership ($u_{\{ij\}}$) of each galaxy x_i to each cluster v_j is calculated using the distance between x_i and v_j. The membership degree is calculated as follows:

$$u_{ij} = \frac{1}{\sum_{k=1}^{c}\left(\frac{d_{ij}}{d_{ik}}\right)^{\frac{2}{(m-1)}}} \quad (1)$$

In the given context, where $u_{\{ij\}}$ is the membership degree of galaxy x_i in cluster v_j, and $d_{\{ij\}}$ is the distance between galaxy x_i and cluster center v_j, typically, the Euclidean distance is as follows:

$$d_{ij} = |x_i - v_j| \quad (2)$$

m is the fuzziness parameter (typically $m = 2$), which determines the level of cluster fuzziness.

3. Cluster Center Update:

Update the cluster centers (v_j) based on the weighted average of the galaxies' features, with weights given by their membership degrees:

$$v_j^{(t+1)} = \frac{\sum_{i=1}^{n} u_{ij}^m x_i}{\sum_{i=1}^{n} u_{ij}^m} \quad (3)$$

In the given context, where $v_j^{(t+1)}$ is the updated cluster center, and u_{ij}^m is the membership degree raised to the power of m.

4. Iteration:

Continue iterating between membership calculation and cluster center update until convergence is achieved. Convergence is typically defined as the point where the changes in membership degrees and cluster centers are below a predefined threshold. Mathematically, this can be represented as follows:

$$|v_j^{(t+1)} - v_j^{(t)}| < \epsilon \text{ and } |u_{ij}^{(t+1)} - u_{ij}^{(t)}| < \epsilon \quad (4)$$

where ϵ is a small positive constant representing the convergence threshold, $v_j^{(t+1)}$ and $v_j^{(t)}$ are the cluster centers at iterations t + 1 and t, respectively, and $u_{ij}^{(t+1)}$, $u_{ij}^{(t)}$ are the membership degrees at iterations t + 1 and t, respectively.

FCM clustering allows galaxies to be grouped based on their morphological characteristics. It allows for overlapping clusters, providing a more flexible and nuanced clustering approach compared to traditional hard clustering methods. By iterating through membership calculations and cluster center updates, FCM can accurately reflect the inherent uncertainty and variability in galaxy data.

3.2.2. Random Forest (RF) for Predictive Modeling

Random Forest (RF) is an ensemble learning method known for its high accuracy and robustness in classification tasks [29]. It operates by building multiple decision trees during training and providing the most frequent class as the output for classification. This method reduces overfitting and improves the generalizability of the model.

RF can effectively manage datasets with many features, which are common in astronomical data. Its ability to manage a diverse set of input features without requiring extensive preprocessing makes it an ideal choice for our problem.

One of the key advantages of RF is its ability to measure the importance of each feature in making predictions. This factor is key for understanding which morphological characteristics of galaxies are most influential in determining their cluster memberships.

RF determines the intrinsic cluster membership of a galaxy based on its features. While FCM assigns degrees of membership across multiple clusters, RF provides a definitive prediction of the most likely cluster for a new galaxy, using its morphological characteristics.

Let X be the feature matrix and y be the response vector. The steps are as follows:

1. Bootstrapping and Bagging: Bootstrapping is a statistical technique in which multiple datasets are generated by sampling with replacement from the original dataset. This method forms the basis for Bagging (Bootstrap Aggregating), where each decision tree is trained on a different bootstrapped dataset. Specifically, B bootstrapped datasets, denoted as (X_b^*, y_b^*), are created from the original dataset (X, y). This approach enhances the stability and accuracy of machine learning models by reducing variance and improving generalization.

2. Tree Construction: At each node, a random subset of features is chosen, and the best possible split is determined from within this subset. The tree continues to grow until a specified stopping condition is reached, such as a maximum depth or a minimum number of samples per leaf. For each bootstrapped dataset (X_b^*, y_b^*), a decision tree T_b is constructed by selecting the best split from a random subset of features at every node.

3. Voting and Prediction: For classification tasks, each tree in the forest casts a vote for the predicted class. The final prediction is the class with the majority vote across all trees. For regression tasks, the overall prediction is derived by averaging the predictions from each tree. Given a new input x, the Random Forest prediction $\hat{y}$ is calculated as follows:

$$\hat{y} = \text{mode}\{T_b(x)\} \text{ for classification} \qquad (5)$$

$$\hat{y} = \frac{1}{B} \sum_{b=1}^{B} T_b(x) \text{ for regression} \qquad (6)$$

The use of RF to predict galaxy cluster membership is justified by its robustness, accuracy, and ability to manage high-dimensional data. Its non-linear nature and feature importance measures further enhance its suitability for this task. The inherent non-interpretability of RF paves the way for the application of XAI techniques, ensuring that we can obtain deterministic information about the model's decision-making process and build confidence in its predictions.

3.3. Interpretable Machine Learning

Explainability techniques are necessary because Random Forest, although robust and accurate, is considered a "black-box" model. This means that, although it can make accurate predictions, the internal decision-making process is not inherently transparent or interpretable. Therefore, Explainable AI (XAI) techniques are used to understand and trust the model's predictions [30].

1. Inherently Interpretable Models [31] are designed to be simple and transparent, making their decision-making process easy to understand. Examples include linear regression, decision trees, and logistic regression. These models are often used when interpretability is crucial, such as in regulatory environments or when stakeholder trust is paramount.
2. Black-box Models [32] are complex, and their internal workings are not easily interpretable. Examples include Random Forest, neural networks, and support vector machines. These models are often favored for their high predictive power, especially when handling large and complex datasets.
3. Model-agnostic Explainability [33] techniques can be applied to any machine learning model, regardless of its internal complexity. Examples include SHAP (SHapley Additive exPlanations) and LIME (Local Interpretable Model-agnostic Explanations), which work by approximating or analyzing the model's output rather than its internal structure.
4. Model-dependent Explainability [34] techniques are specific to certain types of models and leverage their internal structure to provide explanations. Examples include decision tree feature importances and neural network saliency maps. These methods are often more efficient but less generalizable than model-agnostic approaches.

3.3.1. SHAP (Shapley Additive exPlanations)

Derived from cooperative game theory, SHAP values offer a standardized measure of feature importance [35]. They explain the contribution of each feature to the prediction for each individual data point.

Calculating the Shapley value for a feature involves taking the average of its marginal contributions over all possible feature combinations.

Mathematically, it is given by the following:

$$\phi_j = \sum_{S \subseteq F \setminus \{j\}} \frac{|S|!(|F| - |S| - 1)!}{|F|!} \left(f_{S \cup \{j\}} \left(x_{S \cup \{j\}} \right) - f_S(x_S) \right) \quad (7)$$

where F is the set of all features, S is a subset of features excluding j, and $f_S(x_S)$ is the prediction from the model using the feature subset S.

SHAP values help in understanding the overall importance of each morphological feature in determining galaxy cluster memberships. This provides insights into the key factors that drive the clustering process.

3.3.2. LIME (Local Interpretable Model-Agnostic Explanations)

LIME provides local explanations by approximating the model's behavior in the vicinity of a particular instance with a simple interpretable model [36].

For a given instance x, LIME perturbs x to generate a set of new instances $\{x'_i\}$ and obtains predictions $\{f(x'_i)\}$ from the original model.

Weights are assigned to these instances based on their proximity to x:

$$\pi(x_i) = exp\left(-\frac{|x - x_i|^2}{\sigma^2}\right) \quad (8)$$

A simple linear model (g) is then fit to the weighted instances to approximate the complex model's behavior locally:

$$g(z') = \underset{g \in G}{argmin} \sum_i \pi(x_i)\big(f(x_i) - g(z'_i)\big)^2 \tag{9}$$

where z' is the binary vector representing the presence or absence of features in x.

LIME can be used to explain why a particular instance belongs to a certain cluster by showing the contribution of each feature to the prediction. This is achieved by analyzing the weights and coefficients of the locally fitted linear model.

LIME allows for explaining individual predictions, showing why a particular galaxy is assigned to a specific cluster. This helps validate the model's decisions on a case-by-case basis.

3.4. FCM–XAI

The integration of Fuzzy C-Means for clustering and Random Forest for predictive modeling, complemented by SHAP and LIME for interpretability, establishes a comprehensive and robust framework for galaxy classification. This approach harnesses the strengths of both FCM and RF, ensuring that model predictions are transparent and interpretable, thus enhancing confidence and reliability in the classification results.

Moreover, the employed methodology is adaptable to various environments where information from multiple sources is processed. By making classification decisions understandable and explainable, this approach increases the robustness and trustworthiness of the outcomes.

4. FCM-XAI Methodology for Galaxy Classification

4.1. Data Collection, Processing, and Transformation

The methodology presented in this study aids in the development of classification models based on specific characteristics. For our research, the Galaxy Zoo data model was employed. The characteristics detailed in this model are summarized in Table 2.

Table 2. Features of the Galaxy Zoo dataset.

	Features
Class1.1	Probability that the galaxy is smooth (featureless).
Class1.2	Probability that the galaxy has features or a disk.
Class1.3	Probability that the image is a star or an artifact.
Class2.1	Probability that the galaxy is edge-on.
Class2.2	Probability that the galaxy is not edge-on.
Class3.1	Probability that the galaxy has a bar.
Class3.2	Probability that the galaxy does not have a bar.
Class4.1	Probability that the galaxy is spiral.
Class4.2	Probability that the galaxy is not spiral.
Class5.1	Probability that the galaxy has no prominent bulge.
Class5.2	Probability that the galaxy's bulge is just noticeable.
Class5.3	Probability that the galaxy's bulge is obvious.
Class5.4	Probability that the galaxy's bulge is dominant.
Class6.1	Probability that the galaxy has odd features.
Class6.2	Probability that the galaxy does not have odd features.
Class7.1	Probability that the galaxy is completely round.
Class7.2	Probability that the galaxy's shape is in-between.
Class7.3	Probability that the galaxy is cigar-shaped.
Class8.1	Probability that the galaxy has a ring.
Class8.2	Probability that the galaxy has a lens or arc.
Class8.3	Probability that the galaxy is disturbed.
Class8.4	Probability that the galaxy is irregular.
Class8.5	Probability that the galaxy has some other strange feature.
Class8.6	Probability that the galaxy is merging.
Class8.7	Probability that the galaxy has a dust lane.
Class9.1	Probability that the galaxy's bulge is round.

Table 2. *Cont.*

	Features
Class9.2	Probability that the galaxy's bulge is boxy.
Class9.3	Probability that the galaxy has no bulge.
Class10.1	Probability that the galaxy's spiral arms are tightly wound.
Class10.2	Probability that the galaxy's spiral arms are moderately wound.
Class10.3	Probability that the galaxy's spiral arms are loosely wound.
Class11.1	Probability that the galaxy has one spiral arm.
Class11.2	Probability that the galaxy has two spiral arms.
Class11.3	Probability that the galaxy has three spiral arms.
Class11.4	Probability that the galaxy has four spiral arms.
Class11.5	Probability that the galaxy has more than four spiral arms.
Class11.6	Probability that the number of spiral arms cannot be determined.

The selection of the following features from the Galaxy Zoo model was guided by their relevance to galaxy classification based on morphological properties [37]. These features capture essential aspects of a galaxy's shape, structure, and specific characteristics that are determinant for accurate and interpretable classification (Table 3).

Table 3. Selected features of Galaxy Zoo dataset.

	Selected Features
GalaxyID	Unique identifier for each galaxy.
Class1.1	Probability that the galaxy is smooth (featureless).
Class2.1	Probability that the galaxy is edge-on.
Class3.1	Probability that the galaxy has a bar.
Class4.1	Probability that the galaxy is spiral.
Class5.1	Probability that the galaxy has no prominent bulge.
Class5.2	Probability that the galaxy's bulge is just noticeable.
Class5.3	Probability that the galaxy's bulge is obvious.
Class5.4	Probability that the galaxy's bulge is dominant.
Class7.1	Probability that the galaxy is completely round.
Class7.2	Probability that the galaxy's shape is in-between.
Class7.3	Probability that the galaxy is cigar-shaped.
Class8.1	Probability that the galaxy has a ring.

The first step in the process involves conducting an Exploratory Data Analysis (EDA). After selecting the features that will determine the cluster types, a correlation matrix is constructed, as shown in Figure 6.

The conclusions that can be drawn from this correlation matrix are as follows:

- Smooth Galaxies: There is a strong negative correlation between "smooth" and "has_signs_of_spiral" (-0.67) as well as "spiral_barred" (-0.48). This indicates that galaxies classified as smooth are less likely to exhibit spiral characteristics. Additionally, a positive correlation is observed between "smooth" and "completely_round" (0.63), suggesting that smooth galaxies tend to have a round shape.
- Edge-On Galaxies: "On_edge" has a moderate positive correlation with "cigar_shaped" (0.50), indicating that edge-on galaxies are often cigar-shaped. Furthermore, there is a negative correlation between "on_edge" and "completely_round" (-0.33), suggesting that edge-on galaxies are less likely to be completely round.
- Spiral Barred Galaxies: "Spiral_barred" is strongly positively correlated with "has_signs_of_spiral" (0.55), indicating that barred spiral galaxies often show signs of spiral arms. Additionally, the negative correlation with "smooth" (-0.48) suggests that barred spirals are less likely to be smooth.
- Bulge Prominence: A progression of positive correlations is observed from "no_bulge" to "dominant_bulge," through "just_noticeable_bulge" and "obvious_bulge." This indicates a spectrum of bulge prominence, ranging from none to dominant.
- Roundness: "Completely_round" has a strong positive correlation with "smooth" (0.63), supporting the observation that round galaxies are often smooth. Negative correlations with "on_edge" (-0.33) and "cigar_shaped" (-0.34) indicate that round galaxies are less likely to be viewed edge-on or to have a cigar shape.
- Presence of a Ring: "Ring_present" shows a moderate positive correlation with "obvious_bulge" (0.41), suggesting that galaxies with rings often have an obvious bulge. Negative correlations with "smooth" (-0.25) and "on_edge" (-0.09) suggest that ringed galaxies are less likely to be smooth or edge-on.

Once the conclusions are drawn from the Exploratory Data Analysis (EDA), the next step involves scaling the features. This is accomplished using the 'StandardScaler' function from Python. This process involves standardizing the features by removing the mean and scaling to unit variance, which ensures that all features contribute equally to the analysis and are on the same scale.

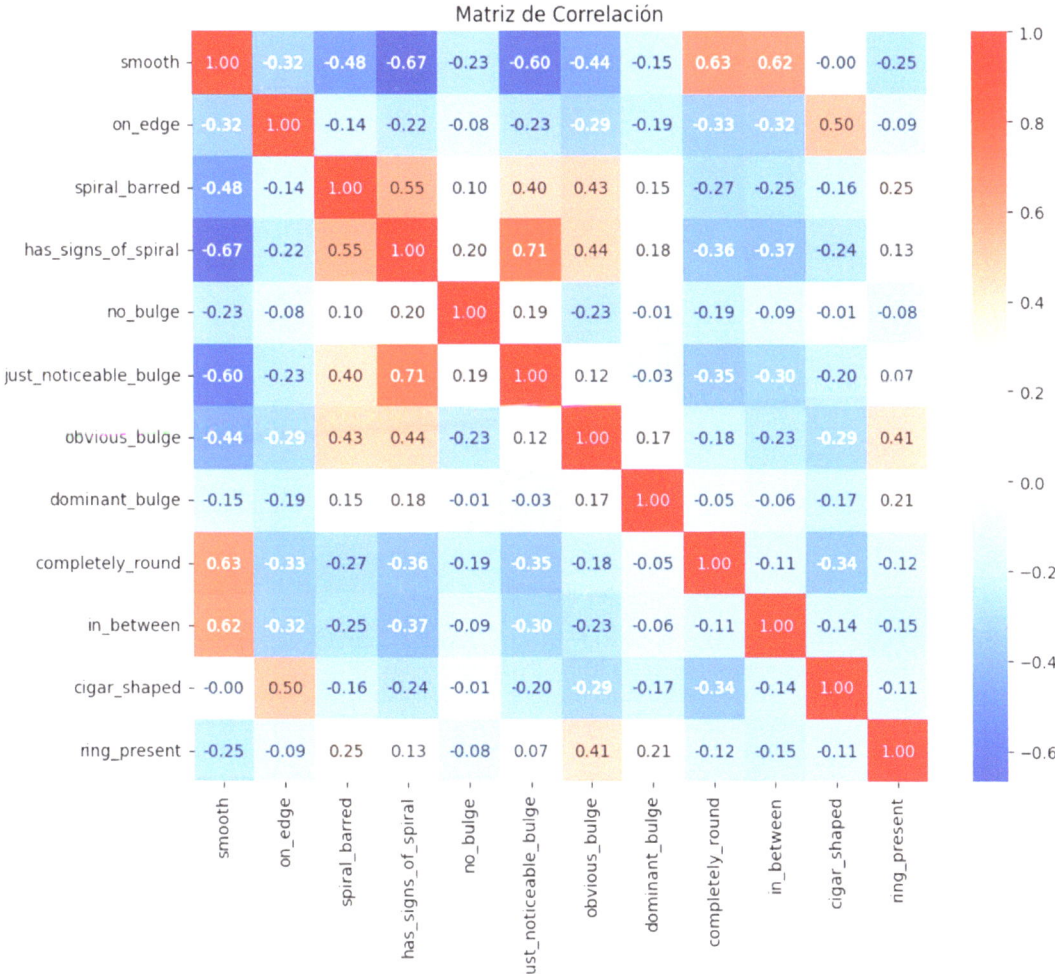

Figure 6. Correlation matrix.

4.2. Galaxy Classification

Before proceeding with the classification of galaxies, the optimal number of clusters must be determined. This step ensures that the clustering algorithm accurately captures the underlying structure of the data. An effective method to achieve this is to use the Silhouette Index.

The Silhouette Index is a widely used metric for assessing the quality of clustering results. It measures how similar an object is to its own cluster (cohesion) compared to other clusters (separation). The silhouette value ranges between -1 and 1, where values close to $+1$ suggest that the object is well matched to its own cluster, a value of 0 indicates that the object is near the boundary between clusters, and values near -1 suggest that the object may be misclassified and better suited to a different cluster.

A range of possible cluster numbers is defined, usually starting from 2 up to a maximum value, which can be 10 or higher depending on the size and complexity of the dataset.

The optimal number of clusters identified is 3, as shown in Figure 7. This number maximizes the average silhouette score across all points in the dataset.

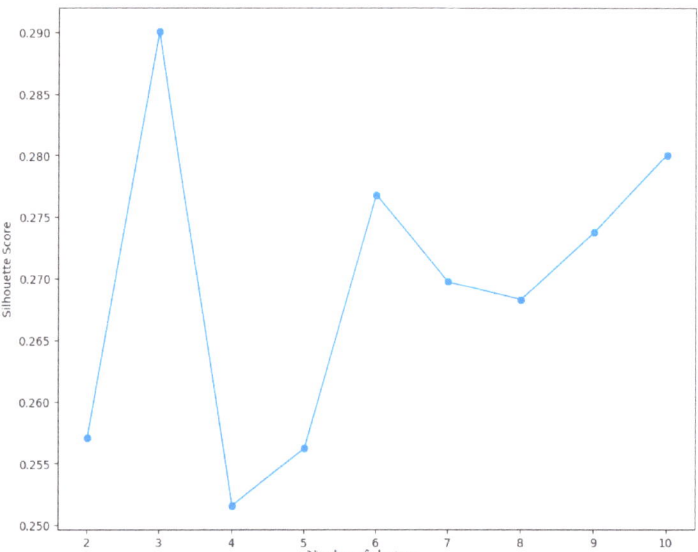

Figure 7. Optimal number of clusters.

The next step involves visualizing the centroids of these clusters to understand the distinguishing features of each cluster (Figure 8).

Based on the resulting centroids, the main characteristics of each cluster can be interpreted. The centroids show the average values of the features for each cluster, allowing for a better understanding of the distinctive characteristics of each group of galaxies.

Cluster 0 is characterized by galaxies that are mostly smooth, with a smooth value of 0.393019. A considerable proportion of galaxies appear edge-on, indicated by an on_edge value of 0.208261. Some galaxies in this cluster have a spiral bar (spiral_barred: 0.074008), and others show signs of spirals (has_signs_of_spiral: 0.128044). Additionally, some galaxies have a just noticeable bulge (just_noticeable_bulge: 0.155726) or an obvious bulge (obvious_bulge: 0.168610). The cluster also contains galaxies with a shape that is intermediate between completely round and cigar-shaped (in_between: 0.201720). Overall, Cluster 0 groups galaxies that are primarily smooth, with a mix of spiral and bulge characteristics.

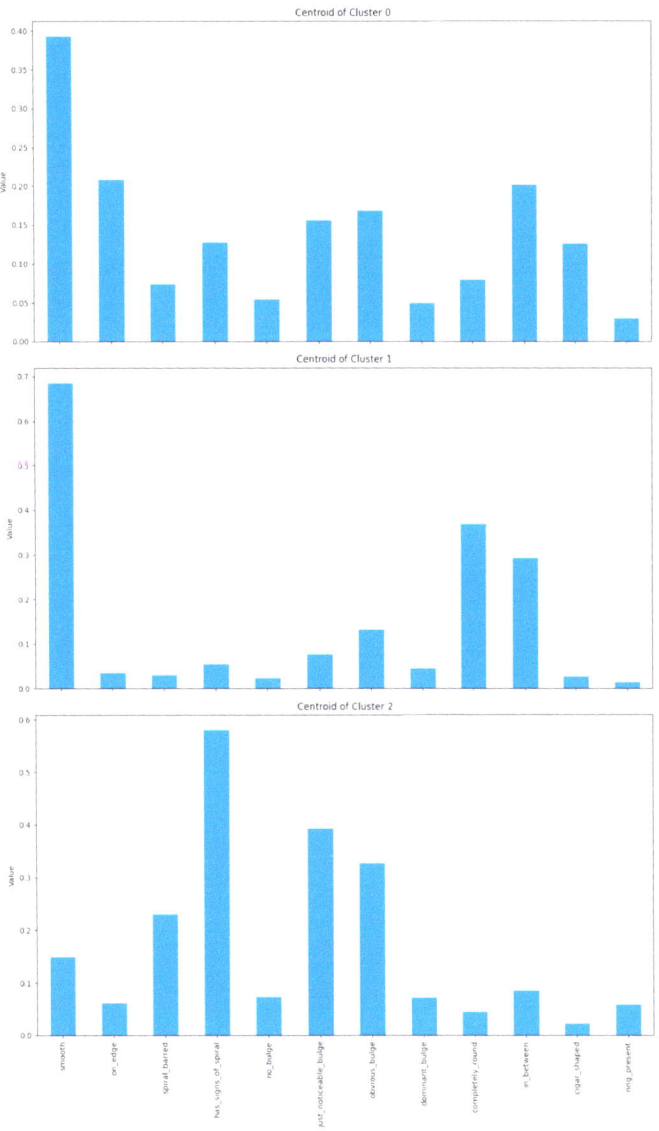

Figure 8. Centroids by cluster.

Cluster 1 is composed mainly of galaxies that are predominantly smooth, with a smooth value of 0.685219. Many galaxies in this cluster are completely round (completely_round: 0.368346), while a significant fraction has an intermediate shape (in_between: 0.292676). Some galaxies have obvious bulges (obvious_bulge: 0.131435), and others have just noticeable bulges (just_noticeable_bulge: 0.076257). Overall, Cluster 1 groups galaxies that are mostly smooth, either completely round or with intermediate shapes, with relatively few complex internal structures, although some have bulges that are either obvious or just noticeable.

Cluster 2 consists of galaxies that frequently exhibit signs of spirals, with a has_signs_of_spiral value of 0.579736. Many galaxies in this cluster have a just noticeable bulge (just_noticeable_bulge: 0.393652), while others have obvious bulges (obvious_bulge:

0.326246). A significant fraction of the galaxies have a spiral bar (spiral_barred: 0.230096), and some galaxies are smooth (smooth: 0.149425). Overall, Cluster 2 groups galaxies that typically display spiral features and bulges, with some galaxies also being smooth.

Figures 9–11 display a sample of galaxy images from the Galaxy Zoo dataset, displaying the cluster memberships as determined by the Fuzzy C-Means clustering algorithm. Each image is annotated with the GalaxyID and the degrees of membership to three different clusters (Cluster 0, Cluster 1, and Cluster 2).

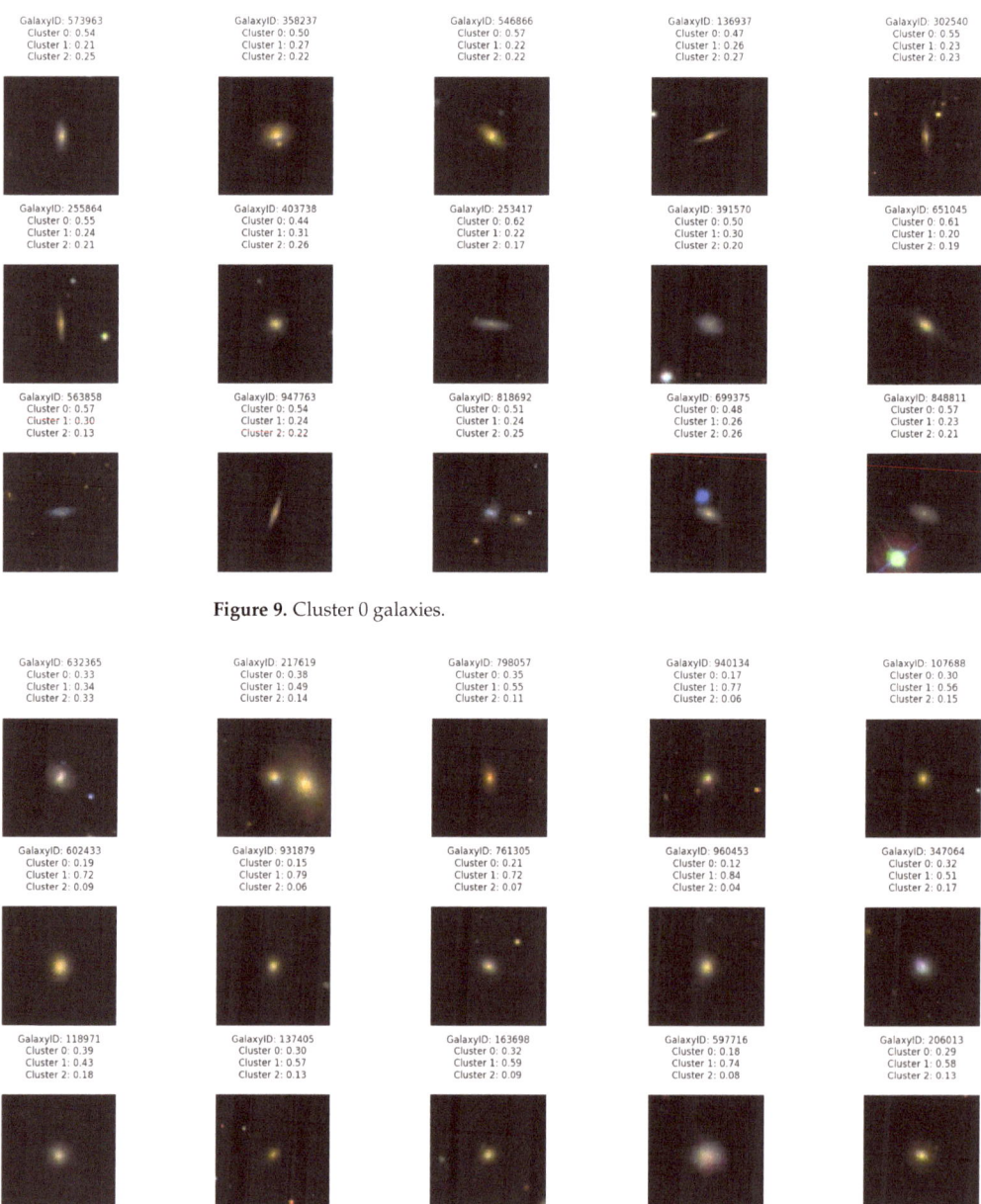

Figure 9. Cluster 0 galaxies.

Figure 10. Cluster 1 galaxies.

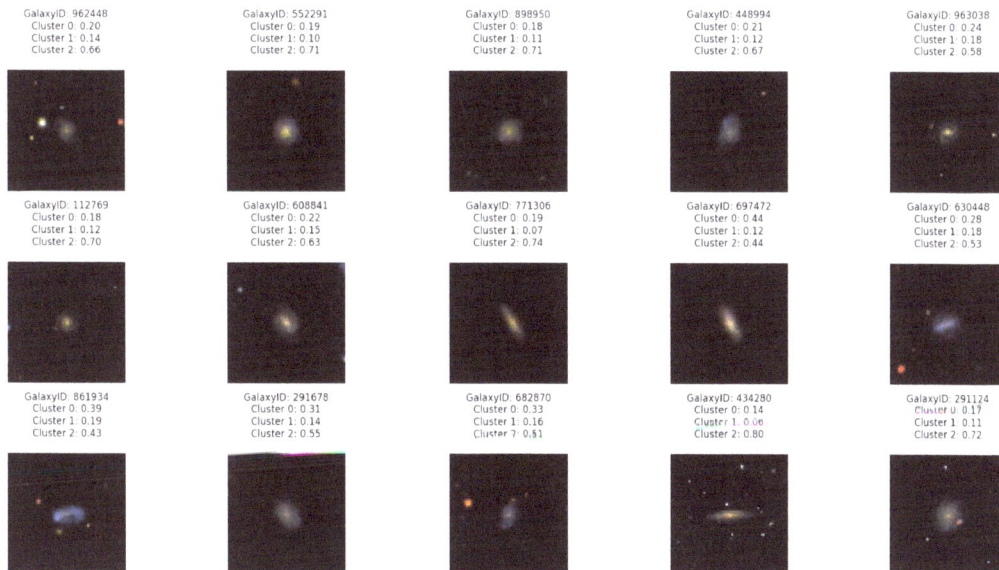

Figure 11. Cluster 2 galaxies.

4.3. Prediction Model Random Forest (RF)

After classifying galaxies based on the previously discussed criteria, the next step involves creating a predictive model to assess the accuracy of our classification and predict the cluster membership for new galaxies. For this purpose, the RF algorithm has been chosen due to its robustness and efficiency in handling complex, non-linear relationships between features.

RF is an ensemble learning method that constructs multiple decision trees during training and outputs the mode of the classes for classification tasks. This approach helps improve the model's accuracy and controls overfitting.

In the data preparation stage, the dataset with labeled clusters from the Fuzzy C-Means clustering is split into training and testing sets. During model training, the Random Forest model is trained on the training set, learning the patterns and relationships between the features (such as smoothness, edge-on probability, presence of spiral structures, etc.) and the cluster labels. Finally, in the model testing phase, the trained model is tested on the testing set to evaluate its predictive performance.

To evaluate the model's performance, a confusion matrix is generated (Figure 12). This matrix provides a detailed breakdown of the model's predictions compared to the actual cluster labels, showing the number of true positives, true negatives, false positives, and false negatives for each cluster.

The Random Forest model can be used to predict the cluster membership of new galaxies based on their morphological characteristics. When a new galaxy is detected, its features are input into the model, which then predicts the probability of the galaxy belonging to each cluster. The cluster with the highest probability is assigned to the new galaxy.

To calculate the precision of the Random Forest model, the following equation is used:

$$\text{Precision} = \frac{\text{True Positives}}{\text{True Positives} + \text{False Positives}} \qquad (10)$$

From the confusion matrix, the precision for each cluster can be calculated as follows:

Cluster 0: (true positive: 6416; false positive: 73 + 107)

$$[\text{Precision}_0 = \frac{6416}{6416 + 73 + 107} = \frac{6416}{6596} \approx 0.973]$$

Cluster 1: (true positive: 6900; false positive: 63 + 0)

$$[\text{Precision}_1 = \frac{6900}{6900 + 63} = \frac{6900}{6963} \approx 0.991]$$

Cluster 2: (true positive: 4850; false positive: 65 + 0)

$$[\text{Precision}_2 = \frac{4850}{4850 + 65} = \frac{4850}{4915} \approx 0.987]$$

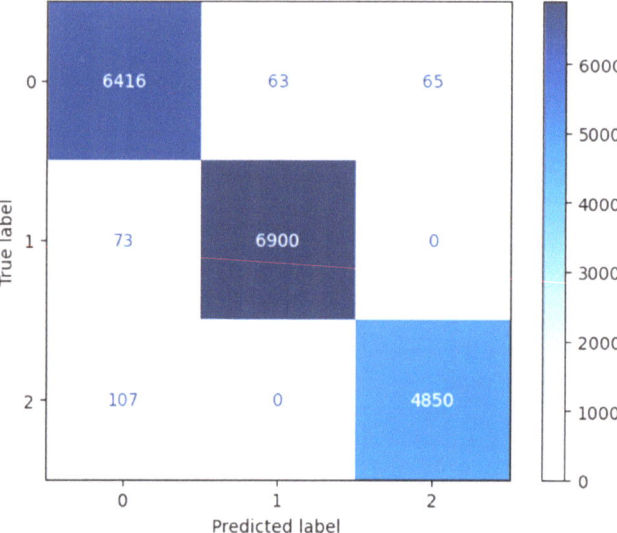

Figure 12. Confusion matrix.

The model demonstrates high precision across all clusters, with Cluster 1 showing the highest precision at 99.1%, followed by Cluster 2 at 98.7%, and Cluster 0 at 97.3%. This indicates that the model is highly effective at correctly classifying galaxies into their respective clusters with minimal false positives.

Each cluster represents a group of galaxies with similar characteristics. Based on the cluster membership, specific actions can be taken:

1. Cluster 0 (Predominantly Smooth Galaxies): For galaxies predicted to belong to this cluster, further investigation into their formation and evolution can be conducted. These galaxies may be less likely to exhibit complex structures, making them ideal candidates for studying early galaxy formation.
2. Cluster 1 (Round and Intermediate Shapes): Galaxies in this cluster often have round or intermediate shapes. Research can focus on understanding the factors contributing to their shape and how they evolve over time. This cluster may also include galaxies in various evolutionary stages.
3. Cluster 2 (Spiral and Bulge Structures): This cluster includes galaxies with prominent spiral features and noticeable bulges. These galaxies can be studied to understand spiral arm formation, star formation rates, and the dynamics of bulge development.

4.4. Interpretable Machine Learning

To ensure the transparency and interpretability of our predictive model, we incorporated explainability techniques into the RF algorithm used for predicting the cluster membership of new galaxies. While RFs are algorithms that excel in accuracy and the ability to manage complex, non-linear relationships between features, they are often considered "black-box" models due to their lack of inherent interpretability.

By employing Explainable AI (XAI) methods such as SHAP, we can gain insights into the decision-making process of the Random Forest model. SHAP values provide a global understanding of feature importance by assigning each feature an importance score based on its contribution to the model's predictions. This allows us to interpret the influence of individual features on the classification of galaxies into specific clusters.

LIME, on the other hand, offers a local perspective by creating interpretable models around individual predictions. This enables us to understand the reasons behind a particular galaxy's assignment to a specific cluster, providing a clearer and more actionable explanation for each decision made by the model.

The integration of these XAI techniques with the RF model ensures that the decisions made by the model are accurate and understandable, thereby increasing the confidence in the results obtained from the classification of new galaxies based on their morphological characteristics.

4.4.1. SHAP

The application of SHAP allows for an understanding of which features are most influential in the model's predictions for each cluster. By analyzing feature importance with SHAP, one can validate that the model behaves consistently with prior knowledge about galaxy classification and detect if the model uses irrelevant or biased features for predictions. Providing explanations about how the model arrives at its predictions increases trust in the system, especially for end-users who may not have deep technical expertise. Furthermore, the SHAP technique helps identify which features might be missing or misinterpreted by the model, allowing for adjustments and improvements in the model's design (Figures 13–15).

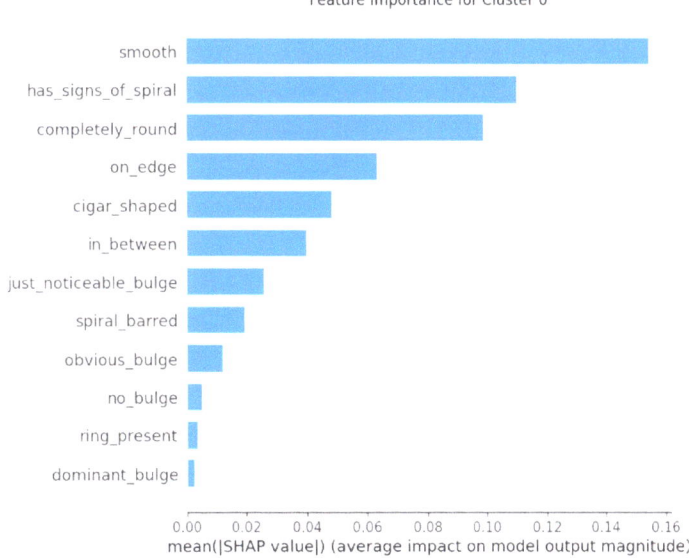

Figure 13. Feature importance for Cluster 0.

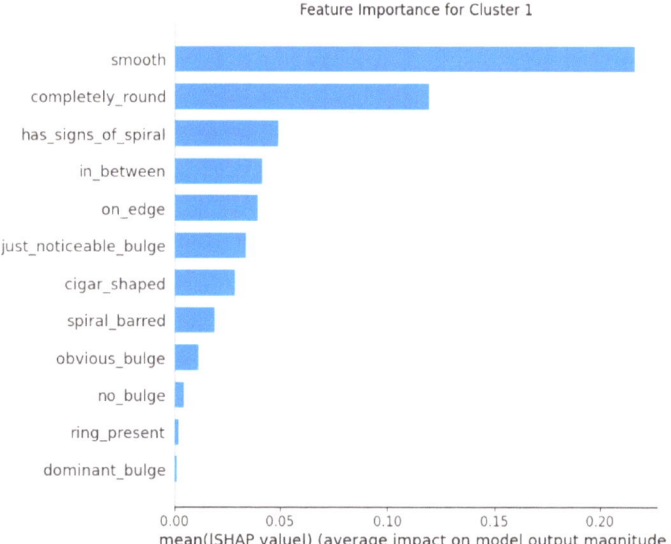

Figure 14. Feature importance for Cluster 1.

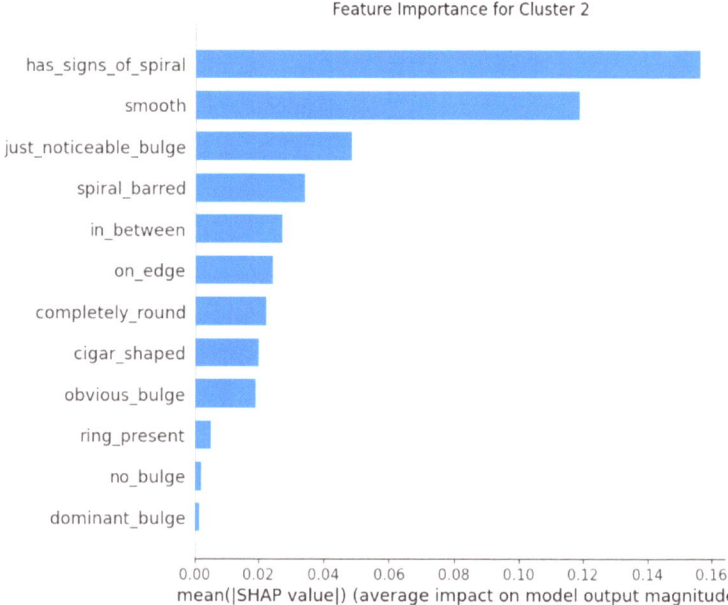

Figure 15. Feature importance for Cluster 2.

It is important to compare the feature importance derived from SHAP with the central features of each cluster obtained previously. This allows us to validate that the Random Forest model aligns with the definitions of the clusters established by Fuzzy C-Means.

1. In Cluster 0, it is observed that the smooth feature is the most influential, aligning with the prior interpretation that galaxies in this cluster are predominantly smooth.
2. In Cluster 1, the "completely_round" feature stands out, consistent with the interpretation that many galaxies in this group are completely round.

3. In Cluster 2, the "has_signs_of_spiral" feature emerges as the most significant, corresponding with the interpretation that galaxies in this cluster frequently display spiral signs.

This correspondence between feature importance and cluster definitions reinforces the validity of the model and provides a deeper understanding of the factors driving the model's decisions.

4.4.2. LIME

LIME operates by creating locally accurate explanations. It takes a specific data point and constructs a dataset of similar but slightly altered instances, then produces a simpler, interpretable surrogate model (such as linear regression) to elucidate the complex model's predictions near the selected data point. This method is particularly useful for comprehending the factors affecting specific predictions, especially when the original model is a "black box" like Random Forest [5].

The LIME result shown in Figure 16 provides an explanation for the classification of a galaxy (Galaxy ID = 553402) into Cluster 2 with a 100% probability.

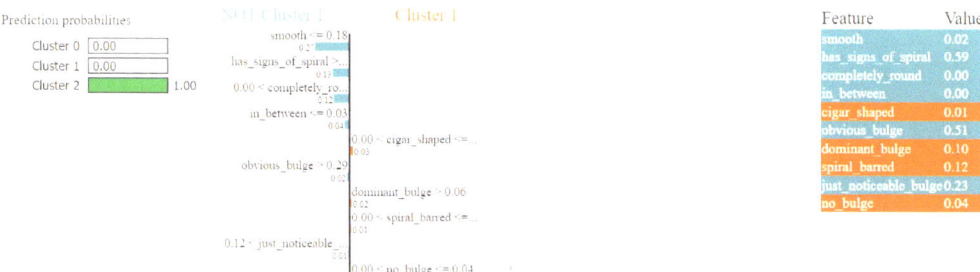

Figure 16. Local cluster prediction (cluster = 2).

The galaxy is classified into Cluster 2 with high certainty due to its morphological characteristics. The noteworthy features influencing this decision include a low "smooth" value, a high "has_signs_of_spiral" value, and the absence of a "completely_round" shape. Specific characteristics, such as low values of "cigar_shaped" and moderate "obvious_bulge", "dominant_bulge", and "just_noticeable_bulge", also contribute to this classification.

The image corresponding to Galaxy ID = 553402 is shown in Figure 17.

Figure 17. Galaxy ID 553402, Cluster 2.

The LIME result shown in Figure 18 provides an explanation for the classification of a galaxy (Galaxy ID = 236126) with a 68% probability for Cluster 0, 0% for Cluster 1, and 32% for Cluster 2.

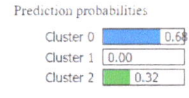 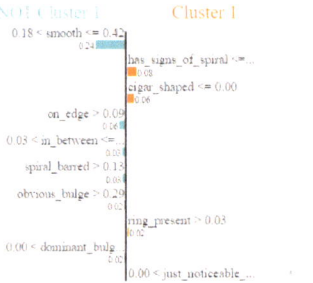

Figure 18. Local cluster prediction (cluster = 0).

The galaxy is primarily classified into Cluster 0 with a 68% probability based on its morphological characteristics. Key features influencing this decision include moderate values for smoothness and the absence of "has_signs_of_spiral" and "cigar_shaped" traits. Additionally, low values for "dominant_bulge" and "just_noticeable_bulge", along with moderate values for "on_edge", "spiral_barred", and "obvious_bulge", further support this classification.

The image corresponding to Galaxy ID = 236126 is shown in Figure 19.

Figure 19. Galaxy ID 236126, Cluster 0.

The LIME result shown in Figure 20 provides an explanation for the classification of a galaxy (Galaxy ID = 113992) with a 100% probability for Cluster 1.

The galaxy is classified into Cluster 1 with a 100% probability, primarily due to its high values for smooth (0.90) and "completely_round" (0.64), which are the most significant contributors to this classification. The absence of spiral characteristics ("has_signs_of_spiral": 0.00), cigar shape ("cigar_shaped": 0.00), and edge-on view ("on_edge": 0.00) also support this result. Additionally, low values for "just_noticeable_bulge" (0.00), "spiral_barred" (0.00), "obvious_bulge" (0.08), and "ring_present" (0.02) further reinforce the classification into Cluster 1. The feature "in_between" has a moderate value (0.25), slightly contradicting the overall classification, but is outweighed by the stronger contributions from other features.

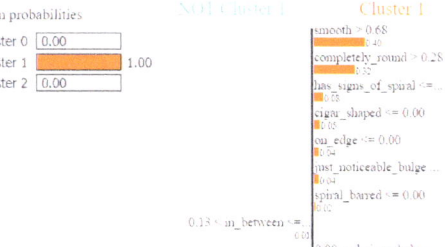

Figure 20. Local cluster prediction (cluster = 1).

The image corresponding to Galaxy ID = 113992 is shown in Figure 21.

Figure 21. Galaxy ID 113992, Cluster 1.

5. Discussion

The present study has developed a comprehensive model for galaxy classification by integrating Fuzzy C-Means (FCM) clustering and predictive modeling, validated through Explainable AI techniques such as SHAP and LIME.

As observed in the related work (Section 2.2), most studies in galaxy classification primarily rely on convolutional neural networks and other deep learning techniques. These methods often achieve high classification accuracy but lack transparency and interpretability. Our study, in contrast, integrates Explainable Artificial Intelligence (XAI) methods, specifically SHAP and LIME, to provide both global and local interpretability, ensuring that the decision-making process behind the classification is understandable, unlike the studies analyzed in Section 2.1 that predominantly use global explanations. Furthermore, while many studies focus exclusively on algorithmic data classification, our work incorporates human-generated data from the Galaxy Zoo project, where non-expert volunteers contribute to the labeling of galaxies. This human interaction introduces an element of uncertainty, which we address through Fuzzy C-Means (FCM) clustering.

The model development and integration process began with employing Fuzzy C-Means (FCM) clustering to identify natural groupings within the galaxy dataset based on their morphological characteristics. This method allowed for the classification of galaxies into overlapping clusters, effectively accommodating the inherent uncertainty and variability in the data. Following this clustering step, a Random Forest model was trained to predict the cluster membership of new galaxies. By leveraging the identified clusters, the Random Forest model provided a robust predictive framework capable of accurately classifying new galaxy data.

Cluster characterization involves examining the centroids of the clusters to identify the key morphological features that define each group. These centroids represent the average values of the features for galaxies within each cluster, providing a clear understanding of the distinguishing characteristics of each cluster. SHAP values were then utilized to quantify the importance of each feature in the model's predictions for each cluster, confirming that the model's behavior aligns with prior knowledge about galaxy classification and ensuring relevant features were appropriately weighted. Additionally, LIME was used to provide local explanations for specific predictions, offering insights into how individual features influenced the model's decisions for particular galaxies.

Model validation and interpretation revealed a strong correspondence between the defined clusters and the morphological types of galaxies through the combination of FCM clustering and SHAP/LIME analysis. This correspondence validates the effectiveness of both the clustering method and the predictive model. The predictive model also demonstrated a high degree of accuracy in assigning new galaxies to the appropriate clusters, extending its application to real-time galaxy classification tasks, thereby enhancing the efficiency and accuracy of astronomical research.

The utility and future improvements of this study highlight the potential for combining human visual analysis with AI-driven image analysis. By leveraging the strengths of both approaches, the model achieves a higher level of precision and reliability in galaxy classification. The enhanced interpretability provided by SHAP and LIME ensures that the model's predictions are transparent and easily understandable, which is important for building trust in AI systems, particularly in scientific research where understanding the rationale behind decisions is critical. Additionally, the methodology developed in this study has broader applications and could be applied to other domains requiring the classification of complex data.

6. Conclusions and Future Work

The developed framework provides a robust and transparent approach to galaxy classification, which provides insight into the complex morphological characteristics of galaxies.

The integration of Fuzzy C-Means clustering with the Random Forest predictive model has proven to be highly effective in classifying galaxies based on their morphological characteristics, particularly when working with the Galaxy Zoo dataset, which is built upon non-expert opinions. The successful clustering of these images, despite the inherent variability introduced by the non-expert input, demonstrates a notable achievement in extracting meaningful clusters from the data.

Moreover, the application of Explainable AI (XAI) techniques such as SHAP and LIME further enhances our ability to assign new galaxy images to the appropriate clusters with high confidence. By utilizing user-generated data from Galaxy Zoo, the model successfully predicts the cluster to which a galaxy belongs, leveraging both the non-expert input and sophisticated clustering and prediction techniques.

The precision achieved in the Random Forest classification for each cluster further reinforces the accuracy of the methodology. The precision values obtained are as follows: Cluster 0 = 0.973; Cluster 1 = 0.991; and Cluster 2 = 0.987.

It can be noted that after applying SHAP to analyze feature importance for each cluster, the results aligned perfectly with the observations made through the Fuzzy C-

Means classification. Additionally, using LIME, we conducted an in-depth analysis of three specific galaxy examples from each cluster, which allowed us to understand clearly why each galaxy was assigned to its respective cluster.

These results highlight the high level of precision achieved in classifying galaxies based on user-labeled data from Galaxy Zoo. The combination of Fuzzy C-Means and Random Forest, supported by XAI, allows for transparent and accurate classification, making it possible to incorporate human-driven insights into automated clustering models with great success.

In future work, it is proposed to advance in the following lines of research:

- Integration with Other Datasets: Future work will focus on integrating this methodology with other astronomical datasets to validate its robustness across several types of galaxy data.
- Refinement of Features: Further refinement of the features used in the model could enhance classification accuracy. Exploring additional morphological and contextual features will be a key area of research.
- Real-Time Classification: Developing a real-time classification system that can process and classify galaxies as new data becomes available will be a significant advancement.
- Expanding Interpretability Techniques: While SHAP and LIME have proven effective, exploring other interpretability techniques could provide deeper insights and improve model transparency further.
- Application to Other Domains: The methodology could be adapted for use in other domains where image classification and interpretability are important, such as medical imaging and remote sensing.

The results of this study demonstrate the potential of combining advanced clustering and predictive techniques with interpretability methods to achieve both high accuracy and transparency in complex classification tasks. This approach sets the stage for future advancements in galaxy classification and other scientific and technical fields requiring robust and Explainable AI models.

Author Contributions: Conceptualization, G.M.D.; methodology, G.M.D.; software, G.M.D.; validation, G.M.D., R.G.M. and J.A.A.J.; formal analysis, G.M.D.; investigation, G.M.D., R.G.M. and J.A.A.J.; resources, G.M.D.; data curation, G.M.D.; writing—original draft preparation, G.M.D.; writing—review and editing, G.M.D., R.G.M. and J.A.A.J.; visualization, G.M.D.; supervision, G.M.D., R.G.M. and J.A.A.J.; project administration, G.M.D. All authors have read and agreed to the published version of the manuscript.

Funding: This research received no external funding.

Data Availability Statement: Publicly accessible datasets were utilized for the analysis in this study. The data source can be accessed via the following link: https://www.zooniverse.org/projects/zookeeper/galaxy-zoo (accessed on 28 July 2024).

Conflicts of Interest: The authors declare no conflicts of interest.

References

1. Zooniverse. Galaxy Zoo. 2004. Available online: https://www.zooniverse.org/projects/zookeeper/galaxy-zoo (accessed on 28 July 2024).
2. Lintott, C.J.; Schawinski, K.; Slosar, A.; Land, K.; Bamford, S.; Thomas, D.; Raddick, M.J.; Nichol, R.C.; Szalay, A.; Andreescu, D.; et al. Galaxy Zoo: Morphologies derived from visual inspection of galaxies from the Sloan Digital Sky Survey. *Mon. Not. R. Astron. Soc.* **2008**, *389*, 1179–1189. [CrossRef]
3. Bezdek, J.C.; Ehrlich, R.; Full, W. FCM: The fuzzy c-means clustering algorithm. *Comput. Geosci.* **1984**, *10*, 191–203. [CrossRef]
4. Lundberg, S.M.; Lee, S.I. A unified approach to interpreting model predictions. In Proceedings of the 31st International Conference on Neural Information Processing Systems, Long Beach, CA, USA, 4–9 December 2017; pp. 4766–4775.
5. Ribeiro, M.T.; Singh, S.; Guestrin, C. "Why Should I Trust You?": Explaining the Predictions of Any Classifier. In Proceedings of the 2016 Conference of the North American Chapter of the Association for Computational Linguistics: Demonstrations, San Francisco, CA, USA, 13–17 August 2016; pp. 97–101. [CrossRef]

6. Zhang, X.F.; Zhang, C.M.; Tang, W.J.; Wei, Z.W. Medical image segmentation using improved FCM. *Sci. China Inf. Sci.* **2012**, *55*, 1052–1061. [CrossRef]
7. Pham, D.L.; Xu, C.; Prince, J.L. Current Methods in Medical Image Segmentation. *Annu. Rev. Biomed. Eng.* **2000**, *2*, 315–337. [CrossRef] [PubMed]
8. Ghosh, A.; Mishra, N.S.; Ghosh, S. Fuzzy clustering algorithms for unsupervised change detection in remote sensing images. *Inf. Sci.* **2011**, *181*, 699–715. [CrossRef]
9. Pal, N.R.; Pal, S.K. A review on image segmentation techniques. *Pattern Recognit.* **1993**, *26*, 1277–1294. [CrossRef]
10. Wright, J.; Yang, A.Y.; Ganesh, A.; Sastry, S.S.; Ma, Y. Robust Face Recognition via Sparse Representation. *IEEE Trans. Pattern Anal. Mach. Intell.* **2009**, *31*, 210–227. [CrossRef]
11. Ho, T.H.; Park, Y.H.; Zhou, Y.P. Incorporating satisfaction into customer value analysis: Optimal investment in lifetime value. *Mark. Sci.* **2006**, *25*, 260–277. [CrossRef]
12. Díaz, G.M.; Carrasco, R.A.; Gómez, D. RFID: A Fuzzy Linguistic Model to Manage Customers from the Perspective of Their Interactions with the Contact Center. *Mathematics* **2021**, *9*, 2362. [CrossRef]
13. Datta, S.; Datta, S. Methods for evaluating clustering algorithms for gene expression data using a reference set of functional classes. *BMC Bioinform.* **2006**, *7*, 397. [CrossRef]
14. Kocak, C.; Egrioglu, E.; Bas, E. A new explainable robust high-order intuitionistic fuzzy time-series method. *Soft Comput.* **2023**, *27*, 1783–1796. [CrossRef]
15. Ghosh, I.; Chaudhuri, T.D.; Sarkar, S.; Mukhopadhyay, S.; Roy, A. Macroeconomic shocks, market uncertainty and speculative bubbles: A decomposition-based predictive model of Indian stock markets. *China Financ. Rev. Int.* **2024**. [CrossRef]
16. Kmita, K.; Kaczmarek-Majer, K.; Hryniewicz, O. Explainable Impact of Partial Supervision in Semi-Supervised Fuzzy Clustering. *IEEE Trans. Fuzzy Syst.* **2024**, *32*, 3189–3198. [CrossRef]
17. Sevas, M.S.; Sharmin, N.; Santona, C.F.T.; Sagor, S.R. Advanced Ensemble Machine-Learning and Explainable AI with Hybridized Clustering for Solar Irradiation Prediction in Bangladesh. *Theor. Appl. Climatol.* **2024**, *155*, 5695–5725. [CrossRef]
18. Sirapangi, M.D.; Gopikrishnan, S. MAIPFE: An Efficient Multimodal Approach Integrating Pre-Emptive Analysis, Personalized Feature Selection, and Explainable AI. *Comput. Mater. Contin.* **2024**, *79*, 2229–2251. [CrossRef]
19. Arabikhan, F.; Gegov, A.; Kaymak, U.; Akbari, N. Fuzzy Networks for Explainable Artificial Intelligence. In Proceedings of the 2023 IEEE Conference on Artificial Intelligence (CAI), Santa Clara, CA, USA, 5–6 June 2023; pp. 199–200. [CrossRef]
20. Priya, C.; Durai Raj Vincent, P.M. An Efficient CSPK-FCM Explainable Artificial Intelligence Model on COVID-19 Data to Predict the Emotion Using Topic Modeling. *J. Adv. Inf. Technol.* **2023**, *14*, 1390–1402. [CrossRef]
21. Akpan, A.G.; Nkubli, F.B.; Ezeano, V.N.; Okwor, A.C.; Ugwuja, M.C.; Offiong, U. XAI for medical image segmentation in medical decision support systems. In *Explainable Artificial Intelligence in Medical Decision Support Systems*; Imoize, A.L., Hemanth, J., Do, D.T., Sur, S.N., Eds.; The Institution of Engineering and Technology: Stevenage, UK, 2022; Volume 50, pp. 137–165.
22. Lin, S.Y.; Wei, J.T.; Weng, C.C.; Wu, H.H. A Case Study of Using Classification and Regression Tree and LRFM Model in a Pediatric Dental Clinic. *Innov. Manag. Serv. Icms* **2011**, *14*, 131–135.
23. Ndung'u, S.; Grobler, T.; Wijnholds, S.J.; Karastoyanova, D.; Azzopardi, G. Advances on the morphological classification of radio galaxies: A review. *New Astron. Rev.* **2023**, *97*, 101685. [CrossRef]
24. Ma, X.; Li, X.; Luo, A.; Zhang, J.; Li, H. Galaxy image classification using hierarchical data learning with weighted sampling and label smoothing. *Mon. Not. R. Astron. Soc.* **2023**, *519*, 4765–4779. [CrossRef]
25. Stoppa, F.; Bhattacharyya, S.; Ruiz de Austri, R.; Vreeswijk, P.; Caron, S.; Zaharijas, G.; Bloemen, S.; Principe, G.; Malyshev, D.; Vodeb, V.; et al. Astrophysics Star-galaxy classification using a convolutional neural network. *Astron. Astrophys.* **2023**, *680*, A109. [CrossRef]
26. Schneider, J.; Stenning, D.C.; Elliott, L.T. Efficient galaxy classification through pretraining. *Front. Astron. Space Sci.* **2023**, *10*, 1197358. [CrossRef]
27. Şenel, F.A. A Hyperparameter Optimization for Galaxy Classification. *Comput. Mater. Contin.* **2023**, *74*, 4587–4600. [CrossRef]
28. Shafique, U.; Qaiser, H. A Comparative Study of Data Mining Process Models (KDD, CRISP-DM and SEMMA). *Int. J. Innov. Sci. Res.* **2014**, *12*, 217–222. Available online: http://www.ijisr.issr-journals.org/ (accessed on 20 July 2024).
29. Parsaei, M.R.; Rostami, S.M.; Javidan, R. A Hybrid Data Mining Approach for Intrusion Detection on Imbalanced NSL-KDD Dataset. *Int. J. Adv. Comput. Sci. Appl.* **2016**, *7*, 20–25. [CrossRef]
30. Molnar, C. *Interpretable Machine Learning: A Guide for Making Black Box Models Explainable*; 2019; p. 247, *Self-published*; Available online: https://christophm.github.io/interpretable-ml-book (accessed on 20 July 2024).
31. Carvalho, D.V.; Pereira, E.M.; Cardoso, J.S. Machine learning interpretability: A survey on methods and metrics. *Electron.* **2019**, *8*, 832. [CrossRef]
32. Monje, L.; Carrasco, R.A.; Rosado, C.; Sánchez-Montañés, M. Deep Learning XAI for Bus Passenger Forecasting: A Use Case in Spain. *Mathematics* **2022**, *10*, 1428. [CrossRef]
33. Ribeiro, M.T.; Singh, S.; Guestrin, C. Model-Agnostic Interpretability of Machine Learning. 2016. Available online: http://arxiv.org/abs/1606.05386 (accessed on 20 July 2024).
34. Lou, Y.; Caruana, R.; Gehrke, J.; Hooker, G. Accurate intelligible models with pairwise interactions. In Proceedings of the 19th ACM SIGKDD International Conference on Knowledge Discovery and Data Mining, Chicago, IL, USA, 11–14 August 2013; Part F1288; pp. 623–631. [CrossRef]

35. Lundberg, S.M.; Erion, G.G.; Lee, S.-I. Consistent Individualized Feature Attribution for Tree Ensembles. *arXiv* **2018**. [CrossRef]
36. Zafar, M.R.; Khan, N.M. DLIME: A Deterministic Local Interpretable Model-Agnostic Explanations Approach for Computer-Aided Diagnosis Systems. *arXiv* **2019**. [CrossRef]
37. Willett, K.W.; Lintott, C.J.; Bamford, S.P.; Masters, K.L.; Simmons, B.D.; Casteels, K.R.V.; Edmondson, E.M.; Fortson, L.F.; Kaviraj, S.; Keel, W.C.; et al. Galaxy zoo 2: Detailed morphological classifications for 304 122 galaxies from the sloan digital sky survey. *Mon. Not. R. Astron. Soc.* **2013**, *435*, 2835–2860. [CrossRef]

Disclaimer/Publisher's Note: The statements, opinions and data contained in all publications are solely those of the individual author(s) and contributor(s) and not of MDPI and/or the editor(s). MDPI and/or the editor(s) disclaim responsibility for any injury to people or property resulting from any ideas, methods, instructions or products referred to in the content.

Article

Neural Network-Based Design of a Buck Zero-Voltage-Switching Quasi-Resonant DC–DC Converter

Nikolay Hinov [1,*] and Bogdan Gilev [2]

[1] Department of Power Electronics, Technical University of Sofia, 1000 Sofia, Bulgaria
[2] Department of Mathematics and Computer Science, University of Transport "Todor Kableshkov", 1574 Sofia, Bulgaria; bgilev@vtu.bg
* Correspondence: hinov@tu-sofia.bg; Tel.: +359-29652569

Abstract: In this paper, a design method using a neural network of a zero-voltage-switching buck quasi-resonant DC–DC converter is presented. The use of this innovative approach is justified because the design of quasi-resonant DC–DC converters is more complex compared to that of classical DC–DC converters. The converter is a piecewise linear system mathematically described by Kirchhoff's laws and represented through switching functions. In this way, a mathematical model is used to generate data on the behavior of the state variables obtained under various design parameters. The obtained data are appropriately normalized, and a neural network is trained with them, which in practice serves as the inverse model of the device. An example is considered to demonstrate how this network can be used to design the converter. The key advantages of the proposed methodology include reducing the development time, improving energy efficiency, and the ability to automatically adapt to different loads and input conditions. This approach offers new opportunities for the design of advanced DC–DC converters in industries with high efficiency and performance requirements, such as the automotive industry and renewable energy sources.

Keywords: quasi-resonant DC–DC converter; neural networks; power electronic device design; zero-voltage switching

MSC: 65K10; 90C31

Citation: Hinov, N.; Gilev, B. Neural Network-Based Design of a Buck Zero-Voltage-Switching Quasi-Resonant DC–DC Converter. *Mathematics* **2024**, *12*, 3305. https://doi.org/10.3390/math12213305

Academic Editor: Francisco Rodrigues Lima Junior

Received: 28 September 2024
Revised: 18 October 2024
Accepted: 19 October 2024
Published: 22 October 2024

Copyright: © 2024 by the authors. Licensee MDPI, Basel, Switzerland. This article is an open access article distributed under the terms and conditions of the Creative Commons Attribution (CC BY) license (https://creativecommons.org/licenses/by/4.0/).

1. Introduction

Power electronic devices play a critical role in the sustainable development of society by supporting the management and conversion of electrical energy in ways that can significantly improve energy efficiency and reduce the environmental footprint [1–3]. These devices are at the heart of many technologies that contribute to environmental sustainability and have applications in a number of key areas, such as the following [4–6]: renewable energy, where they manage and convert electrical energy produced by photovoltaic devices and wind generators into a useful form that can be either used locally or transmitted to the grid; electric transportation, as electric vehicles (EVs) depend on power electronics to manage their batteries and drive their motors, with the efficiency of these systems directly affecting vehicle performance and range; energy management and storage, where the application of energy management systems, including power electronic devices, enables more efficient use of energy, optimization of its consumption, and reduced losses; sustainable production, because power electronics improve the management and efficiency of electrical energy in production processes. After all, power electronic devices are fundamental to the realization of energy-efficient and environmentally friendly technologies that are vital for the sustainable development of society. Their constant improvement and integration into various technological solutions contribute significantly to achieving a greener and more sustainable future. In this respect, there is a constant effort to develop both their circuit engineering and topologies, as well as to apply various methods for their optimal design, prototyping, and operation [7,8].

DC–DC converters are electronic devices that convert direct current (DC) from one voltage to another, finding wide application in various electronic systems and devices. One variety of these converters is the buck converter, which steps down the input voltage to a lower output voltage. As technology advances, there is a need to improve the efficiency and reliability of converters, leading to the development of zero-voltage-switching (ZVS) quasi-resonant topologies [9]. Quasi-resonant DC–DC converters are a specialized type of converter that uses resonant techniques to improve efficiency and reduce electromagnetic interference. They offer various advantages compared to traditional converters, but they also have their own specific disadvantages. The comparative analysis of their strengths and weaknesses is as follows [10,11]:

1. Advantages include increased efficiency, as quasi-resonant converters minimize switching losses by switching transistors at zero voltage (zero-voltage switching—ZVS) or zero current (zero-current switching—ZCS). This leads to lower heat load and higher overall efficiency. Additionally, electromagnetic radiation is reduced due to the resonant mode of operation, which decreases high-frequency interference and is particularly beneficial in applications requiring strict compliance with electromagnetic compatibility (EMC) norms. The converter also generates less noise, as the resonant operation significantly reduces the noise and vibration caused by switching. Finally, reliability is improved, since components are less stressed under ZVS or ZCS conditions, extending their life and enhancing their reliability.

2. Disadvantages include design complexity, as quasi-resonant converters are typically more complex than traditional DC–DC converters. This can make it difficult to achieve an optimal design due to the presence of multiple criteria and constraints, and it may also lead to higher initial development and production costs. Additionally, quasi-resonant converters have a limited operating range, functioning efficiently only within a specific load and voltage range, with significant drops in efficiency and performance outside of this range. Although they can be more efficient, quasi-resonant converters often require larger or more specialized components, such as resonant capacitors and inductors, which can increase the overall size and weight of the device. Lastly, the higher cost of components, due to the need for specialized parts and the associated demand for higher manufacturing precision, can make quasi-resonant converters more expensive to produce compared to standard DC–DC converters.

In conclusion, quasi-resonant DC–DC converters provide significant advantages in terms of efficiency and electromagnetic compatibility, making them particularly suitable for applications with such requirements. However, their more complex design and higher component costs may limit their suitability for all applications. The choice of the appropriate converter type depends on the specific requirements and constraints of the application.

2. Literature Review

Optimal design and performance assurance of power electronic devices require a multi-level approach that combines rigorous engineering design with proven technology and innovation. The key to success lies in detailed planning, quality execution at each stage, and commitment to the continuous improvement and renewal of technologies and processes.

The design of power electronic devices is a critical process that requires a careful balance between performance, reliability, efficiency, and cost. Technological advances and new requirements for energy efficiency and electromagnetic interference minimization have led to the development of various classical and innovative design methods. The following are the most important of them [12–14]:

1. Classic design methods involve several key approaches. First, by using analytical models, the traditional approach calculates parameters such as voltages, currents, and thermal profiles through analytical equations, providing an initial understanding of the device design. Second, prototyping and testing involve creating physical

prototypes that are tested to identify any problems in real working conditions, including thermal management, performance, and reliability testing. Finally, modeling of electrical circuits is conducted using standard software packages such as SPICE and MATLAB/Simulink to simulate electrical circuits and components before their actual creation.

2. Innovative design methods include several advanced approaches. First, computer modeling and simulations are increasingly complex and efficient, allowing for detailed analysis of projects in a virtual environment. These tools can simulate thermal, mechanical, and electromagnetic properties, reducing the need for early physical prototypes. Second, the use of optimization algorithms involves innovative methods such as genetic algorithms and particle swarm optimization to automatically find optimal design solutions, leading to better results and shorter development times. Third, machine learning and artificial intelligence can automate the design process by analyzing design data and past results, helping to identify unforeseen opportunities for optimization. Finally, the integration of modular and standardized components simplifies the engineering, production, and servicing processes by designing devices with a high degree of modularity and using standardized components.

Using a combination of classical and innovative methods can significantly increase the efficiency, reliability, and environmental friendliness of power electronics. By taking advantage of advances in technology and new engineering tools, designers can achieve better results and develop products that meet today's demands for sustainability and innovation.

Neural networks are a powerful tool in the field of machine learning, mimicking the structure and functioning of the human brain. They are composed of layers of neurons (computational units) that process and transmit information through connected weights. The main types of neural networks and their applications are as follows [15,16]:

- Perceptrons: The simplest type of neural network, usually with a single layer of neurons, used for simple classification tasks.
- Multi-layer perceptrons (MLPs): These consist of an input layer, one or more hidden layers, and an output layer. They can model more complex functions and solve a variety of tasks.
- Convolutional neural networks (CNNs): These specialize in image processing, using convolutional layers that automatically and efficiently learn spatial hierarchies of features.
- Recurrent neural networks (RNNs): These are suitable for processing sequential data, such as speech or text, because they have a "memory" that retains information about previous inputs to the network.
- LSTM (long short-term memory) networks: These are a variety of RNN designed to avoid the vanishing gradient problem and retain information for very long periods of time.
- GANs (generative adversarial networks): These are composed of two networks competing with each other: one generates data, while the other tries to distinguish the generated data from the real data.

Neural networks continue to be the subject of intensive research and development, and their application possibilities in various fields are constantly expanding. They find diverse applications in power electronics, improving the performance, efficiency, and reliability of electronic systems. The following are some concrete examples of the application of neural networks in this field:

1. Control of power electronics converters:

Neural networks are used to control various types of converters in power electronics, including DC–DC converters, DC–AC inverters, and AC–DC rectifiers. They help optimize power management, increase efficiency, and reduce harmonic distortion. In [17,18], the authors provide concrete examples of the synthesis and implementation of various algorithms for maintaining the maximum power mode of photovoltaic and wind generators [19,20] by applying various artificial intelligence techniques, including genetic algorithms and artificial neural networks (ANNs), while [21] discusses the synthesis of optimal control and adjustment of the controller for the implementation of robust control for the needs of various applications—electrical means of transport [22], decentralized energy production [23], and the implementation of industrial technologies [24]. An alternative to the synthesis of control with neural networks is the application of model predictive control (MPC). In [25], an MPC technique is proposed that uses a nonlinear inductor (NI) in a DC cascade system to achieve faster and more stable transitions. The proposed controller is based on the dead time principle, which simplifies the prediction process and reduces the complexity of the digital implementation in the control stage. The inclusion of an NI extends the stability range of the control-to-output transient characteristics of the power conversion stage with a CPL (constant power load)-type load. In addition, two Luenberger observers are implemented that can handle load uncertainties and adapt to system parameter changes. Criteria for selecting the observer parameters and their stability are discussed. The proposed MPC controller is easy to implement and operates at a fixed switching frequency. Validation through simulation and experimental results confirms the effectiveness of this technique to improve system stability.

2. Forecasting the charge and discharge of batteries:

Modeling and predicting battery behavior using neural networks can significantly improve energy management in energy storage systems. These models help optimize charging and discharging processes, extending battery life and improving the overall system performance. In particular, these tools are useful for creating models of various energy storage elements, based on which optimal operation can be achieved during the entire life cycle [26,27].

3. Reliability and fault diagnosis:

Neural networks are used to monitor and diagnose the condition of power electronic devices and systems. They can analyze sensor data in real time to detect anomalies, early signs of component failure, or wear. This helps prevent breakdowns and reduces downtime. This is a well-researched area subject to numerous publications related to the evaluation of the thermal load [27] and the related reliability of the circuit elements [28], as well as the derivation of dependencies related to their degradation over time [29].

4. Management of electrical networks:

Against the backdrop of the growing complexity of electrical networks, neural networks can be used to control power transmission networks, optimize energy distribution, and manage renewable energy sources. They also help in the integration of different energy sources and the management of their stability. In practice, the application of various artificial intelligence techniques is the basis of the realization of the intelligent management of energy flows in smart electrical networks [30,31] and smart cities [32].

5. Optimization of power electronics design:

Neural networks have been successfully used for the design and optimization of power electronic devices by analyzing and predicting the behavior of various circuit configurations and components. This results in faster product development and more efficient use of materials. For example, in [33], a method based on an artificial neural network (ANN) surrogate model is considered for the optimal design of a modular electric vehicle (EV) fast charging station. This is a typical combinatorial optimization problem that has no analytical solution because the key parameters used in the design are discrete—for example, the number of charging columns, power electronic converter modules, and switching contactors. On the other hand, in [34,35], a method for the modeling and determination of circuit parameters for DC–DC converters based on artificial intelligence (AI) is proposed. First, a switching loss database was built, and then an artificial neural network (ANN) was trained from the database. Then, with transfer learning (TL), other ANNs were trained for other semiconductor switches requiring the use of fewer training data. Finally, a heuristic optimization algorithm was used to obtain the most efficient and optimal device design. In the same way, neural networks have been used to evaluate various output parameters of power electronic devices [36,37], thereby achieving optimal loading of the power elements.

The present review concludes that the use of neural networks in power electronics enables a more intelligent and flexible management of energy resources, which is particularly important in a world of increasing energy requirements and with an emphasis on sustainability.

The considered artificial intelligence technique deals with the problems related to conditions, such as different loads, switching losses, or tolerances of the components of the power electronic devices, with the following capabilities:

- Neural networks can be trained on datasets that include different load conditions to improve their adaptability. By exposing the network to a wide range of operating scenarios during training, the model can learn to generalize its predictions and maintain optimal performance even when the load varies. In addition, techniques such as online learning can be used, allowing the neural network to dynamically adjust its predictions as the load changes in real time, ensuring stable and efficient operation.
- Switching losses are nonlinear and vary with factors such as frequency and voltage levels, making them difficult to predict. Neural networks can handle this by being trained on experimental data that accurately reflect these losses under various operating conditions. Through this data-driven approach, the neural network can develop a predictive model that accounts for switching losses more effectively than traditional control methods. In addition, reinforcement learning can be used to optimize switching patterns and frequencies, reduce overall energy losses, and improve efficiency.
- Component tolerances introduce variability in system behavior that can affect the accuracy of predictions. Neural networks can be trained to account for these tolerances by incorporating variations in component parameters in the training dataset. For example, the network can be trained using data that simulate variations in capacitors, inductors, and transistors to ensure that it can handle variations in component values. Furthermore, robust optimization techniques can be incorporated into the network architecture to ensure that the predicted results remain accurate even in the presence of component biases.
- To cope with non-ideal conditions, neural networks can use transfer learning or be retrained periodically as new data become available, allowing them to remain accurate as the system ages or as conditions change. Additionally, ensemble learning methods, where multiple neural networks work in tandem, can increase robustness, allowing the system to make more reliable decisions by averaging or combining predictions from different models.
- Neural networks can be integrated with real-time monitoring systems that continuously feed operational data back into the control loop. This enables adaptive control, where the neural network adjusts its predictions and optimizations based on the current state of the system, compensating for any deviation from ideal conditions, such as switching losses or changes in load. This type of feedback is critical to ensure that the system remains robust and performs optimally despite the presence of non-ideal factors.

In summary, neural networks can be trained and optimized to handle non-ideal conditions such as varying loads, switching losses, and component tolerances by using comprehensive datasets, advanced training techniques, and real-time feedback. These approaches improve the network's ability to predict and adapt to a wide range of operational scenarios, providing stable and efficient control of power converters in real-world conditions.

On the other hand, there is a certain gap related to the application of artificial intelligence techniques for the design of neural networks. Certain solutions involve the use of large databases or complex computational procedures that require significant computing resources. This determines the main goal of this paper: to propose a new innovative approach for the design of quasi-resonant DC–DC converters, based on the use of neural networks, to be implemented with standard means and minimal hardware requirements.

3. Mathematical Model of a Zero-Voltage-Switching Buck DC–DC Converter in MATLAB/Simulink

The buck DC–DC converter is a classic step-down converter that uses the following components: a transistor switch, a diode, an inductance, and a capacitor. The basic principle of its operation is based on pulsating control of the input voltage through the transistor switch to achieve a reduction in the output voltage. The disadvantage of conventional buck converters is the loss of energy during switching, especially at high operating frequencies [38].

The ZVS quasi-resonant buck converter uses zero-voltage-switching (ZVS) technology, which greatly reduces switching losses and increases efficiency. The main idea of ZVS is that the switching of the transistor takes place when the voltage on it is close to zero, which minimizes the losses from turning the switch on and off [39].

In the quasi-resonant topology, the inductance and capacitance of the circuit are used to generate resonant oscillations that help achieve ZVS. This is achieved by including additional components such as resonant inductance and resonant capacitors. These form a resonant tank that allows the voltage across the switching transistor to drop to zero before turning it on [4,6]. The schematic of the studied power converter is shown in Figure 1.

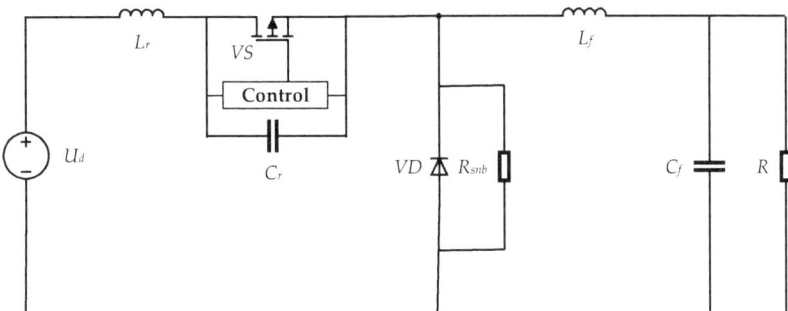

Figure 1. Power schematic in MATLAB/Simulink environment.

The following notations are used for the circuit elements:

U_d—input voltage; VS—semiconductor switch (transistor); VD—semiconductor switch (diode); L_r—resonant inductance; C_r—resonant capacity; L_f—filter inductance; C_f—filter capacity; R—resistance in the load circuit; R_{snb}—the resistance of the snubber circuit of the diode.

Since the zero-voltage-switching buck DC–DC converter is a system with a variable structure, using the switching functions K_0, K_1, and R_{snb}, as well as Kirchhoff's laws, the following mathematical model was obtained:

$$L_r \frac{di_r}{dt} + K_0 \cdot u_{Cr} + R_{snb}(i_{Lr} - i_{Lf}) = U_d$$

$$L_f \frac{di_{Lf}}{dt} = -u_{Cf} + R_{snb}(i_{Lr} - i_{Lf}) \quad (1)$$

$$C_r \frac{du_{Cr}}{dt} = K_0 \cdot i_{Lr} - K_1 \cdot \frac{u_{Cr}}{0.01}$$

$$C_f \frac{du_{Cf}}{dt} + \frac{u_{Cf}}{R} = i_{Lf}$$

where

$$K_0 = \begin{cases} 0, & \text{for external trigger ing of the transistor or } u_{Cr} \leq 0 \\ 1, & \text{for the other cases} \end{cases}$$

$$K_1 = \begin{cases} 1, & \text{for } u_{Cr} < 0 \\ 0, & \text{for the other cases} \end{cases} \quad (2)$$

$$R_{snb} = \begin{cases} 500, & \text{for } i_{Lr} - i_{Lf} \geq 0 \\ 0.01, & \text{for } i_{Lr} - i_{Lf} < 0 \end{cases} \quad (3)$$

In the presented model, the system of differential Equation (1) describes the electromagnetic processes in the circuit, and the logic Equations (2) and (3) describe the change in the configuration of the power circuit during the period of the control frequency.

To verify the adequacy of the modeling, a step-down quasi-resonant DC–DC converter with input voltage of 20 V, output voltage of 10 V, output current of 1 A, and an operating frequency of 1 MHz was designed using the methodology given in [11]. As a result, the following circuit element values were obtained: resonant inductance L_r = 1.6 µH, resonant capacity C_r = 4 nF, filter inductance L_f = 60 µH, filter capacity C_f = 0.35 µF, and load resistance R = 10 Ω. Based on the above Equations (1)–(3), the converter's mathematical model was implemented in a MATLAB environment. This was achieved through the following code (parameters defined during the design of the considered example were entered into the program):

```
function Main

global Lr Cr Lf Cf R f Ud

Ud=20; R=10; f=1000000;
Lr=1.6e-6; Lf=6e-5; Cr=4e-9; Cf=1.5e-7

X0=[0;0;0;0];
options=odeset('MaxStep',1e-7);
[t,x] = ode15s(@SS,[0 1e-4], X0,options);

function dx = SS(t,x)
global R Lr Cr Lf Cf f Ud

% x(1) = ilr; x(2) = iLf; x(3) = uCr; x(4) = uCf;

K=sin(2*pi*f*t);
if (K>0|x(3,1)<-0.1), Co=0; else, Co=1; end
if (x(3,1)<0), Coe=1; else, Coe=0; end
if (x(1,1)-x(2,1))>0, Rsnb=500; else Rsnb=0.01; end

dx = [(Ud-Rsnb*(x(1,1)-x(2,1))-Co*x(3,1))/Lr;...
(-x(4,1)+Rsnb*(x(1,1)-x(2,1)))/Lf;...
(Co*x(1,1)-Coe*x(3,1)/0.01)/Cr;...
(x(2,1)-x(4,1)/R)/Cf];
```

With this model, the behavior of the power circuit was simulated, as represented by the time dependences of the state variables i_{Lr}, i_{Lf}, u_{Cr}, and u_{Cf}. Their form (consequentially, from top to bottom: i_{Lr}, i_{Lf}, u_{Cr}, and u_{Cf}) is given in Figure 2. From the presented results, it can be established that the design task was achieved, and thanks to the model's adequate description of the conditions for working with soft voltage commutations, the DC–DC converter has the properties, i.e., is of the ZVS type.

By itself, this code would not be very useful, because simulations can be conducted directly in a Simulink/MATLAB environment (Figure 1). In the next section, the code is included in a procedure for generating input and reference data on which a neural network used for designing the DC–DC converter is trained.

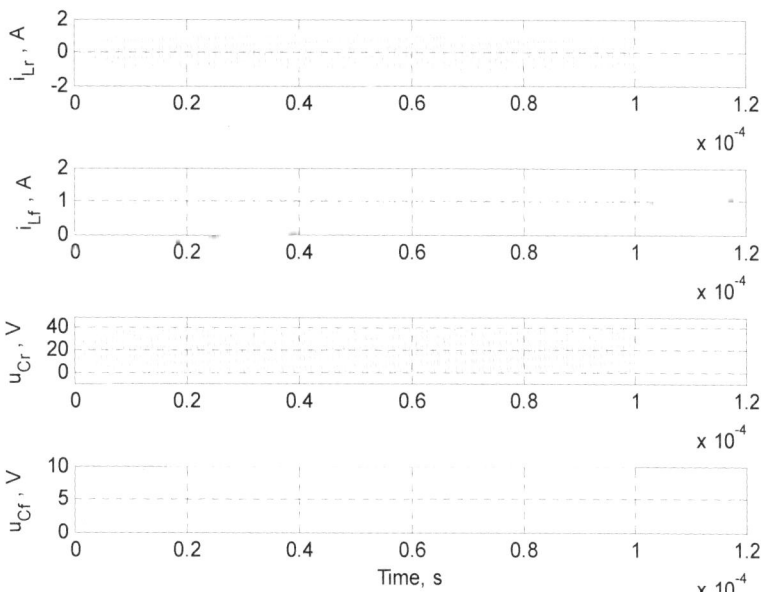

Figure 2. Results of numerical simulations with the model of the DC–DC converter—time diagrams of the state variables.

4. Input and Output (Reference) Data, Type, Structure, and Training of the Neural Network

The procedure of designing the inverter can be considered as a process that is taking place inside a "black box", where characteristics (most often averaged) of the state variables are fed into its inputs i_{Lr}, i_{Lf}, u_{Cr}, and u_{Cf}, and the values of the circuit elements L_r, C_r, L_f, and C_f are obtained at the output.

At the input, averaged values of the four output state variables are provided: u_{aver} and i_{aver} from Equation (4). Averaged input variables (4) are selected because this makes the design task static, which, in turn, allows it to be solved with a feed-forward neural network (NN). This avoids the use of more complex (for initialization and training) recurrent networks, which are more suitable for dynamic modeling.

$$i_{aver} = \left(\sum_{k=1}^{n} i_{Lf,k}\right)/n \text{ and } u_{aver} = \left(\sum_{k=1}^{n} u_{Cf,k}\right)/n \qquad (4)$$

Based on the graphs from Figure 2, it can be seen that the values for the output current and voltage are established over the time from $t \in [0.4 \times 10^{-4}, 1 \times 10^{-4}]$. For this reason, the two average values (4) are calculated specifically within this time interval.

After initializing and training the neural network, it can be found that the choice of inputs (4) allows the appropriately selected and trained neural network to find a reliable relationship between the input (averaged state variables) and the output (values of circuit elements).

In order to generate the data with which the network will be trained, a range of parameter variations for L_r, C_r, L_f, and C_f is selected. To maintain the converter's operating mode (conditions for continuous–continuous output current, resonance in the resonant circuit, etc.), a range of parameter variations is selected, including the design parameters from Section 2 and variations within one order of magnitude, i.e.,

$$L_r \in [0.5 \times 10^{-6}, 5 \times 10^{-6}], \; L_f \in [1 \times 10^{-5}, 10 \times 10^{-5}] \text{ and } C_f \in [0.5 \times 10^{-7}, 5 \times 10^{-7}]$$

The value of the parameter C_r is determined by the resonance condition in the resonant circuit, i.e.,

$$L_r C_r = 6.4 \times 10^{-15} \tag{5}$$

Subsequently, values for the parameters L_r, C_r, L_f, and C_f can be randomly selected (falling within the upper limits), and the mathematical model from Section 2 can be used to find the corresponding values of the averaged output state variables (4). In this way, a database can be created for training the neural network.

Due to a simpler program implementation, the following approach was chosen here, where the values of the parameters L_r, C_r, L_f, and C_f were changed with constant steps $\Delta L_r = 1 \times 10^{-6}$, $\Delta L_f = 2 \times 10^{-5}$, and $\Delta C_f = 1 \times 10^{-7}$, and these values were combined with each other. This was achieved using three nested loops, i.e.,

```
dLr=1e-6;
dLf=2e-5;
dCf=1e-7;

m=1;
Lr=0;
for p=1:5
Lr=Lr+dLr;
Cr=6.4e-15/Lr;
Lf=0;
for q=1:5
Lf=Lf+dLf;
Cf=0;
for u=1:5
Cf=Cf+dCf;
LLr(m)=Lr; LLf(m)=Lf; CCr(m)=Cr; CCf(m)=Cf;
%%%%%%%%%%%%%%%%%%%%%%%%%%%%%%%%%%%%%%%%%%
X0=[0;0;0;0];
options=odeset('MaxStep',1e-7);
[t,x] = ode15s(@SS,[0 1e-4], X0,options);
ti=t;
Ii=x(:,2);
tu=t;
Uu=x(:,4);
%%%%%%%%%%%%%%%%%%%%%%%%%%%%%%%%%%%%%%%%%%
ni=length(ti);
si=0;
for k=1:ni
if ti(k)>0.4e-4, si=si+1;Tipart(si)=ti(k); Ipart(si)=Ii(k); end
end

nu=length(tu);
su=0;
for k=1:nu
if tu(k)>0.4e-4, su=su+1;Tupart(su)=tu(k); Upart(su)=Uu(k); end
end
```

```
            Uaver(m)=sum(Upart(1:su))/length(Upart(1:su));
            Iaver(m)=sum(Ipart(1:si))/length(Ipart(1:si));
            m=m+1;
        end
    end
end

save data.mat LLr LLf CCr CCf Uaver Iaver deltaU deltaI
```

Using the source code provided, the data generated with the model for the values of the searched circuit elements L_r, C_r, L_f, and C_f were generated, as shown in Figure 3.

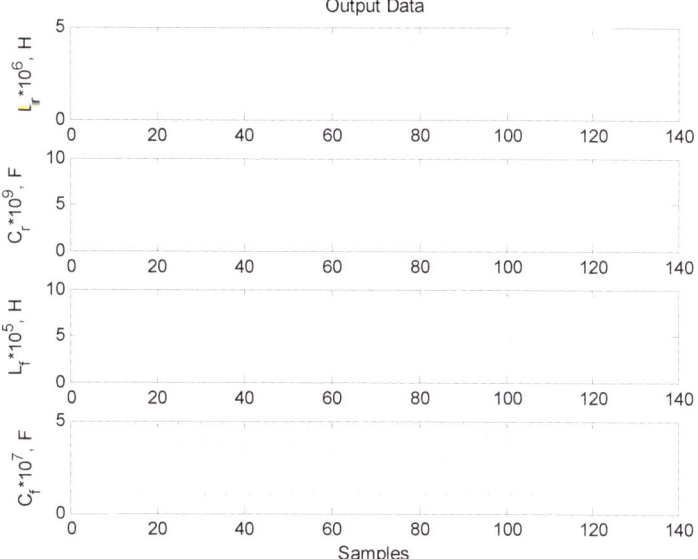

Figure 3. Generated data for the searched circuit parameters L_r, C_r, L_f, and C_f.

As can be seen from Figure 3, these data were multiplied by the scaling factors 10^7, 10^5, 10^9, and 10^6. This is necessary because the data from Figure 3 will serve as the output reference data with which the neural network will be trained. It is desirable for the output data to be commensurate with the input data. Therefore, this multiplication avoids the need for the network to perform alignment between the orders of the input and output data. Additionally, this can further improve the network's performance.

Because we need to keep the resonance condition, the value of C_r determines the value of L_r uniquely, as shown in Figure 3. For the remaining parameters, each is combined with every other value. The data for u_{aver} and i_{aver} were also generated using the same code, as shown in Figure 4. The data in Figure 4 are the input data with which the neural network will be trained.

For the needs of this study, a multi-layer perceptron (MLP) network was used, which consists of an input layer, one hidden layer, and an output layer. The main idea is for the network to predict the optimal parameters of the buck ZVS quasi-resonant DC–DC converter based on training data generated through simulations. The network is trained through a back-propagation process, with the objective being to minimize the loss function (e.g., root-mean-square error). The error is calculated as the difference between the predicted values and the actual target values (labels).

In the case of this particular task, a single hidden layer was chosen because it provides sufficient capacity to capture nonlinear dependencies in the data without overcomplicating the model. Determining the number of hidden layers is most often based on experimental testing. A network with one or two hidden layers is usually sufficient for many applications. If the task is very complex or requires the modeling of more complex dependencies, the number of layers can be increased. In our case, the experiments showed that one hidden layer provides sufficient accuracy without leading to overcomplication or excessive training time.

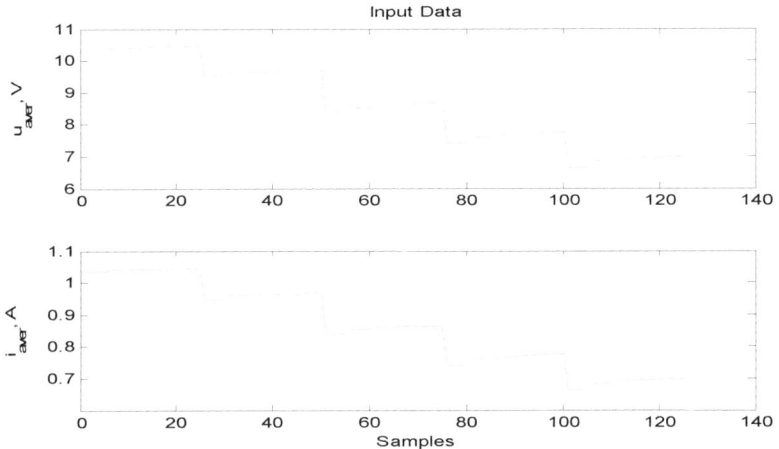

Figure 4. Data for i_{aver} and u_{aver}.

The activation function used in the hidden layer is ReLU (Rectified Linear Unit), which is popular for its advantages, such as computational efficiency and avoiding the vanishing gradient problems common with sigmoid and hyperbolic tangent functions. ReLU provides faster convergence and better performance for regression-related tasks such as the present task. A linear activation function is used in the output layer because the task is regression and the goal is to predict continuous values (optimal transducer parameters).

The accuracy of the network was measured by calculating the mean absolute error (MAE) and mean square error (MSE) on the test dataset. These metrics show how close the predicted values are to the actual targets. The test data were separated from the trainers to test the network's ability to generalize beyond the training set. The results showed a low error, which demonstrates the good generalization ability of the model. Furthermore, to assess the robustness of the model, cross-validation was conducted by training and testing the network on different data splits.

The structure of the neural network chosen for our design needs is shown in Figure 5. As already mentioned, this network has one hidden layer and one output layer. We used the "logsig" activation function in the hidden layers, and the output layers were linear. The neural network was trained using the batch training method. It is a well-known fact that a neural network with this structure is an effective approximator.

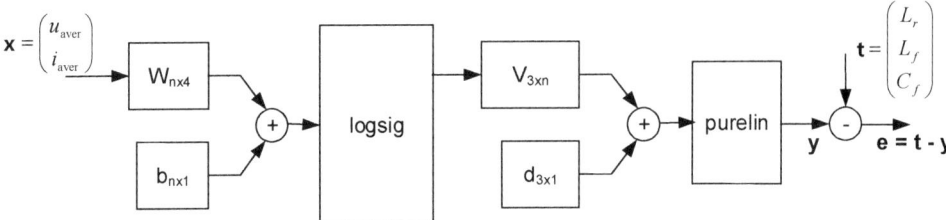

Figure 5. Structure of the neural network used to determine the circuit elements of the DC–DC converter.

In the NN structure shown in Figure 5, n represents the number of neurons in the first layer. The number of these neurons determines the capacity of the network. During the training process, it was found that for a value of $n = 60$, the network is powerful enough. The neural network was trained by the batch training method, where the aim is to minimize the total mean squared error.

$$S(\mathbf{w}) = \frac{1}{2}\sum_{k}(\mathbf{t}(k) - \mathbf{y}(k))^2 \qquad (6)$$

where
$\mathbf{w} = (W, b, V, d)$ is the generalized weight matrix of the network;
$\mathbf{x}(k) = [i_{aver}, u_{aver}]^T$ are the inputs of the network;
$\mathbf{t}(k) = [L_r, L_f, C_f]^T$ are the reference outputs by which the network is trained;
$\mathbf{y}(k) = [L_r, L_f, C_f]^T$ are the implemented outputs of the network, where for the chosen network structure we have $\mathbf{y}(k) = V \cdot \text{logsig}(W\,\mathbf{x}(k) + b) + d$.

The actual training involved numerically solving an optimization problem, which in this case was achieved using the Levenberg–Marquardt method. At the core of initializing and training the network in the MATLAB environment is the following code:

```
load data.mat

t=[LLr*1e+6; LLf*1e+5; CCf*1e+7];
x=[Uaver;Iaver;deltaU;deltaI];

net=newff(minmax(x),[60,3],{'logsig','purelin'});

net=train(net,x,t);

y=sim(net,x);
```

We trained the neural networks in a maximum of 1000 iterations. Error and other optimization parameters are shown in Figures 6 and 7.

After the NN was trained, the output values obtained by the neural network $\mathbf{y}(k) = [L_r, L_f, C_f]^T$ (recall that $C_r = 6.4 \times 10^{-15}/L_r$) were compared with the reference values $\mathbf{t}(k) = [L_r, L_f, C_f]^T$, and the inputs $\mathbf{x}(k) = [i_{aver}, u_{aver}]^T$ were retained. Figure 8 shows that, after the completion of training, we had a very good match between $\mathbf{y}(k)$ and $\mathbf{t}(k)$. It can be seen from Figure 8 that the third coordinate C_f is practically not trained. This shows that the output variables L_r and L_f are determined almost uniquely by the two input variables u_{aver} and i_{aver}, while the output variable C_f, in practice, does not depend on the input variables u_{aver} and i_{aver}. If we want to determine the value of C_f uniquely, it is necessary to introduce another input to the network, such as $\max(u_{Cr})$. In this case, the following problem arises: The quantities u_{aver}, i_{aver}, and $\max(u_{Cr})$ are interdependent and are provided as inputs to the network by the user. Since the dependencies between these variables are not obvious, the user provides values that are chosen arbitrarily (subjectively) and does not consider the dependencies between them. This may worsen the network's performance; therefore, we prefer to keep the two inputs to the network, specifically leaving the inputs u_{aver} and i_{aver}. This will ensure that the two most important parameters u_{aver} and i_{aver} will be designed correctly.

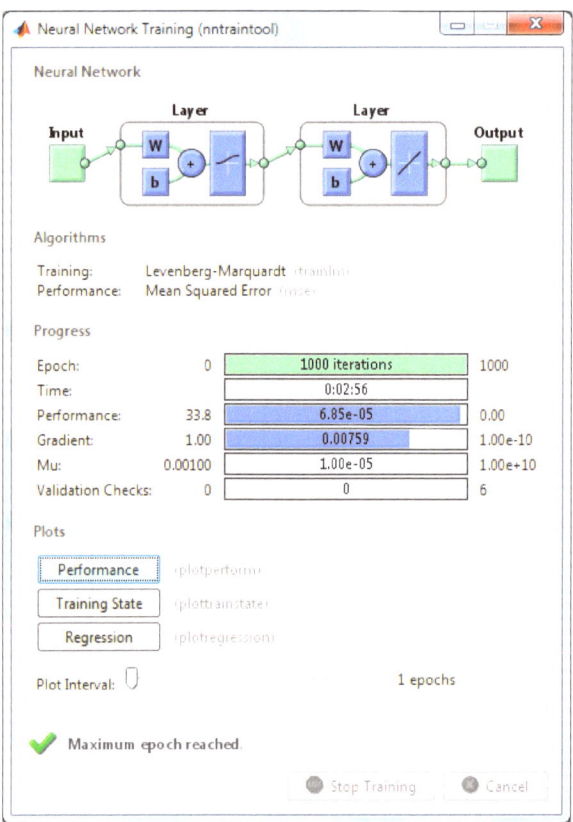

Figure 6. Screen showing the neural network training parameters' setup window.

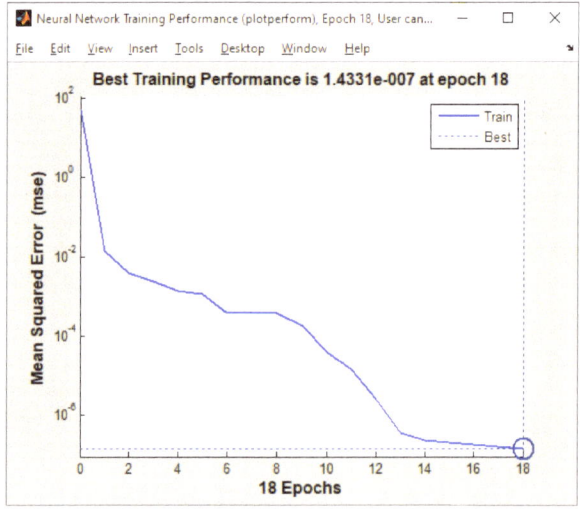

Figure 7. Screen showing the neural network training process using root-mean-square error analysis.

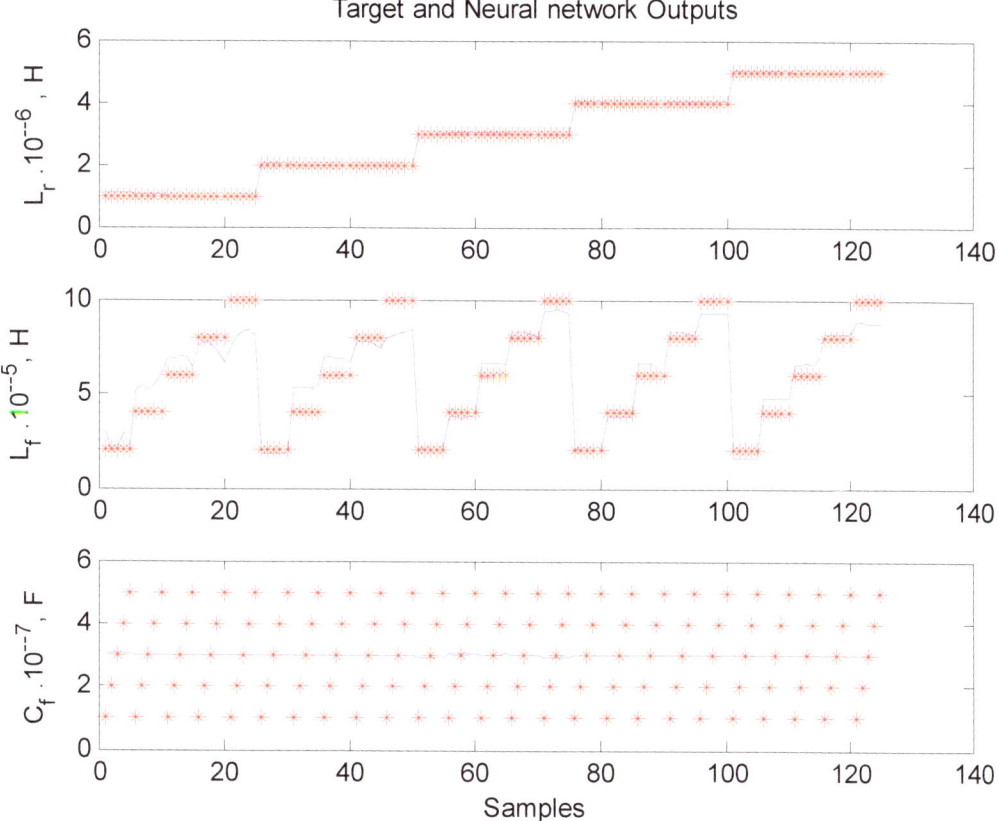

Figure 8. Targets **t**(k) (line) and neural network outputs **y**(k) (stars).

5. Examples of Converter Design

In this section, various computational examples for the design of the investigated type of DC–DC converters are discussed. Specific data for all considered cases are presented in Table 1.

Example 1. *We will evaluate the efficiency of the design procedure through an example. For this purpose, we selected the following values for the inputs of the NN:*

$$\mathbf{x} = [i_{aver} = 0.9, u_{aver} = 9]^T$$

Table 1. Systematized design results of various examples of quasi-resonant DC–DC converters with the trained neural network.

Design Examples	Schematic Element Values
Design assignment: u_{aver} = 9 V; i_{aver} = 0.9 A Design result: u_{aver} = 8.9941 V; i_{aver} = 0.9 A	L_r = 2.6905 µH L_f = 122.81 mH C_f = 41.312 µF C_r = 2.6812 µF
Design assignment: u_{aver} = 12 V; i_{aver} = 1.1 A Design result: u_{aver} = 11.31 V; i_{aver} = 1.1283 A	L_r = 1.1372 µH L_f = 158.99 mH C_f = 18.521 µF C_r = 1.1372 µF

Table 1. *Cont.*

Design Examples	Schematic Element Values
Design assignment: u_{aver} = 12 V; i_{aver} = 1.2 A Design result: u_{aver} = 11.7144 V; i_{aver} = 1.172 A	L_r = 937.09 nH L_f = 219.29 mH C_f = 286.32 nF C_r = 937.1 nF
Design assignment: u_{aver} = 8 V; i_{aver} = 0.8 A Design result: u_{aver} = 8.0562 V; i_{aver} = 0.7944 A	L_r = 3.7405 µH L_f = 89.566 mH C_f = 1.5969 µF C_r = 3.7405 µF
Design assignment: u_{aver} = 7 V; i_{aver} = 0.8 A Design result: u_{aver} = 6.9758 V; i_{aver} = 0.8062 A	L_r = 3.6065 µH L_f = 39.584 mH C_f = 0.30984 µF C_r = 3.6064 µF
Design assignment: u_{aver} = 7 V; i_{aver} = 0.7 A Design result: u_{aver} = 6.9981 V; i_{aver} = 0.7048 A	L_r = 4.9944 µH L_f = 91.647 mH C_f = 0.29975 µF C_r = 4.9945 µF
Design assignment: u_{aver} = 11 V; i_{aver} = 1.1 A Design result: u_{aver} = 11.0793 V; i_{aver} = 1.1091 A	L_r = 1.1950 µH L_f = 215.92 mH C_f = 0.30758 µF C_r = 1.1950 µF

By means of the neural network and the command sim(net,x), we "designed" the converter and obtained an output:

$$L_r = 2.6905 \times 10^{-6} \text{H}, \ L_f = 1.2281 \times 10^{-4} \text{H}, \ C_f = 4.1312 \times 1010^{-7} \text{F},$$

$$C_r = 2.6812 \times 10^{-6} \text{F}.$$

With the obtained values for L_r, C_r, L_f, and C_f, we simulated the mathematical model of the converter. The results of the simulation are shown in Figure 9.

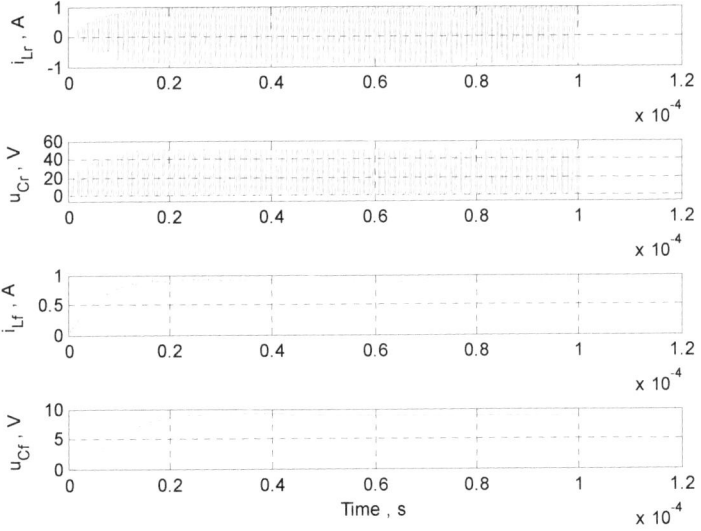

Figure 9. Simulation results of Example 1, regarding the state variables of the DC–DC converter.

For the data obtained in Figure 8 (in state space), the values of i_{aver} and u_{aver} were calculated using Formula (2). The results are as follows:

$$u_{aver} = 8.9941 \text{ V}, i_{aver} = 0.9000 \text{ A}$$

We compared the values obtained in this way for i_{aver} and u_{aver} with the values that were initially chosen as inputs to the neural network at. We observed that they were extremely close.

Example 2. *We select new input values for the neural network:*

$$\mathbf{x} = [i_{aver} = 0.8, u_{aver} = 8]^T.$$

Using the neural network, we obtained

$$L_r = 3.7405 \times 10^{-6} \text{H}, L_f = 89.566 \times 10^{-6} \text{H}, C_f = 1.5969 \times 10^{-6} \text{F},$$

$$C_r = 3.7405 \times 10^{-6} \text{F}.$$

With the values obtained in this way, we simulated the mathematical model and obtained Figure 10.

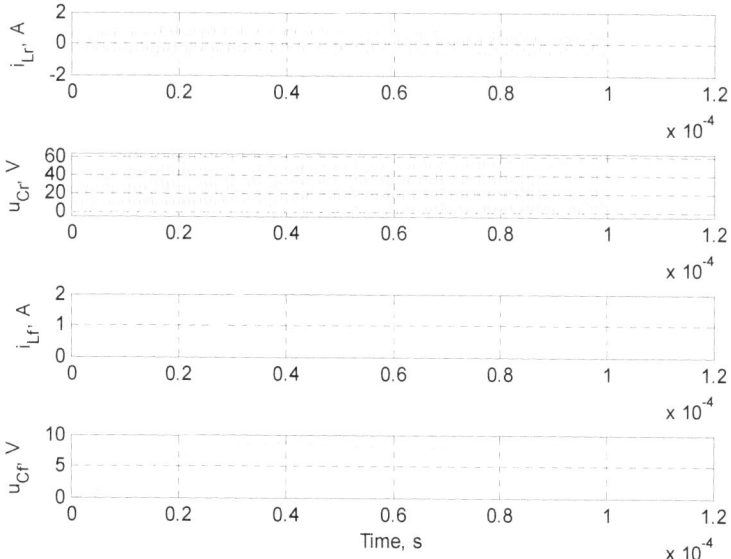

Figure 10. Simulation results of Example 2, regarding the state variables of the DC–DC converter.

For the state variables in Figure 8, the values of i_{aver} and u_{aver} were calculated, and we found:

$$u_{aver} = 8.0562 \text{ V}, i_{aver} = 0.7944 \text{ A}$$

We compared the obtained values of i_{aver} and u_{aver} with the values that were initially chosen as inputs to the neural network, and again, we observed that they were extremely close.

Example 3. *We choose new values for neural network inputs:*

$$\mathbf{x} = [u_{aver} = 11, i_{aver} = 1.1]^T$$

With the help of the neural network we obtained

$$L_r = 1.1950 \times 10^{-6} \text{H}, \ L_f = 2.1592 \times 10^{-4} \text{H}, \ C_f = 3.0758 \times 10^{-7} \text{F and}$$

$$C_r = 1.1950 \times 10^{-6} \text{F}.$$

Using the calculated values, we simulated the mathematical model, and the results are shown in Figure 11.

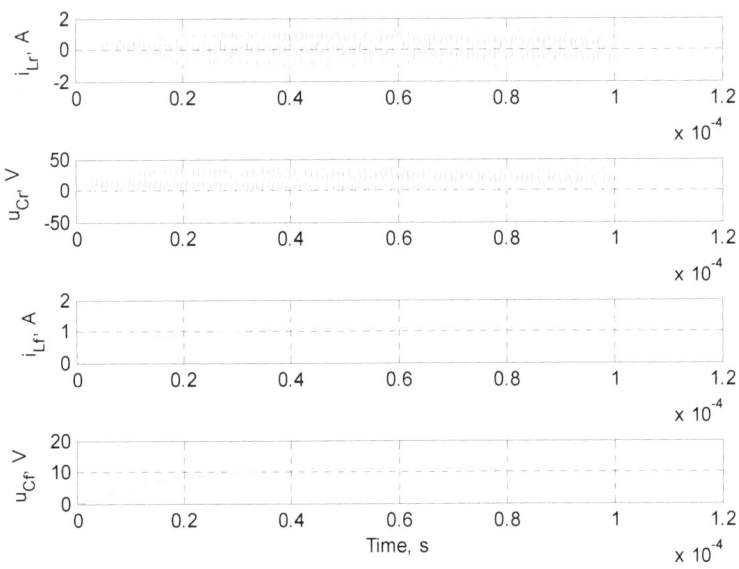

Figure 11. Simulation results of Example 3, regarding the state variables of the DC–DC converter.

The values of the state variables from Figure 11 were calculated. The values of u_{aver} and i_{aver} were as follows:

$$u_{aver} = 11.0793 \text{ V and } i_{aver} = 1.1091 \text{ A}$$

We compared the values obtained in this way with the values that were selected as inputs for the neural network at the beginning of this example, and again we found that they were extremely close. In the selected intervals of variation in the input data (Figure 4), the network thus initialized and trained works as a design medium with astonishing accuracy.

6. Discussion

The analysis of the presented results shows that the neural network is an effective design tool in the field of power electronics. Specifically, the neural network is able to predict the relevant parameters, such as resonant frequency, resonant and filter inductances, and capacitances, with an accuracy that greatly exceeds that of traditional analytical approaches. This makes it easier to achieve ZVS under various loads and operating conditions, resulting in reduced switching losses and increased overall converter efficiency.

One of the main advantages of the neural network-based approach is its ability to deal with multidimensional problems where multiple parameters need to be optimized simultaneously. In the traditional design process, each parameter is usually set manually, which is laborious and often results in performance compromises. In this context, the neural network allows for a more flexible and efficient setup process, while minimizing the risk of unexpected problems during operation.

On the other hand, this approach also has some limitations. Building and training a neural network requires significant computing resources and a large set of training data. Creating these data through simulations or experiments can be an expensive and time-consuming process. Furthermore, the neural network itself is a complex model, which can make the optimization process opaque to engineers. This leads to the need for careful verification of network results through physical experiments or additional simulations to ensure that the obtained parameters meet the reliability and efficiency requirements.

Additionally, one of the main caveats to using neural networks in such applications is related to model generalization. While the trained network demonstrates high accuracy in designing the transducer within the given input parameters, there is a risk of reduced performance under conditions other than those used for training. This means that the network may require additional training or adjustments if new operating conditions are introduced that are significantly different from those used during training. A summary of the proposed approach is given with the algorithm shown in Figure 12.

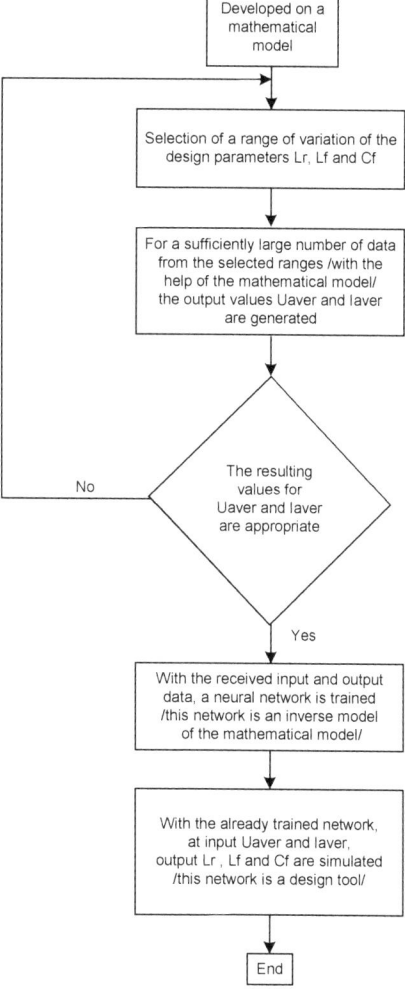

Figure 12. Schematic representation of the steps for designing a DC–DC converter with a neural network.

Despite these challenges, the results of our research clearly show the potential of neural networks in power electronics, especially in the field of quasi-resonant converters. This methodology offers promising opportunities for design automation while improving efficiency and reducing development time. In the future, the integration of such techniques with other optimization methods, such as genetic algorithms or swarm algorithms, may lead to even better results and extend the applicability of this approach to other types of transducers and systems.

Finally, this study highlights the need for continued experiments to improve the stability and reliability of neural networks under different operating conditions. Incorporating more diverse data and creating more complex simulations may allow for better model generalization and more reliable optimization for future industrial applications.

7. Conclusions

This paper presents an innovative design approach for a buck ZVS quasi-resonant DC–DC converter based on neural networks. The use of neural networks to optimize the design of electronic devices offers significant advantages over traditional methods. While traditional design approaches rely on trial and error and extensive theoretical modeling, neural networks enable rapid adaptation and refinement of transducer parameters.

The presented system was able to reduce power losses and improve the efficiency of the converter, which is of key importance for applications related to renewable energy sources and electric vehicles. By training the neural network on real data and simulations, the model demonstrated the ability to predict the optimal operating conditions, adapting to different loads and input voltages.

Furthermore, the integration of the ZVS technique into the converter design contributes to better performance and longer component life by minimizing mechanical and thermal stresses. However, further research is needed to optimize neural network training and to evaluate performance under real-world conditions.

The use of neural networks to design buck ZVS quasi-resonant DC–DC converters represents a significant advance in the field of power electronics. The present study demonstrates that neural networks can effectively predict and optimize converter parameters, ensuring high performance and energy efficiency. The results show that neural networks not only speed up the design process but also offer new opportunities for innovation in electronic systems.

Practical challenges when integrating neural networks into the control or design of quasi-resonant DC–DC converters in high-frequency switching applications include the following:

- Response time and latency: In high-frequency switching applications, the system response time is critical. Neural networks require time to compute their outputs, especially for complex models. This latency can be a significant issue when controlling converters operating at MHz frequencies, where any delay may lead to unwanted oscillations or system instability.
- Model accuracy: Neural networks are often trained based on data collected under limited operating conditions. In high-frequency switching, nonlinear and unpredictable effects (such as parasitic inductances and capacitances) may arise, which are not accounted for in the training dataset. This can lead to a loss of control accuracy in real-world scenarios.
- Training complexity: Training a neural network to effectively control a high-frequency DC–DC converter requires the collection of highly accurate and reliable data. Moreover, applying neural networks to complex nonlinear processes such as quasi-resonant switching may necessitate more advanced training techniques (e.g., reinforcement learning), increasing the time and computational resources needed for model preparation.
- Stability and robustness: Using neural networks to control quasi-resonant converters raises concerns regarding system robustness and stability. Neural networks can react unexpectedly to unforeseen input signals or conditions, potentially leading to

- unpredictable converter behavior. Additional monitoring and protection mechanisms are needed to ensure system stability.
- Integration with traditional control methods: Neural networks often need to work alongside traditional control methods, such as proportional–integral–derivative (PID) controllers. This integration can be challenging, since different methods may have different response times and dynamics, making coordination between the control mechanisms difficult.
- Hardware constraints: Integrating neural networks requires additional hardware resources, such as powerful microcontrollers or processors capable of handling complex real-time computations. In high-frequency switching, this may necessitate the use of specialized hardware (e.g., FPGA or DSP), which increases costs and system complexity.
- Energy efficiency: Neural networks may demand significant computational resources, which, in turn, can increase power consumption. In applications like DC–DC converters, where energy efficiency is critical, this additional load could pose a challenge.

These practical challenges show that despite the potential advantages of neural networks in controlling power electronics systems, successfully integrating them into quasi-resonant DC–DC converters requires overcoming significant technical and engineering barriers.

Future research could focus on extending the methodology to include different topologies and operating conditions, as well as implementing neural networks in real-world applications. Combining traditional engineering approaches with artificial intelligence and machine learning can open new horizons for the design of power electronic devices, which is particularly important in the context of global efforts for energy efficiency and sustainable development.

Author Contributions: N.H. and B.G. were involved in the full process of producing this paper, including conceptualization, methodology, modeling, validation, visualization, and preparing the manuscript. All authors have read and agreed to the published version of the manuscript.

Funding: This research was funded by the Bulgarian National Scientific Fund, grant number КП-06-H57/7/16.11.2021, and the APC was funded by КП-06-H57/7/16.11.2021.

Data Availability Statement: The original contributions presented in the study are included in the article, further inquiries can be directed to the corresponding authors.

Acknowledgments: This research was carried out within the framework of the project "Artificial Intelligence-Based modeling, design, control, and operation of power electronic devices and systems", КП-06-H57/7/16.11.2021, Bulgarian National Scientific Fund.

Conflicts of Interest: The authors declare no conflicts of interest.

References

1. Krishnamoorthy, H.S.; Krein, P.; Zahnstecher, B. From 'Power Electronics Inside' to 'Human-Centered Power Electronics'. *IEEE Power Electron. Mag.* **2023**, *10*, 61–63. [CrossRef]
2. Pollefliet, J. *Power Electronics: Switches and Converters*; Elsevier Inc.: Amsterdam, The Netherlands, 2017. [CrossRef]
3. Tan, D. Emerging System Applications and Technological Trends in Power Electronics: Power electronics is increasingly cutting across traditional boundaries. *IEEE Power Electron. Mag.* **2015**, *2*, 38–47. [CrossRef]
4. Mohan, W.P.R.N.; Undeland, T.M. *Power Electronics: Converters, Applications, and Design*, 3rd ed.; Wiley: Hoboken, NJ, USA, 2007.
5. Bose, B.K. Power Electronics: My Life and Vision for the Future [My View]. *IEEE Ind. Electron. Mag.* **2022**, *16*, 65–72. [CrossRef]
6. Rashid, M.H. *Power Electronics Handbook*; Elsevier Inc.: Amsterdam, The Netherlands, 2023. [CrossRef]
7. Kassakian, J.G.; Perreault, D.J.; Verghese, G.C.; Schlecht, M.F. *Principles of Power Electronics*; Cambridge University Press: Cambridge, UK, 2023. [CrossRef]
8. Hasan, M. Application of power electronics in power systems. In *Handbook of Research on Power and Energy System Optimization*; IGI Global: Hershey, PA, USA, 2018.
9. Rajabi, A.; Marangalu, M.G.; Shahir, F.M.; Sedaghati, R. Power electronics converters-An overview. In *Intelligent Control of Medium and High Power Converters*; Institution of Engineering & Technology: Stevenage, UK, 2023. [CrossRef]
10. Lopusina, I.; Stanojevic, A.; Bouvier, Y.E.; Grbovic, P.J. Comparison Between ZVS and ZCS Series Resonant Balancing Converters. In Proceedings of the 22nd International Symposium on Power Electronics, Ee 2023, Novi Sad, Serbia, 25–28 October 2023. [CrossRef]

11. Kazimierczuk, M.K.; Czarkowski, D. *Resonant Power Converters*; John Wiley & Sons: Hoboken, NJ, USA, 2012.
12. Zhang, J.; Shi, Y.; Zhan, Z.H. Power electronic circuits design: A particle swarm optimization approach. In *Simulated Evolution and Learning*; Lecture Notes in Computer Science (Including Subseries Lecture Notes in Artificial Intelligence and Lecture Notes in Bioinformatics); Springer: Berlin/Heidelberg, Germany, 2008. [CrossRef]
13. Asadi, F. *Simulation of Power Electronics Circuits with MATLAB®/Simulink®: Design, Analyze, and Prototype Power Electronics*; Springer Nature: Dordrecht, The Netherlands, 2022. [CrossRef]
14. Batarseh, I.; Harb, A. *Power Electronics: Circuit Analysis and Design*; Springer: Cham, Switzerland, 2017. [CrossRef]
15. Sitek, W.; Trzaska, J. Practical aspects of the design and use of the artificial neural networks in materials engineering. *Metals* **2021**, *11*, 1832. [CrossRef]
16. Kanwar, N.; Goswami, A.K.; Mishra, S.P. Design Issues in Artificial Neural Network (ANN). In Proceedings of the 2019 4th International Conference on Internet of Things: Smart Innovation and Usages, IoT-SIU 2019, Ghaziabad, India, 18–19 April 2019. [CrossRef]
17. Ibrahim, N.F.; Mahmoud, M.M.; Al Thaiban, A.M.; Barnawi, A.B.; Elbarbary, Z.S.; Omar, A.I.; Abdelfattah, H. Operation of Grid-Connected PV System with ANN-Based MPPT and an Optimized LCL Filter Using GRG Algorithm for Enhanced Power Quality. *IEEE Access* **2023**, *11*, 106859–106876. [CrossRef]
18. Lakhdara, A.; Bahi, T.; Moussaoui, A. MPPT Techniques of the Solar PV under Partial Shading. In Proceedings of the 18th IEEE International Multi-Conference on Systems, Signals and Devices, SSD 2021, Monastir, Tunisia, 22–25 March 2021. [CrossRef]
19. Mohammad, K.; Musa, S.M. Optimization of Solar Energy Using Artificial Neural Network Controller. In Proceedings of the 2022 14th IEEE International Conference on Computational Intelligence and Communication Networks, CICN 2022, Al-Khobar, Saudi Arabia, 4–6 December 2022. [CrossRef]
20. Susmitha, P.; Parventhan, K.; Umamaheswari, S. Artificial Neural Network Control for Solar—Wind Based Micro Grid. In Proceedings of the MysuruCon 2022—2022 IEEE 2nd Mysore Sub Section International Conference, Mysuru, India, 16–17 October 2022. [CrossRef]
21. Gangula, S.D.; Nizami, T.K.; Udumula, R.R.; Chakravarty, A.; Singh, P. Adaptive neural network control of DC–DC power converter. *Expert. Syst. Appl.* **2023**, *229*, 120362. [CrossRef]
22. Nizami, T.K.; Chakravarty, A. Neural network integrated adaptive backstepping control of DC-DC boost converter. *IFAC-PapersOnLine* **2020**, *53*, 549–554. [CrossRef]
23. Hussein, A.I.; Shigdar, B.; Almatrafi, L.; Alaidroos, B.; Alsharif, F.; Aly, R.H.M. Design of a DC/DC Converter with a PID Controller and Backpropagation Neural Network for Electric Vehicles. In Proceedings of the 20th International Learning and Technology Conference, L and T 2023, Jeddah, Saudi Arabia, 26 January 2023. [CrossRef]
24. Siddhartha, V.; Hote, Y.V. Robust PID Controller Design for DC-DC Converters: The Buck Converter. In Proceedings of the 2022 IEEE Electrical Power and Energy Conference, EPEC 2022, Victoria, BC, Canada, 5–7 December 2022. [CrossRef]
25. Lin, H.; Chung, H.S.-H.; Shen, R.; Xiang, Y. Enhancing Stability of DC Cascaded Systems with CPLs Using MPC Combined with NI and Accounting for Parameter Uncertainties. *IEEE Trans. Power Electron.* **2024**, *39*, 5225–5238. [CrossRef]
26. Qiao, Z.; Yang, R.; Liao, W.Q.; Yu, W.; Fang, Y. Impedance modeling, Parameters sensitivity and Stability analysis of hybrid DC ship microgrid. *Electr. Power Syst. Res.* **2024**, *226*, 109901. [CrossRef]
27. Lu, C.; Li, J.; Chen, K.; Zhou, W.; Wu, Q.; Ke, J. System-level Parameters Identification for DC-DC Converters Based on Artificial Neural Network Algorithm. In Proceedings of the 2023 IEEE Energy Conversion Congress and Exposition, ECCE 2023, Nashville, TN, USA, 29 October–2 November 2023. [CrossRef]
28. Zhao, S.; Peng, Y.; Zhang, Y.; Wang, H. Physics-informed Machine Learning for Parameter Estimation of DC-DC Converter. In Proceedings of the IEEE Applied Power Electronics Conference and Exposition—APEC, Houston, TX, USA, 20–24 March 2022. [CrossRef]
29. Chen, S.; Zhang, J.; Wang, S.; Wen, P.; Zhao, S. Circuit Parameter Identification of Degrading DC-DC Converters Based on Physics-informed Neural Network. In Proceedings of the 2022 Prognostics and Health Management Conference, PHM-London 2022, London, UK, 27–29 May 2022. [CrossRef]
30. Kabalci, Y.; Kabalci, E.; Padmanaban, S.; Holm-Nielsen, J.B.; Blaabjerg, F. Internet of things applications as energy internet in smart grids and smart environments. *Electronics* **2019**, *8*, 972. [CrossRef]
31. Padmanaban, S.; Palanisamy, S.; Chenniappan, S.; Holm-Nielsen, J.B. *Artificial Intelligence-Based Smart Power Systems*; Wiley: Hoboken, NJ, USA, 2022. [CrossRef]
32. Haque, A.; Shah, N.; Malik, J.A.; Malik, A. Fundamentals of power electronics in smart cities. In *Smart Cities: Power Electronics, Renewable Energy, and Internet of Things*; CRC Press: Boca Raton, FL, USA, 2024. [CrossRef]
33. Lin, J.; Gebbran, D.; Dragicevic, T. Surrogate-Assisted Combinatorial Optimization of EV Fast Charging Stations. *IEEE Trans. Transp. Electrif.* **2023**, *10*, 2183–2191. [CrossRef]
34. Tian, F.; Cobaleda, D.B.; Martinez, W. Artificial-Intelligence based DC-DC Converter Efficiency Modelling and Parameters Optimization. In Proceedings of the 24th European Conference on Power Electronics and Applications, EPE 2022 ECCE Europe, Hanover, Germany, 5–9 September 2022.
35. Li, X.; Zhang, X.; Lin, F.; Blaabjerg, F. Artificial-Intelligence-Based Design for Circuit Parameters of Power Converters. *IEEE Trans. Ind. Electron.* **2022**, *69*, 11144–11155. [CrossRef]

36. Balci, S.; Kayabasi, A.; Yildiz, B. ANN-based estimation of the voltage ripple according to the load variation of battery chargers. *Int. J. Electron.* **2020**, *107*, 17–27. [CrossRef]
37. Virgili, M.; James, P.; Forsyth, A.J. Black-box model for estimating efficiency curves in DC-DC converters for energy storage systems. In Proceedings of the IEEE Vehicular Technology Conference, Helsinki, Finland, 19–22 June 2022. [CrossRef]
38. Akca, H.; Aktas, A. Examination and experimental comparison of dc/dc buck converter topologies used in wireless electric vehicle charging applications. *Int. J. Optim. Control Theor. Appl.* **2024**, *14*, 81–89. [CrossRef]
39. Vakovsky, D.; Hinov, N. Informational model verification of ZVS Buck quasi-resonant DC-DC converter. *AIP Conf. Proc.* **2016**, *1789*, 060016. [CrossRef]

Disclaimer/Publisher's Note: The statements, opinions and data contained in all publications are solely those of the individual author(s) and contributor(s) and not of MDPI and/or the editor(s). MDPI and/or the editor(s) disclaim responsibility for any injury to people or property resulting from any ideas, methods, instructions or products referred to in the content.

MDPI AG
Grosspeteranlage 5
4052 Basel
Switzerland
Tel.: +41 61 683 77 34

Mathematics Editorial Office
E-mail: mathematics@mdpi.com
www.mdpi.com/journal/mathematics

Disclaimer/Publisher's Note: The title and front matter of this reprint are at the discretion of the Guest Editor. The publisher is not responsible for their content or any associated concerns. The statements, opinions and data contained in all individual articles are solely those of the individual Editor and contributors and not of MDPI. MDPI disclaims responsibility for any injury to people or property resulting from any ideas, methods, instructions or products referred to in the content.